Sonic Writing

Sonic Writing

Technologies of Material, Symbolic, and Signal Inscriptions

Thor Magnusson

BLOOMSBURY ACADEMIC
NEW YORK • LONDON • OXFORD • NEW DELHI • SYDNEY

BLOOMSBURY ACADEMIC
Bloomsbury Publishing Inc
1385 Broadway, New York, NY 10018, USA
50 Bedford Square, London, WC1B 3DP, UK

BLOOMSBURY, BLOOMSBURY ACADEMIC and the Diana logo are trademarks
of Bloomsbury Publishing Plc

First published in the United States of America 2019

Copyright © Thor Magnusson, 2019

For legal purposes the Acknowledgements on pp. xiii–xiv constitute an extension
of this copyright page.

Cover design: Daniel Benneworth-Gray
Cover image: Ryan Ross Smith

All rights reserved. No part of this publication may be reproduced or transmitted
in any form or by any means, electronic or mechanical, including photocopying,
recording, or any information storage or retrieval system, without prior permission
in writing from the publishers.

Bloomsbury Publishing Inc does not have any control over, or responsibility for, any
third-party websites referred to or in this book. All internet addresses given in this
book were correct at the time of going to press. The author and publisher regret any
inconvenience caused if addresses have changed or sites have ceased to exist,
but can accept no responsibility for any such changes.

Whilst every effort has been made to locate copyright holders the publishers would
be grateful to hear from any person(s) not here acknowledged.

A catalog record for this book is available from the Library of Congress.

ISBN: HB: 978-1-5013-1385-1
PB: 978-1-5013-1386-8
ePDF: 978-1-5013-1387-5
eBook: 978-1-5013-1388-2

Typeset by RefineCatch Limited, Bungay, Suffolk

To find out more about our authors and books visit www.bloomsbury.com
and sign up for our newsletters.

Contents

List of Figures	vii
Preface	ix
Acknowledgements	xiii
Introduction: On Objects, Humans, and Machines	1

Part I Material Inscriptions

1	Instrumentality	17
2	New Instruments	33
3	Epistemic Tools	47
4	Digital Organology	59

Part II Symbolic Inscriptions

5	Writing Music	71
6	Printing Music	83
7	New Languages	95
8	Machine Notation	109

Part III Signal Inscriptions

9	Inscribing Sound	123
10	Recording	135
11	Analysing	147
12	Machine Listening	157

Part IV Digital Writing

13	Transductions	167
14	New Notations	179
15	Machine Writing	195
16	Music in Multimedia	207

Part V Conclusion

17 A Future of Music Tech   223
18 Transformation of Tradition   231
19 New Education for New Music   239

Notes   245
Bibliography   256
Index   278

Figures

1.1	The Ecstasy of St Cecilia – with Sts Mary Magdalene, Augustine, John Evangelist, and Paul	26
1.2	An organological illustration from Michael Praetorius's book *De Organographia*	27
2.1	A typical explanatory model of an electronic musical instrument.	36
2.2	Michel Waisvisz performing his Hands instrument, originally built in 1984	37
2.3	The Karlax Instrument made by De Fact in France	40
2.4	The LinnStrument, by Roger Linn	41
3.1	The Physical Laboratory of the Academie des Sciences	49
3.2	A representation of Pythagoras exploring ratios of string length, weight, and size	54
4.1	An Owl Project mLog	64
5.1	An example of Daseian notation in *Musica enchiriadis*, from the ninth century	77
6.1	Ghiselin Danckerts's open score of *Ave Maris Stella* from Pietro Cerone's *El melopeo y maestro: Treatise on the Theory and Practice of Music*	85
6.2	Movable type notational symbols could be arranged onto a grid and thousands of prints made by the press	87
6.3	A page from *Roman de Fauvel*, a fourteenth-century French allegorical poem with early examples of musical notation embedded.	88
7.1	*Mosaica*, a graphical score piece by Milan Adamčiak from 1968	101
8.1	A generative score schemata of *Cybernetics*, a piece by Roland Kayn, released in three parts, from 1966 to 1969	116
8.2	A piece in the making. Susanne Ciani's Buchla 200	119
9.1	A diagram from Ernst Chladni's *Discoveries in the Theory of Sound*, a pioneering eighteenth-century work on acoustics	127
11.1	Charles Seeger working with the Melograph Model B in the late 1960s	150
11.2	A spectrogram of a vibraphone note	152
13.1	The Magnetic Resonator Piano	173
13.2	Halldorophone	174
14.1	Performance on a tangible score by Enrique Tomás	187
14.2	Alexandra Cardenas live coding at the GEIGER Festival in Gothenburg in 2014	190
14.3	A still image from a piano/video piece by Claudia Molitor for pianist Zubin Kanga, called *You touched the twinkle on the helix of my ear* from 2018	192

15.1 A screenshot of three SC Tweets (SuperCollider code) 201
16.1 A screenshot from Tom Betts's *AVSeq*, an interactive musical game from 2010 213
16.2 Picture of Semiconductor's *HALO* installation, interpreting data from the CERN experiment ATLAS on the Large Hadron Collider 217
17.1 A picture from a performance of Robert Henke's *Lumière II* in Italy, 2016 224
18.1 A picture from Marianthi Papalexandri-Alexandri's performance of *Untitled II*, at Festival Con Voce in Lucerne Switzerland in 2014 235
18.2 Marco Donnarumma performing *Hypo Chrysos* at Inspace, University of Edinburgh, 2011 237

Preface

A voice comes to one in the dark. Imagine. A city half-lit by gas lamps, some electric lights. The grating sound of horse-drawn carriage wheels, mixed with the clatter of hooves on the cobbled streets, occasionally interrupted by the loud continuous drone of a motor car. In the distance, the factories are waking up, and the previously silent sky is now vibrant with sound waves. Imagine this birth of the modern age, through new sciences of evolution, light, and sound – relativity. Impregnated with machines and theories that capsized our accepted understanding of the world, modernity introduced a new compass of morality – God was dead! And before the mechanical rumble of the day begins, this voice in the dark whispers a prophecy about a brave new world built on new technologies.

The technological promise of the twenty-first century is alluring, but we have been here before. Luigi Russolo's *Art of Noise* was written in 1913, heralding a new age of noise, machines, industries, and war. The *Futurist Manifesto*, published three years earlier, had called for the destruction of the old; of museums, libraries, and academies, paving the way for new sciences, and "the shock of the new." With combustion engines, machines, electricity, and telecommunication, the early twentieth century was so drastically different from the world just a few decades earlier, that practically all areas of human endeavour had been redefined. In response, Russolo built the *intonarumori*, the instrument of the new age, to proclaim the noise of this new world. The same year Russolo wrote his manifesto, Stravinsky premiered his *Rite of Spring* in Paris. That story has been told many times, but it involves Stravinsky escaping the auditorium because of the loud uproar in the audience; the music of the new century was perhaps too much to take in. New musical ideas required new instruments – and the other way around! Across the Western world, instruments were being invented as modernism emphasised novelty, the never-heard-before, the avant-garde.

A few years later, a Russian inventor, Lev Sergeyevich Termen, presented his theremin instrument, also known as the *aetherophone*, which worked by means of electromagnetic fields – an instrument that did not require the touch of a human hand, but emitted such celestial sonic beauty, so alien and unlike anything ever heard, that it became the mandatory sonic tapestry for all science fiction movies for decades to come. The theremin was not the first electronic instrument, but possibly the first to be mass-produced, affordable to the middle classes, and not too large for people's homes.

The sound of the theremin is familiar today, but imagine again going back a century, to the world described above, and hearing that sonic purity of the two heterodyning oscillators so effortlessly controlled by hands waving in the thin aether. Indeed, there was something electric in the air at the beginning of that century. Modernism in the arts was redefining the role of art itself, and this was supported by developments in the natural sciences, in philosophy, and politics. All this had such an influence on the young composer Edgard Varèse that he began to write music from an extra-musical mindset, drawing inspiration from new technologies and science. In one of his "Liberation of Sound" lectures from 1936, he says: "When new instruments will allow me to write music as I conceive it, taking the place of the linear counterpoint, the movement of sound-masses, of shifting planes, will be clearly perceived" (Varèse 1966: 11). Describing music with the scientific language of modern physics, it was not just Varèse's vocabulary that had changed, but the very theoretical foundation of music as well, to such degree that it had to be redefined. Music "will no longer be the old conception of melody or interplay of melodies" (Varèse 1966: 11), but it will rather be based on colour or timbre, where the new instruments can emit any frequency partials or harmonics, all controlled by the composer working with new machines.

> I am sure that the time will come when the composer, after he has graphically realized his score, will see this score automatically put on a machine which will faithfully transmit the musical content to the listener. As frequencies and new rhythms will have to be indicated on the score, our actual notation will be inadequate. The new notation will probably be seismographic. And here it is curious to note that at the beginning of two eras, the Mediaeval primitive and our own primitive era (for we are at a new primitive stage in music today) we are faced with an identical problem: the problem of finding graphic symbols for the transposition of the composer's thought into sound. (Varèse 1966: 12)

If, nearly a century ago, Varèse was writing at the late dawn of modernism in the arts, we could claim that we are "primitives" no longer. The instruments Varèse envisioned have been built in many ways and forms. Some of them realised with electronic technologies, but others had to wait for the arrival of the digital computer. Indeed, the most common representation of sound on the digital computer is in the form of seismographic notation, as Varèse predicts, the familiar graphic representation that shows amplitude values in time, like a mountain range mirrored in the reflection of a calm lake. However, mechanical machines that could draw waveform onto a physical medium were invented in the late nineteenth century, with reproductive writing following suit. Writing music for machines, via phonographic technologies, was something Stravinsky considered as one way forward in musical composition. New technologies inevitably form a new thinking.

These snapshots of selected events in the twentieth century demonstrate how the new electric and mechanical instruments of sonic writing offered novel ways of composing, performing, and listening. I have introduced them to demonstrate how machines and electricity represented a profound change in human culture, and indeed they did, but so did metallurgy, the printing press, the steam engine, and Jacquard looms. Today we are in the midst of yet another technological revolution represented by highly intelligent, increasingly autonomous, computational systems that learn and adapt to our behaviour. Seen from the historical scope of that revolution, we are still primitives.

This book explores how technology conditions our musical expression. Refraining from the extremes of both technological determinism, which argues that we are pawns of a technological progress that conditions our mode of being, and technological constructivism, which states that humans will always adopt technology to its needs, the book explores the intricate and complex relationships between ideas and materiality, between history and the technological present. The long quote above by Varèse can easily be justified on the first pages of this book, as it inaugurates the transition from this "primitive era" to where we are now. The book will explore these issues by looking at selected genealogies of our instruments, notation, and recording as enabled by current computational technologies. The music technologies of the twenty-first century, or indeed musical culture in its entirety, cannot be apprehended without a referral to the historical past: referencing older writings, instruments, compositions, and recordings in our new work.

Sampling Samuel Beckett in the first two sentences of a book on sonic writing is only appropriate as this master of high modernist prose began to write with sound waves too, composing with tape for the new communication technologies of his day. In *Krapp's Last Tape*, for instance, we hear the recorded voice of Krapp's younger self talking about an "extraordinary silence this evening," and yet it is that very voice which breaks the silence through the aegis of a new form of sonic writing. Krapp is troubled by his younger self, but that is not only because of a weariness of growing old and with his bygone ways of thinking, but also because of a certain fundamental unease with his own voice, expressed in sound, and recorded onto the tape decades earlier. In this work, Beckett embraces new writing technologies, transmuting his fascination with the transmission of information and sound via radio, of a sound recording across time, and of the interpretation an actor can give to the dead letters of the written page into consummate art. And there is a parallel here between the writerly Beckett applying phonographic technologies in his writing and the experiments composers carried out in the recording studios – namely, the study of how the new technologies of phonographic writing afforded worlds of innovative expression, where writing shifted from the symbol to the signal.

Instruments, notation, phonography, and digital systems. Why such a diverse topic in one book? The aims and objectives are to focus on the conditions of digital musical technologies, their technological underpinning, and their ideological constitution. Our digital instruments are not objects emerging from nowhere: they are hybrid systems whose histories, ideologies, embodied performance, musicologies, aesthetics, and styles originate in practices that are pre-digital, often with cultural patterns that can be traced back centuries into our musical past. The objective, then, is to investigate how a distinct form of musical practice is emerging in the twenty-first century, one that applies digital technologies in extremely wide musical contexts. Here, certain forms of musical expression that began in the sixteenth century with movable type printing, and were emboldened by recording practices in the early twentieth century, are transcended with the help of intelligent machines. The book illustrates how this new musical culture, brought forth by computer technologies, is in many ways reminiscent of the cultures of pre-notated musical practices, where music was grounded in theory and not in notated musical works. This book explores selected genealogical threads that support a specific weaving of the current tapestry of musical practices with new technologies, a tapestry in which the musical work is transformed into another mode of existence.

The ontology of music is changing: in twenty-first-century compositional practices music is again seen as a system that is equally embodied in the theory and the technologies that express it, as opposed to former practices of individual pieces written by named composers. It is tempting to add a *k* to the word music when referring to it in the ancient, medieval, and early modern sense of *mousikos* (Greek), *musica* (Latin), and *Musick* (in seventeenth-century English). These forms of sonic writing did not conceive of the musical work as a primary element in the set of musical practices. This is rather *musick* as theorised and practised; as a system or a body of scientific knowledge that is instantiated, rendered, and performed, via instruments that embody the episteme. By using the letter *k*, we also subscribe to Christopher Small's (1998) concept of *musicking*, which enjoins us to see music as a verb, as a social activity that expresses the human condition and forms an important part of how we relate to each other and organise our world. The argument follows, that when we start writing sound with signals instead of symbols, through a process in which our tools become intelligent, we transform our notions of music as a necessary consequence of the new music technologies. The new practices echo, question, and problematise ancient notions that theory (*episteme*) is of more interest and importance in music than skill (*techne*), and this book will explore how our new epistemic tools equally serve as platforms for theory and skill.

Writing a book about sonic writing in the age of digital media may seem like the wrong choice of medium. With computational, networked multimedia, I could have included sounds, pictures, videos, interviews, instruments, and more. The work would not need to be static, but could evolve over time, as new information materialises. Readers could comment, and discuss topics in a dialogue that would doubtless become an essential part of the book. However, there is a quality of the printed format that I value: to constrain, focus, write, rewrite, rehearse, rethink, review, edit and re-edit in an iterative process that involves colleagues, editors, proofreaders, and myself operating in a tradition that extends back to antiquity and is endowed with its own rules, protocols, and formats. Considering the wide scope of the book – tracing the genealogies of instrumental, notational, and phonographic technologies as they get transduced into the processor-based media of the computer – I felt that the format of the book would provide the necessary constraint and focus that I was after.[1] This book is about musical instruments, media, and materialities. It explores how new media formats shape the theory and practice of music. Nietzsche wrote about the typewriter that "our writing tools are also working on our thoughts" (Kittler 1999: 203), but this statement becomes much more relevant and amplified in the real-time and compositional technologies of music making. Let our journey begin!

Acknowledgements

This book is an outcome of a two-year AHRC (UK Arts and Humanities Research Council) fellowship. I am very grateful for the opportunity to research, travel, collaborate, and organise events that directly shaped the ideas here within. With project partners, I organised symposia, workshops, and talks, and I was able to spend time using libraries, studios, and archives, as well as interviewing people. The hospitality and collaborative spirit I experienced at STEIM in Amsterdam, IRCAM in Paris, CNMAT in Berkeley, CMC in Columbia University, New York, and C4DM at Queen Mary University in London was amazing. At these institutions I met some fantastic people working on excellent cutting-edge research which inspired the ideas in this book, but these are too numerous to be listed here. However, I would like to thank especially those who invited me to work at their institutions: Joel Ryan, Kristina Andersen, and Marije Baalman at STEIM; Frédéric Bevilacqua and Norbert Schnell at IRCAM; Ed Campion at CNMAT; Brad Garton at CMC; and Andrew McPherson at C4DM. During this project, I was also invited to a writer's residency at Fondation Jan Michalski pour l'écriture et la littérature in Switzerland, where this work took on a considerable focus. I thank the Foundation for their hospitality and superb attention to a writer's needs.

I have had the fortunate opportunity to interview various people whose practice informs the changes I describe in musical culture. These are: Kristina Andersen, Margaret Birley, Ken Butler, Ed Campion, Suzanne Ciani, Nick Collins, Pierre Couprie, Pete Furniss, Brad Garton, Rama Gottfried, Derek Holzer, Hans Jóhannsson, Shinji Kanki, Roger Linn, James McCartney, Alex McLean, Andrew McPherson, Claudia Molitor, Sarah Nicolls, Godfried-Willem Raes, Christopher Redgate, Ryan Ross Smith, Laetitia Sonami, Laurie Spiegel, Dan Stowell, Bob Sturm, Enrique Tomás, Ge Wang, Anna Xambó, and Pamela Z. I have the feeling that more material will emerge as a result of these interviews.

I owe very much to my colleagues and collaborators at the Sussex Emute Lab (Experimental Music Technologies Lab), with whom I organised the 3rd ICLI (International Conference on Live Interfaces) in June 2016, as part of this research, and are generally some incredibly creative and inspiring people: Dylan Beattie, Danny Bright, Cecile Chevalier, Andrew Duff, Alice Eldridge, Evelin Ficarra, Chris Kiefer, Paul McConnell, Alex Peverett, Thanos Polymeneas-Liontiris, Halldór Úlfarsson, and Joe Watson. Colleagues at the Music Department of Sussex have been very supportive and I have learned much from conversations with Ed Hughes, Martin Butler, Richard Elliott, Mimi Haddon, Tim Hopkins, and Nick Till. I am fortunate to be working at the School of Media, Film and Music at the University of Sussex, as well as being an associate member of the Sussex Humanities Lab, and I'm continually inspired by my colleagues' critical, philosophical, and creative approaches in their research, event organisation, and writing. It would take up too much space to name them all here. The same applies for the fantastic students we have at all educational levels at Sussex.

Four wonderful people joined me on the Sonic Writing research project: Sam Duffy, Alex Peverett, Sally-Jane Norman, and David Hendy. All of them were crucial in diverse parts of the research project, especially planning, event organisation, and data handling, and I thank Alex for his eagle-eyed reading of the book's manuscript. Equally important has been my continuing collaborations with Enrike Hurtado, with whom I founded the ixi software project in 2000, and Alex McLean, who I established the Live Coding Research Network with in 2013, which have been educational, inspiring, and always enjoyable. I don't think this book would have happened without these people.

Writing a book is a long and extensive process. My perception is that of being an antenna for ideas that exist or are emergent in the field, and I've picked them up from innumerable conversations, presentations, and performances at concerts, festivals, and conferences over the years. I feel like Deleuze and Guattari, when they say in *Mille plateaux* that since "each of us was several, there was already quite a crowd." On that account, there have been wave shapers and filters that formed this text into the state it is currently in, static on the printed pages of this book, and I would like to thank the following for their diverse levels of reading, commenting, suggesting, and correcting: Alex Peverett, Halldór Úlfarsson, Nick Till, Chris Kiefer, Enrike Hurtado, Bergsveinn Birgisson, David Bausola, David Hendy, and importantly the superb in-depth engagement by Neal O'Donoghue. I thank the editors and anonymous peer reviewers at Bloomsbury, whose support has been excellent, from the stages of proposal to cover design. I am also very grateful to Ryan Ross Smith, whose score *Study no. 57*, is on the cover of this book – or rather, a still from a dynamic, animated score. It is appropriate to freeze an animated score on the cover of a book that describes the perpetual change in musical technologies: music is a field of human activity that never rests and is never quiet, but rather serves as a looking glass into our possible futures.

Finally, I'd like to thank my wonderful family for their support and understanding, more importantly humour and fun, and I dedicate this book to Birta, Mirra, and Loki.

Introduction: On Objects, Humans, and Machines

Where to begin? Music, instruments, inscription, technology, human activities! Attempting to find a good point of origin, or orientation, in such a diverse field as music, so deeply entrenched as it is in our nature, poses numerous problems. Unlike many other human pursuits, the study of music and musical instruments does not involve exploring activities that begin at some particular point in our history. Quite the opposite: music is the making of humanity, the very thing that makes us human. This is a point musicologist Gary Tomlinson makes in his book *A Million Years of Music* (Tomlinson 2015), where the origins of music are traced back to 800,000 years before the emergence of our species *c.* 200,000 years ago. The argument is convincing: music is an archaic activity written into our psyche, existing beyond the scope of conceptual language or symbolic systems, and connecting with the totality of our intellectual, emotional, and physical being. Originating with early human brain development and tool use, music had a role in the attempt to understand and organise the world via the mimetic function of repeating what is heard, forging the synchrony of rhythmic repetition, understanding time, establishing order, and social cohesion. Human activities or emotions, animals or nature, as well as cosmic cycles, could be represented in the events of musical time. Some authors, for example Rousseau and Darwin, are of the view that music appeared before language, while others claim that both have a common precursor (Mithen 2005). What we do know is that music played a role in training the voice to the degree that human language became possible.

As a cultural technology, music trains our proprioceptive action, interpersonal collaboration, improvisation, creativity, and social cohesion. In the past, it was defined as a mathematical discipline, and current cognitive science does indeed demonstrate that we use the same areas of our brain in maths and music (Cranmore and Tunks 2015). However, even if such general statements can be established on the function of music as one of eternal permanence (with lullabies, ritual, dance, work, or war music), our *ideas* about music and the related cultural *practices* are continually evolving. More importantly for the current study, we observe how our musical technologies have evolved from simple sticks and bone flutes to intelligent, adaptive, and creative systems. It can be argued that music technologies have always been at the cutting edge of human tool use – from simple cave flutes, to harps, organs, mechanical pianos, musical automata, electronic sound, computer software, AI and deep learning in music. The use of technology for expressive purposes is one of the things that distinguishes humanity from other animals, and if there is anything that might best exemplify human nature, it might be our technologies for musical expression. This seems to be the consensus among artists, scientists, politicians, and cultural theorists, and a good example is the Voyager *Golden Record* which has been traversing outer space since 1977. The record contains the key accomplishments of humanity, according to a NASA committee chaired by Carl Sagan, and includes music such as Bach, gamelan, shakuhachi, Stravinsky, and Laurie Spiegel. The record also includes clever symbolic instructions for extraterrestrials to build phonographic playback technology!

The non-origins of music

Music is no one thing. It is a networked mesh of practices, ideas and objects, as Christopher Small so elegantly demonstrates in his book *Musicking: The Meanings of Performing and Listening* (Small 1998). Tracing the etymology of the word "music," we arrive at the ancient Greek *mousike*, or the inspiration and activities we are given by the mythological Muses. The story of the Muses in Greek mythology is well known: they were the nine daughters of Zeus, the principal Olympic god, and Mnemosyne, the god of memory. Their teacher was Apollo, the god of music, art, and poetry. The important status given to music in preliterate societies is well known: how rhyme and metre serve as a mnemotechnic for reciting and passing on long poetic verses. Like the Norse god of poetry Odin, who hung upside down in the tree of knowledge for nine nights and nine days, learning to chant the runes – attaining the power of inscription and its magic – Zeus and Mnemosyne had a love affair for nine nights, the product of which was the nine muses.[1] The domains of the muses include poetry, history, dance, comedy, music, and astronomy, but they were primarily affiliated with Apollo, the god of music. In some earlier mythological sources, the Muses are said to be three: *Aoide* ("song" or "tune"), *Melete* ("practice" or "occasion"), and *Mneme* ("memory"), which is a more music-focused narrative. In early Greek philosophy, for example in Plato's *Republic*, we sometimes find *mousike* as the general domain of the performing arts, which include dance and poetry, but in other instances it is referred to as the theory and practice of something closer to what we understand as music today. The concept of music has certainly evolved over time, and we note that there are many societies that do not make a distinction between music and dance, music and ritual, or music as a part of other social functions.

This book does not seek the origins of music or its technologies: Tomlinson's book would be an excellent source for those interested in the early genesis of human musical practice. My focus is rather to understand how emerging digital music technologies trace their concepts, design, and functionality to practices that precede our cultural epoch. Such a study relates to the field of media archaeology (Parikka 2012), to how the development of material objects shape our social practices, and to how our technologies reveal important aspects of why we make music as we do today. We may begin by searching for historical musical artefacts, but we will always be left with the fact that the human voice is one of the key musical instruments. The voice is clearly an expressive and original instrument, although we cannot reasonably speculate which came first, the use of voice for musical purpose, or, for example, sticks used for banging on objects. It may seem strange to talk about the voice, a human organ, as an instrument, but singers typically define their voice as such, just like dancers see their body as an instrument. In any case, the purpose and function of those early organised sounds, made by a primitively controlled voice, sticks, or rocks, may be far removed from what we would typically call music today. And since the use of those rocks, sticks, or voice might have been accompanied with some other repetitive tasks, such as cracking nuts, we might define these tools as musical instruments, even if applied for another purpose too. Imagine an early tribe working together on a repetitive task with some tools; when a rhythm emerges and activities sync, we have music. From this perspective, the separation of music from other daily activities seems artificial, one that manifests in different ways throughout history and across cultures.

Examples of what we consider early human instruments include bone flutes found in cave dwellings estimated to be around 40,000 years old. The making of these objects by our cave-dwelling ancestors are the earliest forms of what I will define as a form of sonic writing. A bone with regularly distributed holes was found in 1995 in the *Divje Babe* cave in Slovenia, and was for a long time regarded as the oldest musical instrument in the world. The hypothesis was that this was a Neanderthal instrument, implying that Neanderthals had a musical culture. However, prehistorian Francesco d'Errico and his team have argued that these holes may have originated from the teeth of a bear, since similar bones with holes have been found in caves where there were no traces of human occupation (d'Errico et al. 1998; d'Errico et al. 2003). They are careful to mention that even if this bone specimen was not made by a hominoid, it does not exclude the possibility of Neanderthals creating and playing musical instruments (d'Errico et al. 2003: 38). In 2008, some bone flutes made out of bird bones were found in the *Hohle Fels* caves in the Swabian Jura in Germany. One of the flutes is made of vulture bone, 8 millimetres wide and 34 centimetres long, with five finger holes and a V-shaped mouthpiece. These are estimated to be between 37,000 and 43,000 years old (Tomlinson 2015: 251).

Tomlinson argues that these early instruments are technologies that demonstrate important characteristics in the evolution of the human: firstly, the notion of abstracting pitch from signification, or the act of signalling a message, and this happens with the discrete pitches of the flute. This conscious planning of where the holes should be made in the flute is what Bernard Stiegler calls *grammatization* (Stiegler 2010b), or the process of externalising our thoughts using discrete systems, for example with material objects. Stiegler's thesis will be discussed in more detail below, but for Tomlinson, this is one of the first examples of material objects that gain an abstract purpose, removed from practical necessities of function and signification: "Music was from the first, in this sense, *absolute*. Discrete pitch arrays have in fact ever since maintained this distance from signification" (Tomlinson 2015: 258). The location of the holes in a bone flute, or the tuning of the strings in a lyre is abstract, non-semantic, and of no particular function. Secondly, as an effect of making music through action sequences, or patterned gestures, the human understanding of time evolves, perceived through social engagement – in short, as *entrained* repeatable social behaviour. This entrainment of synced action forms temporal hierarchies that we call metre in music.

Mnemotechnics and music

The extension of the human mind and expansion of memory (by means of mnemotechnics) begin when humans store information via symbolic systems, by carving into rocks, bones, and other objects (Leroi-Gourhan 1993; Stiegler 1998). This was a momentous evolutionary step for the cognitive abilities of humans. But what does it mean when we have a material musical object of such durability that it outlives generations of cave dwellers? The bone flute is an object that materialises music theory by dint of a discrete organisation of pitch – an organisation that is by nature musico-theoretical, owing to the discreteness of its tuning and scale. We can define this new object as an epistemic artefact (Kirsh and Maglio 1994) or an epistemic tool (see Magnusson 2009), as it affects, shapes, controls, and directs the way music would be made when the instrument is played. It is an externalisation of musical memory, or what Stiegler would call a mnemotechnic device, that maintains the human in the "stereotypic" form it has evolved into, suggesting that our brain and our external tools are one interdependent system:

> The question then becomes: *where is the memory of the stereotype kept, if not in the material trace of the stereotype in which the preexisting tool itself consists, repeated, duplicated by its "maker" and guiding the latter much more than being guided by him or her?* In this sense, the archaic cortex and equipment are codetermined in a structural coupling of a particular sort. (Stiegler 1998: 158)

This book frames the act of writing in a wider sense than musical notation or sound recording. Sonic writing equally includes material, symbolic, and phonographic inscriptions. It explores how we think *with* and *through* technology; sometimes in a dialogue with technology, other times where it serves as an extension of bodies. Although it is a coinage that derives directly from the Greek concept of *phonography* (and my native Icelandic *hljóðritun*), "sonic writing" refers to a wider area of human expression traceable to certain strands in the philosophy of technology, where the notion of grammatology (Derrida 1997) is used to explain how humans extend their memory and body technics by means of external "supplements."[2] Stiegler developed this notion of grammatology further, and introduced *grammatization* as the technical history of memory, arguing that we cannot separate incorporated memory (*mneme*) and exterior memory (*hypomnesis*) (Stiegler 2010a: 29):

> By grammatization, I mean the process whereby the currents and continuities shaping our lives become discrete elements. The history of human memory is the history of this process. Writing, as the breaking into discrete elements of the flux of

speech (let us invent the word *discretization* for this possibility), is an example of a stage in the process of grammatization. (Stiegler 2010b)

If grammatisation is the task of externalising, of abstracting processes, of categorising, of representing, it is a process that begins at an early stage of human evolution. Tomlinson, in his book on the origin of music, references a Stiegler favourite, the anthropologist André Leroi-Gourhan, whose work *Gesture and Speech* (1993) speculates how early human Palaeolithic technics form the condition for language. From this perspective, repetition is what makes identity possible, and through repetitive human movements, such as cutting meat with a sharp rock, a patterned and rhythmic activity emerges. Subsequently, from this bottom-up, embodied activity based on environmental affordances (of the sharp flint), the human mind is formed. Indeed, it "is not that stone tools are proxies of mind but something closer to the reverse: mind as an outgrowth of the body-stone interface" (Tomlinson 2015: 68). The account of human evolution we find emerging here is one of the development of human skills through technology, not the other way around: so instead of the individual human mind exhibiting a top-down invention of new artefacts, we find a bottom-up discovery of uses by a social group of actors of entrained collaboration. This view fits very well with the way deconstruction describes the non-origin of things, the role of the supplement and of technology in general. Art historian Whitney Davis gives laconic expression to this view when he says: "the tool writes the toolmaker as much as the toolmaker writes the tool" (Davis 1996: 116).

Technics are an externalisation of the mind, but for Stiegler the technical is not merely a tool or a machine – it constitutes the *invention* of the human. "Is it *tekhnē* that arises from *logos*, or the reverse? Or rather, is it not that *logos* and *tekhnē* are modalities of the same being-outside-oneself?" (Stiegler 1998: 193). What characterises the rational human is the transmission of memory via technical artefacts. We have a transgenerational accumulation of memory that is preserved in the technological totality: the people of the cave die and new generations are born, but they are born into a cave equipped with tools, one of which is a bone flute. The experience of the human culture gets written into the tool – and it becomes cumulative and transmissible. This means, however, that the human being is an animal whose knowledge acquisition is not transferred by one individual to its offspring via genes, but via mimesis. This is knowledge that needs to be learned repeatedly by every new generation, and these cycles of iteration will result in difference, in misunderstandings, and in reappropriation in new technocultural contexts.

Musical instruments are key to understanding such mimetic activity, as much in their making as in their use; they are built of materials and, by way of ingenious design, serve as surfaces of musical inscription. The musical theory of each musical culture is written into the functional body of the instrument itself. The instrument is concretised music theory. It becomes what Stiegler calls a "techno-logic," which then, in turn, becomes our external memory: "A tool is, before anything else, memory: if this were not the case, it could never function as a reference of significance" (Stiegler 1998: 254). Before the writing technologies of clay, papyrus, scroll, and codex, we wrote our cultural memory into our material artefacts, such as the caveman's flute which would pass through generations of players, all of whom would automatically align to the musical potential and theory written into the surface of that instrument. To be human is to inherit the already-there of history, in an environment of designed objects into which we are "thrown," to use Heidegger's vocabulary. We, therefore, get access to a past "that is mine but that I have nevertheless not lived" (Stiegler 1998: 159).

Three music technological epistemes

This book is a study of how the new technologies of digital music-making trace their design to the practices of material, symbolic, and signal inscriptions, or, in other words, to the design of instruments, notation, and recording technologies. We need to study the genealogy of digital musical media in their acoustic and analogue roots, through historical periods in which these instruments, with affiliated performance, compositional and engineering techniques, emerge, and are established. The book focuses on contemporary digital technologies that issue from the Western musical tradition, and I will trace their lineage through documentations of ancient Greek medieval musical practices, as well as early modern writings, in order to demonstrate how our current musical practices did not drastically change with the digital, but rather trace their design, ideas, and methods to practices that have existed in our culture for millennia. For a concise framing of the problem we are dealing with when understanding the roots of digital systems, I identify three chosen historical periods to serve as an abstraction for general developments in technological infrastructures, conceptual epistemes, production networks, and performative paradigms. They represent epochal shifts, or epistemes, that can be roughly defined as the instrumental paradigms of the nineteenth, twentieth, and twenty-first centuries, here framed as representations of practices with acoustic, electronic, and digital technologies (Magnusson 2009).

In the nineteenth-century episteme of acoustic instruments, we find sophisticated technological objects, such as pianos, violins, and saxophones. The technological context was one of standardisation in instrument production, in conjunction with mass production where instruments, such as the piano, began to be made in factories with production methods that derived from the Fordist ideology of the conveyor belt (see Petersen 2013). This episteme relates to the modernist emphasis on the ocular, where we see the vibrating instrument and we feel its feedback and workings in and through our bodies, which have held the particular instrument for thousands of hours. A unique relationship between the human and the instrument is forged, namely one of care and love. In this production network music is primarily communicated through symbolic representation, as musical scores. The written score is here a file format that can be mass-produced, disseminated, commodified, and repeatedly performed. To listen to music means performing it; there are no phonographic technologies. Virtuosity in the nineteenth century is about the instrument as a prosthesis of the instrumentalist's body, an extension through which inner states can be expressed with technical excellence. Musical automata, such as player pianos, are part of this episteme. For a period (1880s–1920s), the virtuosic become increasingly machinic, but with the development of phonographic technologies the machinic becomes electronic rather than mechanical.

In the electronic episteme of the twentieth century, the instruments are physical and clearly visible, but we cannot inspect their internal workings. There are no visible vibrating objects, no hammers, bows, strings, or membranes that move. However, we touch them, connect them into a sound-producing assemblage, and manipulate their workings through knobs, sliders, and buttons. The twentieth century is the age of the electronic synthesizer, the pickup, the microphone, the amplifier, and the loudspeaker. The physical energy of the performer is amplified via electronic forces. Unlike acoustic instruments, electronic instruments come with manuals and tutorials, perhaps the electronic analogue to nineteenth-century *etudes*. These are instruments that draw into the spectral domain, and musicologists such as Robert Erickson (1975) and Tellef Kvifte (2011) have argued that in twentieth-century music, the focus shifted from musical notes written as black blobs on staff paper to timbral composition of sonic spectra – a shift, in other words, from the symbol to the signal. With the signal technologies of synthesis and projection, a new focus on acoustics also emerges (van Eck 2017). The commercial production network that characterises this episteme is analogue representation, where sound is inscribed on recorded media as waveforms that are isomorphic to the molecular movement in the air of the sound itself. Unlike musical notation, this is a format that does not enable repeated performances, but rather repeated plays of one performance. The idea of virtuosity changes in the twentieth century, often shifting the focus from the instrumentalist to the instrument itself; the instrument "lands on the operating table" (Craenen 2014: 129), as a found object which is to be investigated and explored through skills that derive from a deep understanding of both musical history and instrumental acoustics.

The third episteme is that of digital instruments. These obviously existed in the twentieth century, but they were largely mimetic screen-based versions of acoustic and electric instruments and technologies, remediated in the mouse-keyboard-screen interface combo of the personal computer. It is really at the cusp of this century that we begin to explore the unique qualities of digital technologies for what they are: namely algorithmic, computational, archival, and subject to artificial intelligence. In the digital episteme of the twenty-first century, the innards of the instruments are hidden – including in live coding, which presents a high-level language representation of the control of switching electric gates. In the digital computer, physical energy is transduced via digital mappings. Any bodily gesture can be mapped to any sound and there is no natural paradigm at play that we can relate to. In this context, Jussi Parikka points out how Elsaesser talks about "ruptures" in our epistemological frameworks, in which the digital is portrayed to be beyond the threshold of the perceptible (Parikka 2012: 23). With digital media we operate in a production network that is of computational representation – that is, the work is digitally notated and interpreted by computer processors, either as linear data or code. The processor-based nature of the digital media would allow us to talk about a format of repeated performances, as if the processor is performing a score (and indeed each DSP algorithm interprets the binary data of the CD, or the MP3 file, as it does with code). If we are tempted still to operate with the concept of virtuosity in twenty-first-century musical culture, we need to explore how the role of the musician (equally composers and performers) has now expanded to include instrument design, composition, stage performance, and all kinds of multimedia productions. The skills are manifold and heterogeneous, deriving from fields as diverse as engineering and HCI to sociology and ethnomusicology. Clearly, the pristine forms of virtuosity we find in musical practice, in forms of music such as classical, pop, rock and jazz, have not disappeared with the new additions: this is a process of additive practices.

Unstable objects

The above paragraphs have pointed the lens at our musical *technologies*, but music is a multiplicity of practices that includes music theory, instrument design, composition, publishing, pedagogy, communicating cultures, media, economies, science, and engineering. In past decades researchers in the humanities, as well as the sciences, have increasingly focused on the role of material objects in human and animal culture. The part played by technology is finally acknowledged as an essential extension of the human body and mind, used to achieve various tasks, from lighting fire, carrying water, or calculating, to playing music, conducting business, or studying the nature of atoms. This shift has often been called the "material turn" and is strongly inspired by Heidegger's philosophy of technology. Wider than Heidegger's scope, the material turn ranges from studying how instruments become part of our perception (Ihde 1979), to the role of physical objects in knowledge generation (Baird 2004), from actor-network theory (Law 1991; Latour 2005), and the social construction of technology (Pinch et al. 1987; Pinch and Trocco 2002), to the more recent object-oriented ontology (Harman 2002; Bogost 2012; Morton 2013). Musicologist Emely Dolan suggests that we are witnessing a material turn in musicology as well: "One recent trend seeks to achieve this by collapsing 'music' back into the bodies, instruments, and machines that produce it. From the disciplinary point of view, this can be liberating: material objects, unlike musical works, do not drag two centuries of musicological literature behind them" (Dolan 2015: 88).

In grounded musical practice we have always been obsessed by materiality: by the nature of wood, the feel of the varnish, hide glue, the strings, horse hair, and, more recently, by oscillators, filters, and knobs, as well as sound cards, physical interfaces, and software algorithms. However, the "turn" within musicology in general, and organology in particular, is certainly one that is inspired by Science and Technology Studies (STS), Actor-Network Theory (ANT), Social Construction of Technology (SCOT), and many other disciplines that study the role of technology and interfaces in shaping our work, thought, and social patterns. For Jonathan Sterne, the material turn is fine, "so long as materiality refers to both physical things and the irreducibly relational character of reality" (Sterne 2014: 121). What is at play are multiple dimensions of instrumentality, where instrumental qualities, compositional ideas, performance excellence, and audience reception cannot be reduced to any one thing, constituting as they do a complex network of relations that are developing in tandem as a techno-social evolution. This approach is supported by the understanding that thinking is grounded in social activity and supported by technological infrastructure, ranging from language to sensory extensions. In the words of a Sussex colleague, Kate O'Riordan, objects "orientate people, knowledge and worlds" (O'Riordan 2017: 1); and for Davis Baird, things bear or "express knowledge both through the representational possibilities that materials offer and through the instrumental functions they deploy" (Baird 2004: 131). If our objects serve as instruments of thinking, as epistemic tools, the millennia-old distinction so clearly articulated by Plato between *episteme* (knowledge) and *techne* (skill) breaks down, and we need to investigate how our tools of grammatisation, of externalising our thoughts into systems of discrete elements, play a crucial role in our musical practice.

While objects are instrumental in the analysis of how music technologies manifest in the twenty-first century, we must not forget that they are the results of movement, changes, development, evolution. Objects are "things in the making, mediated, unstable, not quite given, constantly deferred" (O'Riordan 2017: 2). Musical instruments are themselves objects *of* process; unlike paintings on the wall, they are used in the process of creating other art.[3] Since Aristotle, Western thought has emphasised form over process (Oyama 1985), and the analysis of objects in archaeology, anthropology, science studies, etc., tends to focus on the character of objects as they appear at a specific point in time: this has been a synchronic, as opposed to a diachronic, study. This book sets the conditions for an attentive focus on processes and how musical objects evolve over time: how they carry ideas, and the evolution of ideas, through time. We are faced with an ontological dichotomy when studying the evolution of musical instruments: either we focus on the objects themselves, their qualities and use, or train our gaze on the evolution of the instruments, the effect of material change on musical practice, and the development of musical ideas. These are views that could be imputed to the early Greek philosopher Parmenides, who believed that everything was static and that change is an illusion, and Heraclitus, who believed that things were in constant flux. The former focuses on objects, the latter on events. For programmers, it is like the distinction between an object-oriented programming language (like C++, Java, or SmallTalk) versus a functional programming language (like Lisp, Scheme, or Haskell). From the longitudinal perspective of human evolution, a technological object, such as a musical instrument, is but a slow event. The Buddhist principle of *Pratītyasamutpāda*, or dependent arising, explains this well: things only exist as dependencies of other things, which in turn depend on yet others, and all this is contingent on a fleeting moment. Nothing is permanent or substantial. In contemporary Western philosophy, we find these views in Whitehead, Deleuze, Hansen, perhaps also in Gilbert Simondon's concept of *transindividuation*; all thinkers who have influenced the approach taken in this work.

Ergodynamics

What are the conceptual tools available for analysing the transduction process when ideas, techniques, and methods from established instruments and music technologies are translated and implemented in new digital instruments? I use the word "transduction" in the general sense signifying a conversion of energy flow from one systemic form to another. To freeze water is a process of transduction, and so is the function of the analogue-digital converter (ADC). A more nuanced sense of the transduction process in media studies can be found in the work of Simondon (2017) and Adrian Mackenzie (2002), both of whom analyse transduction as a process of transforming constitutive structures. It involves the study of "how things become what they are rather than what they are" (Mackenzie 2002: 16), thus emphasising the processual nature of change, rather than the static ontology of the object as it is subjected to this recasting. We can ask to what degree we are simulating, borrowing, and remediating (Bolter and Grusin 1999), when we implement an older medium within a new medium. Consider an electric guitar VST-plugin or a MIDI saxophone. Our problem is one of language: what name should we give to this transduction of musical instruments? How should we describe the borrowed elements, the translated ideas, the design process, the support for trained motor movements, and finally, the music theory, feel, and playability of the new musical object? These questions require us to take a neo-Renaissance approach, due to the interdisciplinary skills required to work with materialities so complex that any thoroughgoing analysis of them involve physics, engineering, computer science, materials, ergonomics, aesthetics, expression, performance, community, ideation, musicology, and, of course, art.

Stiegler's concepts of *epiphylogenesis* and *tertiary memory* (technological memory) are useful in explaining how technology is constitutive of human thinking (Stiegler 1998), but they are less useful in explaining the transmission process and the evolutionary mechanics of design. For our analysis we need to emphasise the socio-technical appropriation and continuation (passing on) of ideas, techniques, methods, and technologies. Instead of technology as our tertiary memory (the first being genetic and second epigenetic memory), or a repertoire of our culture, I'm interested in a concept that focuses on action and movements, and especially how kinetic memory is supported over a lineage of new technological inventions. The Greek word *ergon* stands for work, task, or function, so we might as well call this *ergogenetic* memory for now – that is, the incorporated memory of how to *use* an object. This is not Stiegler's tertiary memory, which is written into the technological structure itself; we are thinking here of human work-memory (possibly called *ergomnemonics*), which can be trained, transmitted, and translated into the use of new objects. A bone with holes in it isn't a flute if the cave dweller has never heard (of) a flute. The actions affiliated with technological objects are of the objects, but they can be borrowed and applied in other performative contexts.

Therefore, to establish a language for the analysis of how new digital instruments emerge after a process of transduction, we might consider calling the application of work processes from one domain to another *ergomimesis*. We mime and imitate actions and processes of one area and we implement the same patterns in a different one. Intrinsic to the concept of ergomimesis is the fact that any repetition, copying, or translation is a new event in itself, involving noise, errors, misunderstandings, abstractions, and new affordances. This noise in the translation is clearly the source of many creative solutions, adaptations, and innovations. Musical transmission itself, across geographical areas and time periods, is characterised by such misunderstandings or appropriations – choose your preferred term. The field of *ergography* would study how technological design builds on previous actions and processes. This involves classifying key musical gestures (plucking, hitting, fingering, stroking, blowing, etc.) and trace how a particular behaviour, movement, or design trope carries over to new instruments (the Greek term *organon*, for instrument, is etymologically related to *ergon*). This product of transduction or "carrying over" might therefore be called an *ergophor* (on the pattern of metaphor), as it "carries" the trope, the embodied inscribed pattern of motoric memory, over to a new physical object, a new instrument. Finally, in a musical instrument we have infinite dimensions for expression. The instrument has both apparent and latent potential, some of it directly perceivable as interface affordances, others more hidden and discoverable as expressive constraints (Magnusson 2010). From an enactivist perspective, focusing on how we play the instrument and become one with it, the way the instrument becomes part of our "body schema" suggests a view of its potential as an "enactive landscape," or the "set of possibilities that can in principle be brought into being" (Kirsh 2013: 3, 11). There is a spectrum of organological analysis available that ranges from the musical object and its affordances through the enactive and embodied potential of the instrument to aesthetic and musico-theoretical considerations. Jonathan De Souza, for example, talks about coinciding sounding and kinaesthetic patterns (De Souza 2017: 15). This multi-dimensional discovery of an instrument is a dynamic process; it happens in time, and through it we find the object's power and potential (*potens, dynamis*) which, in turn, affects our ideas and embodied skills. We seem also to be lacking a term that signifies the expressive power and depth of an instrument, where we might say "this instrument has an interesting x," as a ludologist might describe a computer game having a "great gameplay."

In Greek, the power or potential of a word can be called *dynamis*, which also signifies ability, skill, and value. I therefore think that we can benefit from the concept of *ergodynamics* when analysing musical instruments. The concept signifies the instrument's potential for expression, what lies in it, its discoverability, mystery and magic. Ergodynamics is not necessarily about directly perceivable qualities (like affordances) or perceived limits (like constraints); it is a general concept that signifies how the instrument can be played, a historical awareness of its play, its material and virtual qualities, and what potential the instrument has for integration in wider musical contexts (which should be read, in the spirit of Smalley (1997), as "musickal" concerns). Ergodynamics is for music technologies (and perhaps any interface) what the concept of "gameplay" is for game developers and ludologists: gameplay is no singular thing, but pertains to the way a game is played, the feel of interaction, the experience of the narrative, and the potential for "zoning in," which is important in the concept of gameplay (Salen and Zimmerman 2003). The conception of ergodynamics derives from the object, yet it is also subjective and cultural. Ergodynamics are not always designed: they are also what we discover and define as of value, and this differs amongst players and musical cultures. Anyone who plays a musical instrument will be familiar with the special moment when a new instrument is picked up and its ergodynamics studied through play.[4] This experience of ergodynamics recognises that an instrument is an object that never rests, or enters a period of stasis: that every time we pick it up there are new things to discover, new patterns our fingers know, because we have changed, the instrument has changed, and so has the world itself – the general performance context.

To take a concrete HCI example of such an ergography, we could look at the *swipe* ergophor. This movement is familiar to us as we turn the pages of a book or a newspaper, or operate with other layered objects, such as a deck of cards. For the HCI designer who wanted to represent stacked information, the swipe is therefore an ergomimetic design implementation of a well-known human action. We could then talk of the ergodynamics of a PDF reader mobile app, as it supports well-known actions from reading books, but it also supports things such as zooming into the text, copying it, highlighting, and so on. With this constellation of concepts that have to do with the way we engage with the functions of objects, how movements and objects relate in design, and the investigative depth afforded by instruments, I am interested in conceiving of instruments and their players as units that are in constant development, each changing the other through a continuous dialogue. A musical instrument is not a dead object, it is a slow event transpiring through time, materialising and metamorphosing through the hands of makers, players, and destroyers (such as fire or worms in the soil). The instrument has a dynamic *ergotype* (*tupos* for "model") that evolves through history and in different musical cultures.[5] Ergographic analysis emphasises the evolution of the object rather than the object itself, but such a shift is not easy, as the object's evolution is not visible in the object itself. Analysing instruments from this perspective, the object is merely a materialised snapshot of what we could call an evolutionary *ergogenesis* (the origin of practical work patterns, an embodied movement and process-focused evolution). Extending Dolan's described material turn and the call for a musicology of instruments, we might also add the *ergographic* view to our musicological tools, or how embodied musical gestures evolve and are designed into new musical instruments. We engage with all of the above topics throughout this book.

Book structure

This book argues that we can trace the ideas, techniques, aesthetics, and methods of our current music technologies far back in time, as supported by historical texts, and that, in multiple ways, emerging musical practices have more in common with those that existed before the advent of printed notation and recording of music. We are entering an age of musical practice that we might define as post-linear or post-recording (implying that, as with postmodernity or postdigital practices, the previous period is surpassed by a new view, a new focus, but not rejected or abandoned – for example, Newtonian physics is still useful after Einstein). From one perspective the move of musical practice to use electronic and digital technologies might seem like a drastic rupture, while from another it appears to be a natural evolution. However, this book looks at technological modes as signifiers for different musical practices, defined by specific eras or epistemes. The function of such a generalisation is to abstract an argument, for the sake of analysis and discussion, in the domain of language; having done that, however, we should return to the dirty world of wood and glue, oscillators and knobs, code and chips, and continue our hybrid technical practices where things connect, plug, destruct at material levels beyond language and analysis.

Sonic Writing is divided into four parts, representing different types of musical inscriptions: the first part – *Material Inscriptions* – engages with musical instruments as musical systems or inventions into which musical ideologies are written. Chapter 1, *Instrumentality*, questions the nature of musical instruments, considering mythical, philosophical, and historical accounts of musical instruments as social agents. Chapter 2, *New Instruments*, looks at how the advent of electricity and digital computers has changed our notions of instruments, how innovation and design have sped up during this period. The third chapter, *Epistemic Tools*, studies the epistemic nature of instrument design; how digital technologies require us to work at a different level as designers than with previous materials, and how this affects both the design and use of instruments. The fourth chapter, *Digital Organology*, asks how we can understand our new musical technologies in the context of older musicological organologies: how do digital instruments fit the evolution of musical tools?

The second part – *Symbolic Inscriptions* – looks at how the evolution of our written practices inform current digital technologies. How does musical notation work in computational media that can analyse, understand, and recreate the musical material? Chapter 5, *Writing Music*, therefore studies notated music and how changes in media technologies have influenced musical practices: the way music is thought, composed, performed, and consumed. The sixth chapter, *Printing Music*, traces how the further evolution of musical notation technologies brought us the work-concept, a particular ontological conception of the musical work, and one that I argue is being surpassed in current musical practices. The seventh chapter, *New Languages*, discussed how twentieth-century developments in avant-garde and experimental music serve as conceptual foundations for our current practices. This is done through studying open works, graphic notations, improvisation, and the rejection of *Werktreue*, or the worshipping of the work-concept. Chapter 8, *Machine Notation*, then asks how computational technologies can find their directions in the world of notated music, how machines can notate, and we look at centuries of practices in which music has been conceived of as results of computational or combinatorial systems.

The third part – *Signal Inscriptions* – is about the mechanical notation of sound. What happens when our machines begin to write sound, and sonic writing shifts from symbols (theories and notes) to signals (grooves, magnetism, digits). The ninth chapter, *Inscribing Sound*, presents a short history of mechanical writing, from early acoustic experiments to the advent of the phonograph. Chapter 10, *Recording*, looks at the advent of phonography proper, and of the advent of the recording studio. In the eleventh chapter, *Analysing*, we explore how the digital computer is able to read and analyse musical signals: what can we find in them, and how can we apply the computer for an improved understanding of our immaterial sonic medium? Chapter 12, *Machine Listening*, continues this train of thought and presents how machine listening opens up new domains for understanding, and recreation, of musical systems.

The fourth part – *Digital Writing* – sums up how the topics of the previous parts manifest within the domain of digital musical technologies. How do we practice instrument making, musical composition, and sound recording with intelligent and networked computational technologies? The thirteenth chapter, *Transductions*, defines digital instruments as models of music theory, as written material entities that outline a musical scope, and yet are strongly dependent on their technological affordances and origin as a technical assemblage. The fourteenth chapter, *New Notations*, looks at how we conceive of musical composition in new technologies. How do we notate for machines and how do we engage with them as partners in collaborative music making? Chapter 15, *Machine Writing*, focuses on machines making music. Computers can clearly be creative on their own, but what kind of creativity is that? And are machines not primarily our instruments, so they will serve us, as always, as tools? In the sixteenth chapter, *Music in Multimedia*, we study what happens when our technologies for music making are the same equipment as that used for photography, film, animation, games, virtual reality, and installations. Why would we limit musical practice to sound only if the distinctions between the art forms are only historically conditioned?

The three chapters in the final part – *A Future of Music Tech, Transformation of Tradition,* and *New Education* – look at what the future might bestow upon those of us who practice, think, and teach music. How are musical traditions changing with new technologies and how does this affect musical education? The point is not, in a Futurist spirit, to destroy old practices – I have no problem with the opera – but rather to set it straight that musical practices in the twenty-first century will be drastically different from those of the nineteenth and twentieth centuries, and this will affect policies in the music industry, education, and cultural funding. Music is a bottom-up activity that emerges in any human culture, but its control, support, and role in society is eventually decided by governments, funders, and legislators – as evidenced by historical accounts ranging from Plato banning certain musical scales to current European governments cracking down on illegal raves – so the question of what music means to us, and how we, the global people of the twenty-first century, can give it the support it needs, is as pertinent as ever.

Part I

Material Inscriptions

1

Instrumentality

What is a musical instrument? We might agree on a basic definition that a musical instrument is an object that produces sound. This would include human-made as well as natural objects. Any found object could be turned into an instrument, so this also includes objects made for other purposes, say wine glasses, sirens, firecrackers, or typewriters. Such a definition supports Wittgenstein's description of how words gain their meaning through their *use* (Wittgenstein 1968), and we can extend that to say that things also gain their function from their use. In Cockney slang, "tea leaf" stands for a "thief" and a "fatboy slim" for a "gym," and likewise a beer bottle can transform into an ashtray, a flute, or a bottleneck slide for a guitar.

In 1933, musicologist Eric von Hornbostel proposed a broad definition of musical instruments: "For the purposes of research everything must count as a musical instrument with which sound can be produced intentionally" (Hornbostel 1933: 129). This is a good definition, wide and inclusive, the only problem being that musical ideas have changed since the 1930s, and now intentionality is not a necessary prerequisite when making music. The music of John Cage from the latter half of the twentieth century corroborates that. We might also question whether musical instruments necessarily need to generate sound. There are conceptual instruments, as Douglas Kahn has shown (Kahn 2014); there are instruments in the form of ideas, as the concept artists of the 1970s can demonstrate; and then there are imaginary instruments, or "fictophones," as curated in Deirdre Loughridge and Thomas Patteson's *Museum of Imaginary Instruments*.[1] Put simply: anything can be a musical instrument if framed as such. In *The Tangible in Music*, Marko Aho provides a definition of musical instruments based on Bielawski's notion of instruments as transformers of movement into sound, here transducing bodily gestures in physical time and space into musical gestures in musical time and space (Aho 2016: 29). Music and movement have always been inseparable, and in some cultures there is no distinction between music and dance: it is one and the same thing. But we need to be careful, because if human movement transduced into sound becomes central in the definition of instrumentality we need to consider less physical modes of performance, say Alvin Lucier's *Music for Solo Performer,* which uses a brain interface to generate the sound (where the only performative movement is brain activity), or live coding performances, where code becomes the body of the instrument, and the primary visible movement, or gesture–sound relationship, are dancing letters on the projected screen. There is clearly body and movement in both of these examples, and our definition needs to encompass such practices too.

Musical organa

Having established that a hammer might be a musical instrument when used for the purpose of generating sound in a musical context, we might then ask wherein lies the difference between an instrument and a tool? Ignoring the etymological roots of the two words (one being Latin and the other Germanic), we can, of course, say that the guitar is a tool for creating sounds, but there is a universal agreement that for nuanced expression we use the term "instrument." Instruments are tools for delicate work or precise perception. They extend our bodily capacity through their power of amplifying and reifying both impression and expression. Scientific instruments are typically designed to extend our senses, for example through the telescope or microscope, or with tools that measure weight, analyse materials, or sense magnetism. Economic instruments involve the fine-tuning of monetary flow, exemplified by changing interest rates, taxes, prices of services and goods, etc. Similarly, musical instruments tune our emotional moods, inspire, and vitalise, as they extend our capacity to express ourselves beyond language, in sound, as material extensions or prostheses of our bodies. Unlike tools that serve as a medium for making a particular task easier, such as hammering a nail (which can be done with a rock or any solid object), musical instruments are more than mere media for the transmission of a message or achieving an end goal: they are an end in themselves. The harp, the bagpipe, or the synthesizer are not there to make certain tasks simpler for us, nor are they substitutes for anything else; they constitute the meaning of their play through their expressive nature.

If instruments are primarily seen as technological extensions of our bodies, an etymological investigation into the origin of the word "instrument" should be helpful. The Latin term *instrumentum* derives from *instruere*, which means to arrange, furnish, equip, or instruct something. To set into place. The instrument is a means, a medium, for action, a tool. The Latin term has the Greek equivalent in *organon*, which signifies equally a bodily organ, a tool, and a musical instrument. Deriving from *ergon* (from which the English word "work" can be traced, and which we explored in the introduction) as "to do" or affect something, the *organon* is a part of the body, and in the form of technical instruments, extensions of the body: thus prosthetic. For Aristotle, in the *Physics*, the organon is that which mediates the motion between the mover (the doer) and the moved (the deed). For the Greeks, musical instruments were *organa*, amongst other tools and implements. The word *organon* was also used in Latin, and Galileo, for example, used the word "organum" for his newly invented telescope. We might, therefore, entertain a definition of musical instruments as any sonic system with which we extend mind and bodies, requiring the practise of refined movements, with the purpose of making music. To complete this etymological survey here, the word "organology" thus means the study of musical instruments. It has been used since the late nineteenth century for this purpose, appearing during a period when colonialists began to bring exotic instruments to Europe, which, mixed with a nascent museum and scientific culture, resulted in the analysis, categorisation, and preservation of musical objects from around the world.

Here is an experiment: close your eyes and think of musical instruments. And again: close your eyes and think of music technology. Having done this experiment with various people in different countries in the past two years, the results tend to be the same everywhere: musical instruments include guitars, violins, clarinets, trumpets, pianos, and percussion. And music technologies include mixers, screens, software, and all kinds of plastic controllers with knobs, sliders, and rubber pads. We can ask ourselves: why it is that the results are so binary and distinctive? The piano is a sophisticated piece of music technology, and a MIDI controller can form part of an expressive musical instrument. The answer is clearly cultural and deeply embedded in practice, but it involves the standardisation of the technological elements of acoustic instruments to serve mass production. Most acoustic instruments gained their final form in the late nineteenth century and have evolved very little since. They are the result of an industrialisation and professionalisation process that also cemented the roles of the composer, conductor, instrumentalist, instrument maker, recording engineer, record label, publisher, concert organisers, venues, technicians, reviewers, media publishers, radio and TV programmes. Related to this formal definition of roles, the evolution of musical instruments stagnated in the nineteenth century, a process Simondon calls "concretisation," or when a technology stabilises and becomes a standard (Simondon 2017). This also relates to how digital music technologies belong to a much larger techno-social change in which large swathes of human activities are transformed to fit work-processes introduced by the digital computer. Another contributing factor would be how, since the 1980s, the word "technology" has practically become a synonym for everything "digital".

Instruments in myth and philosophy

The function and meaning of musical instruments change over time in the diverse musical cultures of the world. Since this book focuses on digital musical technology, there is not much space for ethnomusicological organology, or how instrumental technologies and practices have been borrowed and appropriated across cultures and continents. Digital music technologies are largely tools of Western musical culture; to explain why this is so would be to probe complex reasons that are beyond the scope of this book. The agenda here is to understand what instruments mean to us currently as a culture, where they come from, and how they manifest in practice in the digital age.

As always, it is a good idea to start with the ancient Greeks. A quick rollercoaster narrative would trace instruments from being the extended organs of the gods in Greek mythology, central to poetry, drama, and dance; through serving as minimal support for the voice in medieval chants, and being effectively banned in some factions of the Christian Reformation; to a more central position in the scientific culture of the Enlightenment. It is evident that musical instruments have always been a particular node of power in our cultural perceptions. If we read mythology, ancient history, and philosophy, or study fine art and photography, instruments are omnipresent in all human cultures at all times. They have served as a vehicle of communication with the gods, as tools for spiritual experiences and revelations, or as precise appliances that give expression to complex inner emotions. They are instruments of the soul, technologies in ritualistic activities, at times freeing people, loosening inhibitions and constraints, especially when connected with the food and drink of public festivities, or with mind-altering substances and trance. Musical instruments have, therefore, often been considered dangerous for conservative powers of all sorts, including the Church in medieval times, Protestant sects, and the contemporary Taliban. There are striking similarities in Plato's rejection of Dionysian music in the fourth century BCE and the arguments of governments attempting to regulate and control illegal raves in European cities during the 1990s.

The notion that instruments are powerful tools to express an inner life beyond rational discourse, and possess a dangerous potential to disrupt public harmony, can be traced back to ancient Greek mythology. Here instruments are no mere tools, but are an intrinsic part of the cosmology and the correct order of things. While string instruments, such as the *kithara*, the harp, and the lyre, were precisely tuned to mathematical ratios, representing the heavenly bodies, wind instruments, most famously the *aulos* (a reed pipe wind instrument), were seen as being of the body, of breath, instinct, and animality. In contrast to the rational instruments of inspiration and of cosmic harmony, we find instruments of trance, drunkenness, and the earth. According to myth, the aulos was invented by the goddess Athena from a bone, and she played it at one of the gods' banquets. After she had been told that her face got distorted whilst playing it, she threw it away and placed upon it a curse that would inflict punishment upon whoever would play it. The satyr Marsyas found the flute, and the instrument began playing Athena's music by itself. Marsyas practised his instrument obsessively and eventually became considered a finer musician than Apollo, the god of music, who played the lyre. Upon hearing rumours of Marsyas's skill, Apollo challenged him to a musical duel before a jury of the Muses, in which Marsyas lost; Marsyas was subsequently flayed alive as a warning to mortals not to challenge the gods. The nymphs, satyrs, and shepherds cried, and a river formed from their tears. A related myth tells of a similar joust between Apollo and the god Pan, a performer of wind instruments. Pan had horns, a goat's beard, and hooves; he was so ugly that even his mother abandoned him at birth. Pan grew up with Dionysus (also known as Bacchus) and also competed with Apollo in the sport of music, but he was more fortunate than Marsyas, as his life was spared. The tension between the rational harp (or *kithara*) and the irrational flute (or *aulos*) is also expressed in the myth of Orpheus, who was the principal musician of the Greek myths. Orpheus played a lyre given to him by Apollo, and he studied its art with the Muses. Orpheus was a poet, of lyrical music. However, during a war fought by the aulos-playing Dionysus, Orpheus was killed, and Apollo installed his lyre as the Lyre star constellation. In general, Greek mythology frames the aulos as an instrument of trance, imbued with earthly, sexual connotations, ritual madness, and ecstasy; it was a symbol of fertility, birth, and renewal.

As an early philosopher, with one foot in mythology and the other in philosophy, Plato was adamant in maintaining the harmony of the state with appropriate music that reflected the order, harmony, and goodness of the world (and these ideas were considerably more real to him than to us in the twenty-first century). To maintain harmony, Plato was against innovation in music as well as certain musical instruments: he preferred "Apollo and the instruments of Apollo to Marsyas and his instruments" (Republic 399e). Aristotle was equally sceptical of the aulos, as it distorted the face of the performer (a reference to the myth of Athena), but also because it was both violently exciting and affective. For Aristotle, the problem with using the flute in an educational setting was that the performer is not able to speak when playing it, and is thus of no use in instruction or singing. Indeed, it is beguiling to consider how the flute was rejected in the ancient Greek context because it removes the power of language from the organ of rationality, the mouth, occupying it instead with a physical extension that projects abstract sounds through breath. If breath signifies the soul (see the etymology of the Latin *spiritus*, or the Greek *psyche*) the flute has a very attuned relation to the inner life of the player, as opposed to the rational execution of plucking a string on the harp, an instrument played at a greater distance to the body, and we lose control of the sound after plucking the strings.[2] The sceptical attitude towards the aulos can also be explained from a philosophical perspective, as the early Greeks spent much energy arguing about their systems of scales and modes. For example, in Book III of the *Republic*, Plato banned all musical modes except the Dorian and the Phrygian. Whilst the lyre had fixed string tunings, the aulos, which often had up to fifteen holes, could be retuned in the middle of a performance by sliding metal rings up or down over unused finger holes, thus sliding from one mode to another and shaking the foundations of what was considered the mimesis of pure cosmic harmony and theoretical structure (Bélis 2001). Aristotle was also a keen theorist of music and generally recommended its practice, but with the caveat that it was not to be pursued to the level of professionalism: "professional musicians we speak of as vulgar people, and indeed we think it not manly to perform music, except when drunk or for fun" (*Politics* 1339b).

Instruments in medieval thought

The tense relationship with instruments and music, evident in Greek culture, continued throughout the Middle Ages. The early Church Fathers were influenced by Neoplatonic views and only adopted a positive attitude towards music when it was considered an aid to salvation. Secular music was judged to be dangerous, with too much potential for evil, debauchery, and diversion from the path of God. Unlike the Greeks, whose culture was more inclusive, diverse, and tolerant of Dionysian rituals, the Church had a more totalitarian approach to music in that it largely rejected instruments in favour of the voice. Medieval theorists maintained the theory, attributed to Pythagoras, that music was a physical expression of the world's harmony: it was the *logos* materialised. A musical scale was a representation of the ratios between the planets orbiting the Earth. This ancient Greek worldview survived, with some variation on the themes, down into medieval Europe, but it is only with the Renaissance that musical instruments begin to develop and take on standardised forms, and gradually losing their mythological and religious reference for a more open and malleable relationship to the world with the Enlightenment. It should be noted that during the Middle Ages, when Greek philosophy and scholarship was largely preserved and continued by Arabic scholars, the Islamic attitude towards music was more liberal and scientific than that of the Church. And unlike ancient Greek and Neoplatonic writings, the Arabic scholars emphasised the practice of music through education and training. We find, for example, a considerable portion of Al-Farabi's *Kitab l-Musiqi al Kabir* (Great Book on Music) dedicated to the *practice* of music, in contradistinction to music as a primarily theoretical discipline.

Although documented medieval music is primarily focused on religious chanting, various medieval writers do comment on instrumental music, for example Boethius, in his *Fundamentals of Music (De Institutione Musica)*, from 520 CE. Basing his work on the earlier musical writings of Nichomachus of Gerasa (60–120 CE) and Ptolemy (100–170 CE), Boethius faithfully transmitted the music theory of Greek antiquity into the Middle Ages. The *Fundamentals of Music* was the key work of the Middle Ages on music theory, a common source for other medieval writing on music and harmony. In it, Boethius does not focus on the *practice* of music; his work is on *theory*, largely based on the theory of Pythagoras and grounded in Neoplatonic metaphysics. Music was seen as a reflection of cosmic harmony, and it was therefore vital that the instruments were correctly tuned and the scales well defined. Musical notation or composition was not considered important and Boethius does not deal with those as such. In *Musica Naturalis*, Jeserich quotes Roger Dragonetti on the peculiar nature of medieval musicology: "Musical metaphysics teaches us not to reflect on music but rather to reflect in accord with it" (Jeserich 2013: 22). The goal with musical performance is to re-enact music as a "transcendent order" of the world. Music and its instruments were a physical manifestation of the mathematical harmony of the world. It is no coincidence, then, that Pythagoras's monochord is called *canon* (κανών) in ancient Greek – appearing later as the Arabic instrument *qanun* – a word that also means "rule." The instrument embodies its rule.[3]

The Greek notion of the *music of the spheres* lasted through the Middle Ages, galvanised as it was by Boethius's tripartite definition of music: "There are three kinds: the first, the music of the universe; the second, human music; the third, instrumental music, as that of the cithara or the tibia or the other instruments which serve for melody" (Boethius in Strunk 1998b: 30). Here we find a hierarchical division into *musica mundana* (of the harmony of the universe), *musica humana* (harmony of the body and soul), and *musica instrumentalis* (instrumental performance), a division that became the generally accepted theory of music for centuries. The music of the universe involved the movement of the heavenly bodies, the seasons, and the elements, one that was beyond the perception of the human ear, but represented the world in its symmetrical beauty. Human music was the music of the voice, inspiring poets and singers; an idea well depicted throughout history, exemplified by how medieval paintings of Gregory the Great often depict him with the Holy Spirit in the form of a dove, singing into his ear. Boethius then says of instrumental music: "The third kind of music is that which is described as *residing in certain instruments*. [*Musica . . . quae in quibusdam consistit instrumentis*] This is produced by tension, as in strings, or by blowing, as in the tibia or those instruments activated by water [e.g., water organs], or by some kind of percussion, as instruments where one beats upon a bronze concavity; by such means various sounds are produced" (Boethius in Strunk 1998b: 31, my italics). For Boethius, music "resides" in instruments, as if they contain a musical agency that we encounter when playing them. Boethius subsequently defines three classes of musicians: at the bottom are the instrumentalists, mere servants to their instruments and skilled craftsmen. Singers and poets are above the instrumentalists, as they are concerned with their poetry, and yet, following Plato's explanation of the art of the poets, they do not really know what they are doing, as their art comes from inspiration rather than true knowledge. The third class is the true musician (*musicus*), the learned theorist who can judge the work of instruments and songs, applying theoretical musical knowledge that relates to the other quadrivium subjects of arithmetic, astronomy, and geometry. The musicus did not necessarily play music.

A few centuries after Boethius, in Odo of Cluny's ninth-century dialogue on theoretical matters between a disciple (D) and a master (M), we find an interesting account of the instrument as a solid reference, as a memory inscribed, and something more reliable than human memory or vocal skills:

(D) How can it be true that a string teaches better than a man?
(M) A man sings as he will or can, but the string is divided with such art by very learned men, using the aforesaid letters, that if it is diligently observed and considered, it cannot mislead. (Strunk 1998b: 91)

They discuss and study further, and after the master has taught his student instrument making, tuning, and scales, and explained pitch theory in some detail, the disciple says of his instrument:

(D) Indeed, I must say that you have given me a wonderful master, who, made by me, teaches me, and teaching me, knows nothing by himself. Indeed, I fervently embrace him for his patience and obedience; he will sing to me whenever I wish, and he will never torment me with blows or abuse when provoked by the slowness of my sense.
(M) He is a good master, but he demands a diligent listener. (Strunk 1998b: 94)

In the ancient Greek and medieval worldview, the world was perfect in its mathematical harmony, and musical instruments (especially string instruments) embodied this theory. Boethius's view of the cosmos resonated down throughout the centuries, for example, in Regino of Prüm's early tenth-century *Epistola de armonica institutione* there is a division of music into *musica naturalis* and *musica artificialis*, the latter using "artificial instruments of sound." Regino writes: "Thus, natural music is that which produces sound with no musical instrument, no touch of the fingers, no human striking or touch, but breathed upon by God" (Jeserich 2013: 178). As a result, instruments produce artificial music: "That music is called artificial [*artificialis musica*] that was conceived and invented by human art and ingenuity, and is based in certain instruments [*in quibusdam consistit instrumentis*]" (Jeserich 2013: 179). Note how the music of the voice can be inspired by God, in consequence of the logic of language, while instrumental music derives from the instrument itself. Centuries after Boethius we still harbour the same scepticism towards instruments, a scepticism we find throughout Western history in relation to technology in general, and that derives from the Platonic dichotomy between *episteme* and *techne*. This is a common theme throughout Western musical history: in medieval times the human voice was raised up against musical instruments; in the eighteenth and nineteenth centuries we see a division in musical performance with a "soul" on the one hand and that of player pianos and musical automata on the other; with twentieth-century synthesizers we find criticism of artificial sound and the lack of personality, as well as how recording technologies kill the music; and with contemporary AI people abhor the idea of machine creativity. Such views are clearly culturally conditioned, and we might take Japan as an example of a culture that has a completely different relationship to technology, material agency, and artificial intelligence. Some have attributed the ease with robots in Japanese culture to the fact that in Shinto animism all objects have a spirit (Kitano 2007).[4]

Instruments of modernity

Medieval instruments were often made with anthropomorphic or zoomorphic designs, with faces, bodies, and limbs of humans or animals, possibly indicating the agency of the instrument and its bespoke character and soul. Foucault argues that symbolic meaning in the Renaissance episteme was characterised by resemblance, or rather mimesis. Eliot Bates, in his research on the social life of musical instruments, describes how some musical instruments gain agency through their anthropocentric signification while others are attributed less (Bates 2012). However, if there was one instrument that epitomises the music of the Middle Ages through its central position, it would have to be the organ; first as a hydraulic organ (with water regulating the air supply to the pipes), and then as a tubular-pneumatic organ. Practically every medieval church had an organ, and there was a robust industry in making and maintaining organs. By the end of the Middle Ages, musical practice diversified, and so did ownership of instruments. In the fourteenth and fifteenth centuries the functions of music proliferated into more diverse parts of society, and the ownership and mastery of musical instruments became more common, which in turn boosted the craftsmanship and trade of instrument making. Artisans in all types of domains of practice increasingly explored mathematical relationships in art, patterns, decorations. Marquetry in musical instruments, as well as interest in mathematical forms, resulted in developments of instrument shapes, sound hole shapes, and tuning techniques. The art of the *intarsiatori* emphasised geometric ornamentation, which in turn influenced instrument design. In the holistic style derived from Renaissance thought, instrumental shapes, such as that of the violin and the cello, took form and inspiration from mathematics, buildings, and nature, with design seeking pure forms wherever possible. More often than not the evolution of an instrument's form was not because of sound quality or ergonomic concerns, but rather as a consequence of ornamental purity.

The European Renaissance brought with it a new concept of the world, namely that of humanism, where the power of the Church dwindled and its dogmas were reinterpreted by Protestant thinkers. Along with renewed knowledge of ancient Greek texts – often translated from Arabic – and the arts in general (painting, architecture, skill, printing), the sixteenth century was a prolific period for instrument makers. The economy had improved and the aristocracy increasingly began collecting musical instruments, which were seen as symbols of wealth and education. However, this increased interest in musical instruments largely focused on plucked string instruments, such as the lute, the lyre, the harp, and later the harpsichord. We find this in the poetry and the painting of the time, which also illustrate how instruments began to shed their anthropomorphic features (e.g., the disappearance of human faces and other bodily references on the instrument bodies). Musical instrumentation and instrument production of the period marginalised flutes, reed instruments, trumpets, and percussive instruments, which can be seen as a prolongation of the Platonic divide between the Apollonian and Dionysian instruments of the lyre and the aulos respectively. The tension between ecclesial and instrumental or folk music continued under Protestantism, and in some denominations and sects, there were real problems with the prohibition and banishment of music and musical instruments, for example with Calvin's banning of instrumental music in church practice.

Boethius's theory of music, in which music is divided into the music of the world, human music, and the music of instruments, was later adopted by Athanasius Kircher, whose 1650 *Musurgia Universalis* became a key musicological treatise, influencing composers such as Bach and Beethoven (Devlin 2002). In her book *Curious and Modern Inventions*, Rebecca Cypess (2016) presents what she terms a "paradox of instrumentality" in the seventeenth century, namely how instruments were placed as an opposition in the tension between material instrumentality and mechanical artifice, on the one hand, and internal emotions or ephemeral *affetti* (passion), on the other. In the seventeenth century, the fine art iconography of the *vanitas*, a category of symbolic art, emphasised the worldly, ephemeral, and fleeting nature of human experience. A type of moralist painting, the symbols used to illustrate the terrestrial and mortal sphere were hour glasses, skulls, and musical instruments, often accompanied by empty wine glasses, and burnt oil or wax candles. Here, the ancient symbolism of reed wind instruments versus stringed harps was also applied. The tension between string and wind instruments can also be found in Shakespeare's *Othello*, where the rough and wild sound of the bagpipe is played against the sophisticated quality of stringed instruments. In the art of the time, the harp is often depicted as belonging to the heavens; indeed, it is the chosen instrument of the angels, while the flute is of the earth and often depicted as broken, impure, and played as part of hedonistic scenes. In Raphael's painting of St Cecilia, the patron saint of music, the saint is portrayed with broken instruments lying on the ground, and although she is holding a small pipe organ, she is looking upwards, into the

Figure 1.1 The Ecstasy of St Cecilia – with Sts Mary Magdalene, Augustine, John Evangelist, and Paul. Print by Philippe Thomassin, in 1617, after Raphael. Engraving. © The Trustees of the British Museum.

clouds where angels sing from books, promoting the view that music is divine, godly, and primarily vocal.

In the Renaissance, the focus shifted from music as belonging to the cosmos to music as having its place and provenance in the human domain – in other words, a shift from the realm of God to that of human musicians and their instruments. Indeed, Kircher based his work on the writings of Gioseffo Zarlino, a sixteenth-century composer, instrument maker, and musicologist who designed an influential and comprehensive classificatory scheme of musical instruments. Zarlino divided instruments into natural and artificial categories, depending on whether they related to the movement of the heavenly bodies, but also into *instromenti mobili* (for variable pitches, such as violin or trombone) or *instromenti stabili* (for fixed pitches, such as

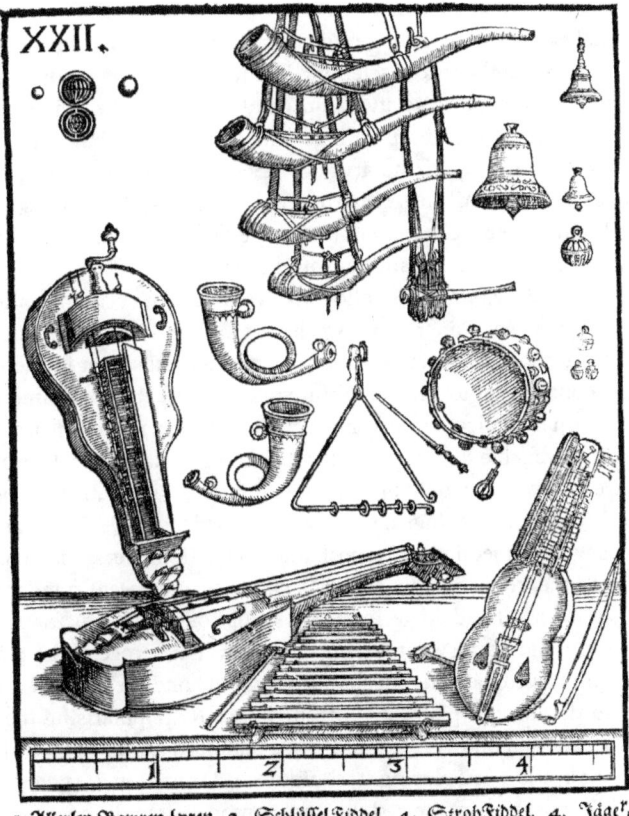

Figure 1.2 An organological illustration from Michael Praetorius's book *De Organographia*, included in his *Syntagma musicum* from 1691.

harp or a flute). However, it was with Michael Praetorius's *De Organographia* of 1619, a work dedicated to musicians and instrument makers, that we find the first modern systematic organological approach (Restle 2008: 259). Praetorius's book was beautifully illustrated, containing a division of instruments divided into wind, strings, and percussion. The book was very influential with composers and performers of the period, and its popularity was not superseded until Hector Berlioz published his *Treatise of Instrumentation* in 1855.

The scientific revolution of the seventeenth century, and the European Enlightenment that followed, brought about new methods of studying and understanding the world: a natural philosophy that applied scientific instruments to measure, weigh, and observe natural phenomena. A new notion of instruments as epistemic tools emerged, both in terms of scientific and musical instruments. Instruments were understood to equally contain and reveal theory, demonstrating a particular perspective or understanding of the world. Indeed, the modern distinction between science and music would have been much less pronounced in the world view of a seventeenth-century practitioner of experimental science. In the scientific revolution of early modernity, musical instruments were integral parts of scientific laboratories, where studies in harmonics, acoustics, waves, mathematical relationships, and sound and hearing in general were conducted. There are numerous paintings and illustrations that depict the role of musical instruments as part of the array of scientific instruments, for example *The Ambassadors* by Holbein. All these instruments, such as scales, prisms, monochords, and camera obscuras, were seen as tools that helped us to think and sense the world, to augment and hone our cognitive and sensorial capacities.

These cultural developments following changes in economy, science, and technology, together with increased wealth resulting from colonialist practices, improved craftsmanship, material techniques, and mechanisation in the instruments. These changes contributed to the establishment of a market of instruments, resulting in mass production in dedicated workshops, which, in turn, made instruments cheaper and more readily available to a larger portion of the population. The wealthier classes had more spare time, which would often be spent on musical practise and performance. Instruments became technologically more sophisticated during the nineteenth century: oboes were embedded with mechanical keys (a universal standard had been developed, called the "conservatoire model"); clarinets underwent standardisation and were kitted out with refined fingerings; the saxophone came out of Adolphe Sax's clarinet workshop and underwent a complex innovation process; the piano became equipped with improved hammer mechanisms and the one-piece metal frame became a standard; and the upright piano was invented, greatly popularising the instrument (Campbell et al. 2004). Following the increased standardisation of instrument design came a standardisation in playing technique and notational syntax. Performers became less idiosyncratic in their style, and composers could count on how their pieces were interpreted. The standardisation of interpretation also influenced the establishment of the orchestra, with the emergent idea of the orchestra as an instrument played by the conductor. This idea was expressed by Berlioz in his 1855 *Treatise of Instrumentation* (Berlioz 1948), in which the conductor should "transmit his feelings to the players" and "his emotion communicates itself to those whom he conducts" (Berlioz 1948: 410).

> The performers of all sorts, constituting together the orchestra, are, so to speak, its strings, tubes, pipes, sounding boards machines endowed with intelligence, but subject to the action of an immense keyboard played by the conductor under the direction of the composer. (Berlioz 1948: 406)

There is a certain musical ontology at work in these lines that characterises Romanticism, which becomes universally accepted in modern music theory; a metaphysic which states that the ideas of the musical work are to be channelled through virtuosic composers, conductors, and instrumentalists. In the following chapters we will investigate how new musical practices deconstruct this ontology by means of their technological nature and systems design – and how this coincides with different conceptions of musical performance. We will question the Romantic notion of the composer, the instrument maker, and the performer, which has lingered in our culture right up to the twenty-first century. This is largely explainable by the fact that most families of musical instruments gained a stabilisation or equilibrium, in terms of technical development, during the nineteenth century, and only the finer evolution of their technical body has happened since. Most people are familiar with the romantic, perhaps even fetishist yearning for older instruments, and many experts proclaim that violin craftsmanship has never surpassed that of Stradivari at the end of the seventeenth century or Guarneri at the beginning of the eighteenth. This is a somewhat problematic account, since as early as 1817 there have been blind listening tests conducted between these instruments and newer instruments (Beament 1997) and in some of these tests, the newer violins have been preferred over the Stradivari (Fritz et al. 2012). Furthermore, and contrary to what many believe due to the strong standardisation of the violin, its development has not stopped, as we see in the work of Carleen Hutchins in the middle of the twentieth century, who attempted a standardisation by creating a family of instruments she called the "violin octet," or, for example, in the work of luthier Hans Jóhannsson (see www.hansjohannsson.com) and his collaborators Signal Wizard Systems on the *Othar* violin.[5]

Instrumental materialities

In order to understand the music of our time, we need to engage with musical technologies and explore the material objects that underpin its practice. Here, the methods of designing, making, and using the instruments are important, as well as their relationship with our musical ideas: who makes the instruments and how do we compose for them? How do we play, collaborate, teach, and learn? This a particularly complex set of questions as the nature of musical instruments has transformed many times in the past three centuries. In the introduction to this book, a rough generalisation was made between the epochs or epistemes of acoustic, electronic, and digital instruments. These should not be conceived of as clear-cut exclusive categories: the electric guitar is both acoustic and electronic, modular synths often contain digital chips, and many digital instruments involve acoustic elements. These periods function rather as particular abstractions to generalise the stages of musical expression. During the nineteenth century, acoustic instruments reached their perfection, became saturated, and gained standardised and streamlined production processes. The widespread conformity in the design and production of instruments was combined with standardisation in musical notation, and also in music as part of cultural and economic activities. We also see the appearance of new instruments like the saxophone. The twentieth century was characterised first by the mechanisation of acoustic instruments, such as automata, player pianos, pianolas, and music boxes (see Patteson 2016), and then by the advent of electronic instruments, such as theremins, sirens, and synthesizers. Many of these were emulated in digital technologies later in the century, but those tended to be *simulations* of the acoustic and the electronic – that is, remediations of the known into the digital medium. Considering that real-time audio was practically impossible in personal computers until the final years of the twentieth century, we might declare the twenty-first century as the century of the digital instrument. With the availability of cheap microchips, sensor interfaces, and user-friendly programming environments, it is really at the beginning of the twenty-first century that we start to see musical instruments that go beyond a remediation of known instrumental patterns, where the properties of the digital are approached and explored in their own right. It is no coincidence then, that the NIME (New Interfaces for Musical Expression – see www.nime.org) conference, which focuses on new technologies for musical expression, began in 2001 as a workshop at the CHI (Human Factors in Computing Systems conference), emerging as an independent conference the following year.

Indeed, in a recent NIME keynote, Kvifte (2011) pointed out how the production of music changed from the nineteenth-century focus on discrete instrumental notes, played from a standard form of musical notation and manipulated as symbols on a page, to the twentieth-century emphasis on timbre. This could be exemplified by pointing to spectralist music (such as Grisey, Murail, Saariaho, and Harvey), drone music (Radigue, Young, Oliveros, Niblock, Conrad), noise (Russolo, Reed, Einstürzende Neubauten, Merzbow), and so on. The instruments created in the twentieth century were embedded with a much stronger and complex control of sonic parameters, and even if discrete note producing interfaces were attached to them (such as the keyboard on the Moog modular synthesizer), the fascination with exploring the timbre of the instruments often overwrote note-based compositions. Even composers writing in the traditional musical notational language for traditional instruments increasingly began thinking of their compositions in terms of timbre and texture (as opposed to the previous emphasis on melody and harmony), as explored by Erickson (1975). Leigh Landy has characterised this as a move from note-based to sound-based music (Landy 2009). If the use of digital computers in the twenty-first century is beginning to express its unique character, it would be through their algorithmic nature, the ease of automation, dynamic mappability, and not least, the profound potential of machine learning, as substantiated by emerging deep learning technologies.

Conclusion

This chapter has moved swiftly from mythology through philosophy and medieval thought, to the techno-sociology of musical instruments. This provides a necessary background to what will later be presented as epistemic tools, instruments of knowledge and agency, and how current ideas of music, the musical work, and musical instruments are becoming reminiscent of earlier musical practices. We have seen how these core technologies of human expression – musical instruments – equally serve as prosthetics of the human body, and as tools for understanding the world, both the objective physical world, and the subjective world of the mind, if such a rough dichotomy makes sense. A discourse on instruments in early mythology, philosophy, and politics was presented, pointing at how instruments have always been social agents, at times creating a frenzy and ecstasy, playing on their own, changing people's shapes and appearances, silencing or inspiring them, while at other times reflecting theory, mathematics, and cosmic harmony. The journey from ancient Greek and medieval texts is important for a key argument of this book – namely, that instruments contain music, theory, and culture. In this section, I have sought to illustrate that such a conception of the intelligence embedded in technological artefacts is not a new idea, but is one that might have temporarily subsided under the weight of the Romantic genius, where the instrument began to be conceived as a neutral medium for virtuosic expression.

2

New Instruments

What does the digital afford to new musical practices? Without diving into an in-depth analysis of "the digital," a topic of considerable complexity, our aim is to study how digital technologies alter the technological conditions of music production, demanding or enabling us (depending on viewpoint) to operate differently in practically all areas of musical practice: from making instruments, composing, and performing to marketing or listening. While new digital musical instruments (DMIs) have become omnipresent in today's music, the focus here is on the *unique qualities* of the digital, as opposed to the digital *simulation* of previous technologies. As an example, many fruitful insights can be gained by analysing the difference between an acoustic, electric, or digital piano, and yet, for the argument put forward here, it is of lesser importance whether Elton John or Herbie Hancock play an acoustic, electric, or digital piano on stage: their playing is what it is, and the nature of the sound generation is irrelevant in this context. Similarly, music recording and production software in the form of digital audio workstations (DAWs) is designed as a simulation of the traditional tape studio, including all its outboard gear, with additional design tropes from the musical score and the piano roll. When the term "digital instruments" is used in this book, I am thinking primarily of computational instruments; digital media are computational media. However, "computational instruments" is an awkward term that perhaps also overly emphasises the computational. While "digital" is not a perfect denominator, it is probably the least bad word. It also references the fingers (digits), the human hand, and thus brings with it the connotation of embodiment, which the computational does not have. In other words, for the argument of this book, it is not of key importance whether a signal is acoustic, analogue, or digital (other work looks into that); what is interesting for us are the affordances, expressive scope, and theoretical potential offered by the instruments – their ergodynamics. Therefore, it is not so riveting for us to study whether someone plays an acoustic or digital piano if "playing the piano" is all they are doing. However, it becomes interesting if suddenly the digital piano responds, changes tuning, morphs between sounds, suggests future paths, or provides an accompaniment to what is played. That is where the digital exhibits its nature as being computational.

As any ethnographer, ethnomethdologist (or better "technomethodologist" – see Button and Dourish 1996), or participatory design researcher would point out, a direct translation of technologies from the analogue to the digital domain is never possible. By moving practices from the real world to the computer, objects, relationships, and work processes are abstracted, quantified, classified, and arranged into an ontology that supports the operational principles of the software. The DAW is a *representation* of the studio, and through subscribing to the work practices laid out by software designers (e.g., their decisions on what appears at the first surface level of the interface versus its back layers), new practices will emerge and others will disappear. What interests me is when the DAW goes beyond its simulative functionality and begins to operate in ways that are uniquely digital, for example by new signal processing techniques, innovations in interface and interaction design, and new AI that enables the software to learn, adopt, suggest, and generally form a dialogue with the musician, where the software becomes more of a partner than a tool. Technical transitions such as those that move from the acoustic, to the electronic, to the digital are transformative *transductive* processes that do not simply change the underlying media functioning: they transform our musical ideas and practices too (Mackenzie 2002).

Instruments and alien objects

In his book *Alien Phenomenology*, Ian Bogost (2012) asks how the world is perceived by things like a bat, a hookah, or a cantaloupe. This is proposed as a new methodology to understand the thing in its context, an object-oriented ontology (Harman 2002). Such an exercise in non-human phenomenology is, of course, impossible, but it is an interesting proposal nevertheless. Let us accept Bogost's challenge and perform an alien phenomenology on musical instruments: acoustic, electronic, and digital. Imagine being an acoustic instrument. You lie on your back in the dark, hoping to be picked up and touched with trained hands that pluck, stroke, and activate your body, exciting sounds from your primary organs. You can feel the room resonate as a result of your own movement and that, in turn, affects how you behave. You feel connected to the room and the performer, as if you, the performer, and the room were one system. You realise how every pair of hands is unique and how every piece of music makes you feel different. But you live a complete existence: you are always there, waiting to be activated, but unlike other artefacts of art, such as an oil painting, it is through human manipulation that you gain a complete existence, pregnant with meaning and function.

As an electronic instrument, you typically lie dormant – half dead – perhaps never to become fully alive again, because without electricity feeding your body, you are nothing. With electrons flowing through your wire-veins, you become functional even if there is no sound. Unlike acoustic instruments that need to be activated for every sound, you fulfil your nature with the injection of electricity, even if the volume is down or there are no speakers in the room. This orgasm of electricity is tantamount to a junkie getting a fix, and when you are up and working, you perform perfectly. The hands of your performer do not really play you, they control your functionality – which keys to press, knobs to turn, and sliders to move; which jacks to plug into which sockets. The player's physical energy is not proportional to the sonic energy you express. This sound is generated by electrons and output as a voltage-current boosted by an amplifier that drives the speaker cone movement – in and out. Analogous to the movement of a tuning fork – in and out. The difference is that the sound does not come from you; it emerges from another location in the room, but there is a clear trace and translation of your movements to the movements of the speaker.

Incarnated as a digital instrument, you might seem of the same nature as the electronic instrument, say a synthesizer, but there is a big difference. All your behaviour can be redefined using a language of algorithms that can be written and rewritten to change your nature. Indeed, you might not feel that you have a nature as such, as a new software upgrade might change your behaviour so completely that it does not remind you of anything you've done before. You turn schizophrenic, polyfunctional, and meta-dimensional. The coupling between the user's touch and what you output depends on the program applied at the particular moment, so you don't really know your user's touch. Unlike the acoustic instrument that knows its user very well, your user behaves differently every time you are played, and that is possibly because you are never the same either! You like being mysterious, conversational, definable, and yet you direct the user in what is possible. What makes you really excited is when you are given the opportunity to learn about your user and establish a relationship. You memorise what has been played, you analyse their performance, and you can respond, suggest, adapt, reject, serve, or tease as you like. This is where you find meaning in your existence.

This playful thought experiment in the alien phenomenology of musical instruments could, of course, be much longer and more detailed, involving all the actors ranging from the instrument maker to the listener, but suffice it to mention that I have previously analysed the qualities of the three types of instruments – acoustic, electronic, digital – in more detail in a journal article (Magnusson 2009). In that text also I exaggerated, for the sake of argumentation, the differences over similarities and continuities, in order to tease out what the ergodynamic character of each type of instrument holds, and acknowledged as much in the conclusion: "This paper has focused on differences at the cost of similarities, and divided into distinct groups phenomena that are best placed on a continuum" (Magnusson 2009: 175). The second-person accounts above are clearly written in jest, but there may be grains of truth in there that relate to the distinct ontological conditions of each instrumental type and the reader is encouraged to meditate further, applying Bogost's methodology, by conducting an in-depth imaginative exploration of how it feels to be acoustic, electronic, or digital.[1]

Interfacing sound

The most obvious difference between the acoustic instrument and its electric and digital counterparts involves the concept of the *interface*, a topic that is of key importance in the design and critique of digital technologies (Galloway 2012; Andersen and Pold 2018). Electronic instruments have designed interfaces that connect to the black box of their functionality. The instrument designer has decided to "expose" certain sound parameters through user control; others lie fixed in the darkness of the box. In general, we can state that the electronic or digital instrument *has* an interface, whereas the acoustic instrument *is* the interface. The term "interface" is not used much

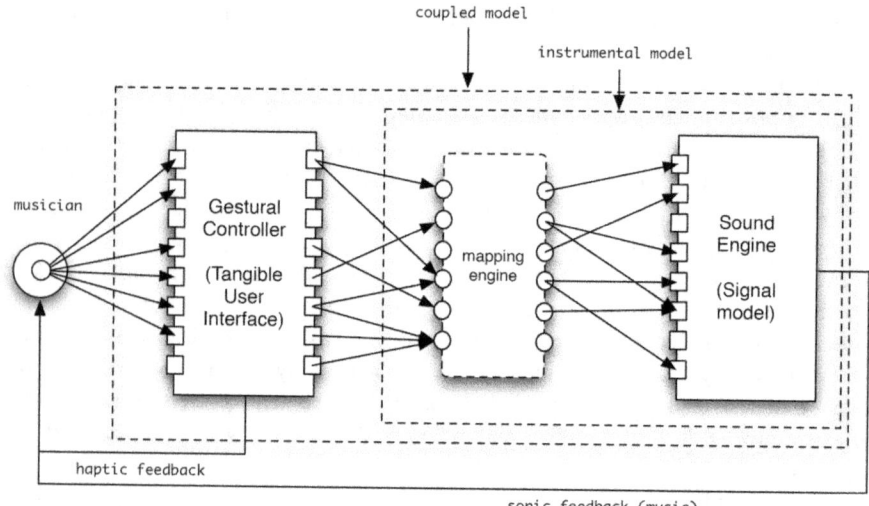

Figure 2.1 A typical explanatory model of an electronic musical instrument (see Wanderley 2000; Leman 2008; Wessel and Wright 2001). New musical instruments typically consist of these three elements. © Thor Magnusson.

in the musical education involving acoustic instruments. Most of Hugh Davis's electro-acoustic musical instruments (Mooney 2017), for example, do not *have* interfaces: they are their sound source, even if amplified and processed electronically. This is because there is nothing that is out of our control when playing acoustic instruments, nothing that does not respond to the energy we put into the instrument, albeit with some exceptions, such as the church organ. This applies to many electronic instruments too. The typical instrumental model, explaining how interface relates to the innards of digital instruments, is often presented with diagrams as in Figure 2.1.

We notice how the instrument in its totality can be seen as a designed and direct coupling between the physical interface, the mapping engine, and the sound engine. Any of these could be swapped out for another design at any point, and yet it would be considered the same instrument.[2] This has been termed the "mapping problem" in the NIME literature (see, for example, Maes et al. 2010), one that is crucial to our understanding of digital instruments, and necessarily belongs to the domain of HCI (Human–Computer Interaction), where instrument designers have attempted to come up with principles (Cook 2001) or criteria (Fels 2004) for the design and evaluation (O'Modhrain 2011) of new musical instruments.

Again, what does the digital bring to the domain of instrument design? Which features and practices are carried over (via the process of ergomimesis) and which are left behind? The long history of digital instruments extends over half a century, but we might take Michel Waisvisz's instrument, "The Hands," as an example of a well-known instrument with a body of musical work written and performed. The Hands were developed in 1984 at STEIM (Studio for Electro-Instrumental Music) in Amsterdam,[3] just after the MIDI protocol specification had been released, and implemented in the Yamaha DX7 synthesizer. Waisvisz and colleagues built a controller that would afford gestural hand movements, moving about on the stage, and dance. The controller had ultrasound sensors (sensing the distance between hands), buttons, switches, bend-sensors, and accelerometers. The controller itself is not an instrument; it is only when coupled to a sound engine that we get the instrumental quality, and it is appropriate here to reference the instrumental model presented above, where the digital instrument is thought of as consisting of an interface, a mapping layer, and a sound engine. The assembled totality of the three elements makes it an instrument. However, the distinction between a musical composition, a mapping, and a sound engine blurs here, which is why some authors have recommended focusing on "mak[ing] a piece, not an instrument or a controller" (Cook 2001). Waisvisz was adamant in not changing the mapping and sound engine when he had hit upon an instrumental configuration that he liked. He wanted to develop a deep relationship with the instrument, in ways similar to the relationship acoustic instrumentalists have with their instruments. Waisvisz merged the notions of an instrument and a composition, typically "freezing" the instrument for a couple of years, in order to practise and perform the piece. Most

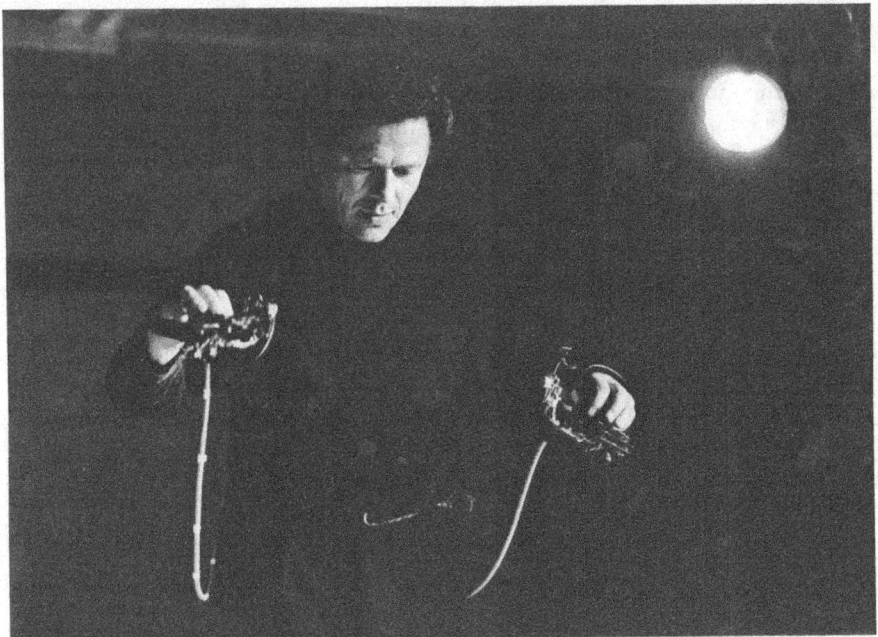

Figure 2.2 Michel Waisvisz performing his Hands instrument, originally built in 1984. The instrument is highly expressive, affording a variety of musical gestures. © Michel Waisvisz Archive.

performers of electronic and digital systems know how tempting it is to change parameters, fix, and develop further, but Waisvisz's approach was unique in that he was able to restrain himself, and not continually change the settings in the attempt to improve the piece just minutes before going on stage, like many of us are guilty of.

Waisvisz was the director of STEIM, a Dutch institute with decades of history in conducting and supporting research on musical performance technologies, offering spaces for artists to work, hosting workshops, and welcoming artists for residencies. One of the artists who has collaborated intensively with STEIM is Laetitia Sonami, whose "Ladygloves" performance system has been an inspiration for a generation of musicians. Sonami works with concrete sounds which she eloquently shapes through the performance with her gloves. The body is extended but there is a direct connection between the performer's movements and the piece itself, as if the sounds are touched and shaped with the hands. However, the lack of physical objects to manipulate does not remove the need for careful mapping and design. Indeed, it is perhaps harder to play non-physical instruments, such as the theremin, the Ladyglove, or motion capture instruments (e.g., made with the Kinect), because they lack the tactile and often haptic feedback of the physical device. In interfaces, the designed affordances, such as pads or knobs, have a dual function in that they serve as a control channel, as well as a reminder of what is possible through visual and tactile cues. The Ladygloves have inspired various projects, such as Imogen Heap's "Mimi gloves," a technology that reached popular awareness during a recent concert tour by pop singer Ariana Grande. The Mimi gloves are currently developed for commercial release; and there is an undercurrent in popular culture of using new instruments, as innovatory gadgets to liven up stage performances. Examples include pop band Coldplay's recent use of the Reactable – an instrument used by Björk on her world tour in 2007 – almost exclusively for visual effect (see Tomás 2016).

The question has been posed: will new digital instruments become part of the current musico-industrial framework with composers, publishers, producers, sound engineers, performers, concert halls, media, critics, audience, etc., or do they herald a new age of distinct musical practice? This is a straightforward question with very complex answers that deserve to be analysed at diverse tiers of musical practice. The availability of new programming languages for audio (such as SuperCollider, Max/MSP, Pure Data or Kyma) and cheap hardware (Arduino, Raspberry Pi, or Bela) has resulted in cultures of new instrument designers in hack labs, conservatories, and universities. These new instruments are used in club gigs and concert halls, are written about in the media and in conference papers, and shared on social media and online video channels, where links spread very fast. Some of these instruments become subject to more comprehensive innovation processes, where the instrument is developed, user tested, branded, and marketed. The innovation of new instruments has been studied, for example, in the field of Social Construction of Technology (Pinch and Bijsterveld 2004). With the increased documentation and data, introduced by new media behaviours, we gain further information about how an instrument emerges in public consciousness. Consider, for example, the difference between the innovation of the saxophone, the Theremin, and the Reactable table synthesizer. What emerges when looking at this field from an innovation perspective is that there is a problem with identity in new digital instruments, perhaps even a crisis. What is a new digital instrument? How do we play it? Who composes for it? Where does it fit in our culture? And is it a sustainable thing? Below I provide two case studies for discussion: the Karlax, and the LinnStrument. Both are marketed, mass-produced, and readily available for buyers.[4] Their inventors operate in a manner that resembles any established technology manufacturer, with publication materials, logos, sales office, customer relations, and a website with user accounts, technical specs, and more.

The Karlax

The Karlax is an apposite case study for a new digital instrument. It was launched in 2010 and promoted as an instrument of high expressivity, with fifty-five independent sensor parameters which can be used for triggering musical events (e.g., playing notes) or controlling a synthesis engine. The instrument is beautifully made, of aluminium and strong plastic, shipped in a leather case, but the price is high: about 3,500 euros. Compared with acoustic instruments, this is in the price range of a good guitar or a clarinet, but the key difference is the instrument's lack of establishment – its individuation and concretisation as part of our musical culture. For an instrument to become established, a variety of factors must conjoin, for example, composers and performers using the instrument, the production becoming streamlined, affordable price of the instrument, the market being responsive, and so on. Remi Dury, the inventor of the Karlax and founder of the Da Fact company that produces it, engaged in commercial promotion of the instrument between 2010 and 2014, but as a busy professor at the Conservatory of Music in Bourges, his focus is now on the educational aspects of new musical interfaces and the company is developing two new instruments – Zil and Bop – which are, together with the Karlax, integrated into the education at the conservatory. This need for an educational infrastructure to support the instrument is also reflected in Karlax workshops given to ten- to eighteen-year-old pupils at the Conservatoire de Vincennes.[5] Indeed, there is a parallel here with how Adolphe Sax considered conservatory tuition of the saxophone as an essential element in establishing his instrument as part of general musical culture. For Sax, it was critical that composers would begin to write for the saxophone, something both Debussy and Berlioz did and which established the reputation of the instrument (see Liley 1998; Horwood 1983).

Similarly, composers of electronic music have embraced the Karlax as an exciting new instrument with strong potential. In a NIME 2014 paper, Tom Mays and Francis Faber discuss their compositional strategies as well as the development of new notations for the Karlax in order to establish a repertoire for the instrument (Mays and Faber 2014). They write that with the Karlax, they see "an opportunity to go beyond the composer/performer/programmer model and start to write pieces for DMIs that could be performed by others – repeatable and shareable" (Mays and Faber 2014: 553). For them, composing for the Karlax involves creating a stable software environment

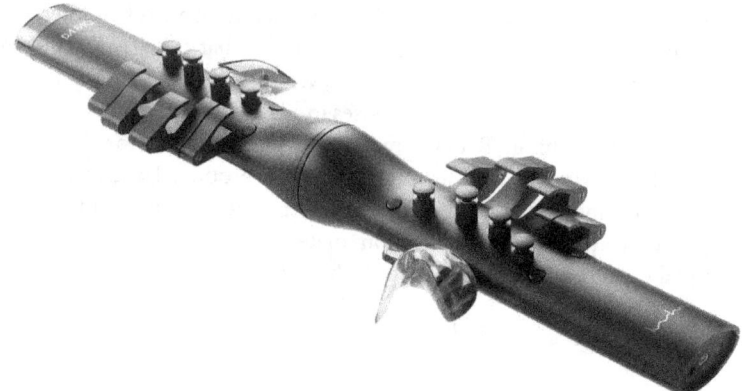

Figure 2.3 The Karlax. A new instrument with fifty-five individual parameter controls, manufactured by the Da Fact company in France.

(presumably technically consistent and sustainable for future use), coherent mapping strategies, and a bespoke system of musical notation. More importantly, this also means establishing certain performance practices and methods of training. During a symposium on composing for the Karlax, a participant expressed the view that when designing an instrumental mapping for the instrument, the composition has to be kept in mind ("En construisant l'instrument, il a déjà la composition en tête"), reflecting Perry Cook's (2001) imperative of composing a piece, not an instrument. Regarding notation, the discussions included whether to write in the form of tablature (where gestural mapping is depicted) or sonic end-result notation of some sort (not necessarily depicting pitch and note lengths as in traditional notation, as the controller is so diverse in function). Mays and Faber's paper describes their Karlax notation system in good detail and they claim that their system is "functional, expressive and readable," albeit there is room for improvement – as in any system of musical notation. They do, however, project that if a standard repertoire of expressive instrumental pieces for the instrument exists (e.g., Max/MSP patches), and if this repertoire is expressed through an idiomatic system of notation, attractive conditions arise for the establishment of a repertoire for the Karlax controller.[6]

The LinnStrument

The LinnStrument is an excellent example of a new musical controller that offers novel modes of expression. Developed by a veteran inventor of influential music technologies, Roger Linn, the interface has a grid of 200 note pads (twenty-five pads on the horizontal axis, and eight on the vertical). The controller's output uses the MIDI protocol by default, but it can be programmed in different ways on the software side and the firmware is open source, so any microtonal or alternative tuning system can be written for it, for example, in the OSC format. Adjacent pads increase by a halftone, but the row above is tuned up a fourth, like on the guitar or the bass, which makes it easy for players to apply their skills and embodied knowledge of chords, scales, and arpeggios on the LinnStrument. This makes the instrument attractive to a large range of performers, and Linn points out on his website that this layout of fourths is becoming a standard for grid button controllers like Ableton Push, Roli Blocks, and diverse mobile apps. The LinnStrument pads have lights (variable degrees of red, green, and blue), they are touch sensitive, a feature which is typically mapped to the amplitude of the note played, and then fingers can be moved on the horizontal and vertical axes, from the key pressed over to the adjacent keys, for example in mapped gestures that control pitch and timbre.

Above, I called the LinnStrument a controller, as it does not have a sound engine or a clear mapping between gesture and sound, but one understands why Linn is adamant in calling his invention an instrument. The LinnStrument is a fine musical object, beautiful in design, feel, and touch. It offers depth and space for exploration, enticing the performer into dimensions of possibilities that can be practised and embodied. Multi-touch, the instrument allows for chords and melodies to be played simultaneously, just as a pianist would play, but with more sophisticated control over each note. The lights in the pads offer ways in which the instrument can begin to communicate back to the performer, suggesting possibilities, or tracing past actions. This is an instrument to learn from. From a sociological and marketing perspective vis-à-vis innovation, it is interesting to observe how the LinnStrument is being introduced to the popular music culture, as opposed to the classical context in which the Karlax controller operates. The musical background of the inventors can partially explain the perceived markets of the instruments, as Linn is a rock guitarist and the inventor of the LinnDrum drum machine, whilst Dury is a conservatory-trained composer working in academia. Such things matter, as any sociologist of music would confirm. And in terms of the longevity of the instrument, from the perspective of marketing and business, the popular culture of a NAMM instrument trade show, where Linn

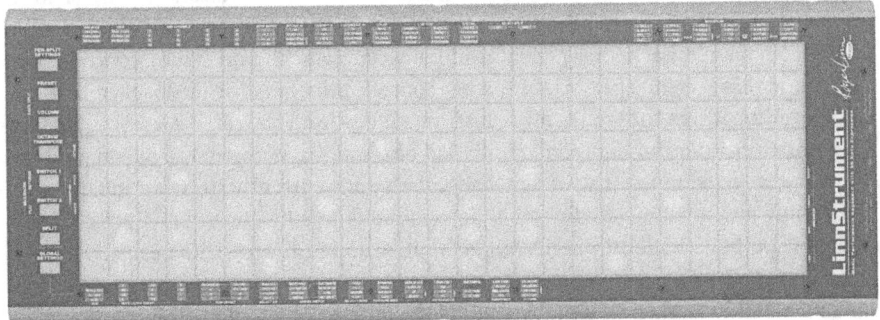

Figure 2.4 The LinnStrument, by Roger Linn. Relating to the tuning of string instruments, and offering the affordances of pressing, sliding, vibrating through finger movements on a 3D sensor (x, y, and pressure). © Roger Linn.

can typically be found, is clearly a better place for business than attending the NIME conference, where people tend to make their own instruments as opposed to buying them.

Resonating acoustic instruments

A different approach is taken by Andrew McPherson and Halldór Úlfarsson, the respective creators of the Magnetic Resonator Piano (MRP) (McPherson 2012) and the halldorophone (Ulfarsson 2018). Not wanting to sacrifice the sonic expressiveness of the acoustic instrument or the trained skills of performers, the MRP constitutes a modification of the grand piano, by adding controlled electromagnets into the body of the instrument above the strings, giving it new expressive possibilities, for example the sounding of a string without an attack. The string slowly comes to life through magnetic activation, with a smooth dynamic envelope, as opposed to the hard attack of the piano hammer, thereby giving the MRP an additional feature that makes it a separate instrument from the regular grand piano. In terms of further compositional opportunities, the angle of the pressed key is also used to emphasise different harmonics of the string, through a delicate tremolo on the key surface. The string's timbre can be changed by pressing into the base of the keys (aftertouch), and pitch bend is achieved by holding one key while lightly touching a neighbouring key. The instrument is augmented; there are no speakers, no microphones, but simply actuators that create a magnetic field that excites the piano string, like a magnet pulling its opposite pole, which is a technique we know from a guitarist eBow. The magnetic actuator of each string has the frequency of that string, or its harmonics (multiples of whole numbers). This enables the performer to "tune into" different harmonic qualities of the string.

Similarly, the halldorophone is an actuated resonating instrument, based on feedback as an integral ergodynamic feature. The halldorophone is modelled on the cello, although it looks slightly different in its modernist design. The instrument has a speaker cone at the back of the sound box that feeds vibrations into its body, typically the sound of its own strings. The feedback emerges when the vibrating body yields a resonating action on the strings, only to be fed back into the instrument's body by the individual pickups. Each string has its unique pickup whose gain can be controlled by sliders and this gives the performer further control in shaping the feedback, unlike the electric guitar where the sound of each string comes from the same pickup. There is a space for electronic and digital manipulation of the signal between the pickup and the speaker cone fixed to the instrument, and Úlfarsson is currently studying the different use in how an electronically fitted halldorophone differs in use and character from a digitally equipped halldorophone. Since the instrument borrows its design from the cello, performers can recycle their knowledge of the cello for performance, although the understanding and control of feedback is a new area of learning. This "recycling" of skills in new instruments is a notion that interests both McPherson and Úlfarsson. Recently, other resonating string instruments have appeared with auxiliary equipment added onto the stringed instrument, such as Alice Eldridge and Chris Kiefer's "Feedback

Cellos" (Eldridge and Kiefer 2017), Tom Davis's "Feral Cello" (Davis 2017), and Thanos Polymeneas-Liontiris's "Feedback Double Bass" (2018).

Both McPherson and Úlfarsson have created opportunities for composers to write for their instruments. They have lent their instruments to performers, sold a few copies, and run workshops where composers have supervised access to the instrument over a period of time, often involving performers too.[7] Both instrument makers have refrained from suggesting a musical notation system for their instruments, as they are interested in musicians exploring the instrument from a neutral mindset; to discover what they find interesting to play with, and eventually to come up with their own idiosyncratic notation. The danger of defining a language for the instrument through symbolic notation is that this would concretisise certain compositional ideas and performer actions; arguably this should be the realm of individual composers who compose their pieces in their own notation, based on what they have discovered through exploring the ergodynamics of the instrument.

Digital affordances

The above examples involve lab productions that require skill, knowledge, and financial means. However, with programmable mobile devices, such as the ubiquitous mobile phone or tablet, musical instruments can be built in the form of apps that exploit the technology of the device itself. This was the approach taken by Ge Wang and collaborators when they created the "Ocarina" instrument in 2004. At the time, when inspecting the recently released iPhone, Wang thought that it would be interesting to design an instrument that would make use of all the interface affordances of the phone: the multi-touch screen, the microphone, the speaker, the gyroscope, the GPS, and so on. The result was the Ocarina, an instrument based on the traditional Mesoamerican clay flute, but with the digital platform affording the design of new features such as embedded musical scores, recording of songs, and communication with other players of the Ocarina around the world (Wang 2014). The instrument sports ergomimetic design features from the real flute, where the sound is controlled by blowing into the phone's microphone (translating the microphone noise into a control signal), and the keys are pressed on the multitouch screen. The Ocarina is a good example of an instrument that has shipped thousands of copies, used by laypeople and professional performers alike, and generally serving as an enjoyable, fun and uplifting musical object that exists in people's pockets, available whenever the urge to play music crops up. The instrument itself supports the knowledge and expertise that people have gained on this age-old instrument, yet offers engaging and novice-friendly entry for new players. It is indeed an excellent example of how learning can be supported by alternative means, such as embedded scores in tablature notation and its game-like functionality. The community aspect is important too, where users share compositions and videos of their work. The Ocarina is amongst the top 20 downloaded apps of all time, which indicates how powerfully it connects with people's general love of music and experimentation. The app is also a good example of how digital instruments can democratise music-making by enabling simpler entry levels. A good source of information for exploring this and related work is Ge Wang's *Artful Design* book (Wang 2018).

Instrument or controller?

In this chapter we have discussed how the digital instrument *has* an interface, whereas the acoustic instrument *is* an interface. We do not typically use the term "interface" when describing acoustic instruments, and the term was not used much until the advent of electronic technologies; it was certainly not a concept used by instrument makers of acoustic instruments.[8] We can, of course, talk about an interface in acoustic instruments – for example, the church organ has quite a sophisticated and complex interface. However, like the piano, the organ has an interface that is mechanically coupled to the instrumental functionality via physical law. In digital instruments, their computational nature and arbitrarily mapped control elements result in technologies that feel thin, yet powerful. Unlike acoustic instruments, whose bodies and play are thick, there is nothing that *necessitates* the design of the digital instrument: it is all a matter of design. The difference between an instrument and a controller also reflects this: we expect controllers to be easy, like a button on a coffee machine, a car radio, or a train ticket kiosk. Instruments, on the other hand, have depth, character, resistance, and individuality. We don't want them to be controller-interfaces: we want mystery and magic, discoverability and surprise. The notion of ergodynamics unveils how the first encounter of a digital musical instrument typically involves exploring its affordances and then a further study in the instrument's constraints. For this reason, digital luthiers often make use of complexity and non-linearity to make the instrument perceptually interesting to play.

The above argument might contradict the ideology of musicians affiliated with what is often called "controllerism" (as in "turntablism"), but here DJs and music producers focus on the performative and expressive use of their software controllers. "Controllerism is the art of manipulating sounds and creating music live, using controllers and software" (Moldover in Golden 2007). Typically using MIDI controllers with rubber pads, plastic buttons, knobs and sliders, this approach attempts to frame the performance on a controller like that of a musical instrument. Although the differences are many and profound, it is impressive to see the expertise and ingenuity demonstrated by controllerists. This is further supported by music software houses such as Ableton and Native Instruments, and equipment manufacturers like Akai or Roland, who are increasingly beginning to present their music technology products as something that equally belongs to the stage as well as the recording studio.

A question arises: how does controllerism differ from the development of new DMIs as we see in the NIME research community? The most prominent difference is the status of a user versus designer of technology. Controllerists are creative users of hardware and software, but their primary focus is working within established musical genres, often with clear criteria for the music's functionality, such as getting people onto the dance floor. In contrast, NIME researchers tend to think more critically about the technology, often designing the instrument from scratch, and questioning what it does, how it works, and what kind of music it encapsulates, to the degree that the distinction between the instrument and the musical piece disappears. Here, building a musical instrument becomes indistinguishable from designing a music-theoretical framework; the musical instrument is a theory of music, and it equally contributes to genres, styles, and musical scenes. Another profound difference is the conception people have of their performance: the controllerist performs their music via an interface, whereas in the NIME performance the instrument itself constitutes the music: without the particular instrument the music would be different.

It is unwise to generalise too much here, as NIME is a broad field of investigation and practice, an inordinately interdisciplinary community populated by engineers, computer scientists, psychologists, musicologists, composers, performers, philosophers, tech innovators, and more. Research topics range from usability (how can this technology best support creative work, is the experience of using it good?), ergonomics (is the technology well designed for the human body, does it support learning and mastery?), human–computer interaction (is the device understandable, well set up, does it communicate its function?), and design (how does it function, what are the materials, is it sustainable?), to aesthetics (how does it look, is it an inspiring object?), music theory (what kind of musical knowledge is inscribed into the instrument, what musicality does it contain?), performance studies (how does it work in live situations, is it open and flexible, fast and controllable?), and audience studies (is the instrument understandable, does it communicate human intention?). This interdisciplinarity is what makes the field so interesting and rewarding: it operates at the most immediate and intense interface, or meeting point, between humans and technology, in one of the most ancient and popular cultural domains: music. Musical instruments present a tech-intense area of interface design for real-time performance, and the design of ergodynamic objects serve as boundary objects (Star and Griesemer 1989), due to the diverse expert knowledge required to build these objects. Other such design areas include the design of interfaces for flying, sailing, driving; surgery or dentistry; playing computer games, martial arts, cooking, or sports. Many of these interfaces are not crucially dependent on the real-time performance aspect, as most are not about the performance itself but about the product, where the critical focus is more likely to be on the aesthetics than strictly technical skill.

Conclusion

From the perspective of innovation and design, as well as composition and performance, the key difference in acoustic and digital organology is that digital musical instruments develop at the speed of general computer technologies. This is a multidimensional field that moves faster than music, so music is borne along by our new technologies. That is a very different situation from what we are used to with acoustic instruments where we have a more considered, slow, and grounded dialogue between the instrument maker, the composer, and the performer in what might be a useful addition or change to an acoustic instrument. The new technologies bring with them practices, ideas and ideologies, and methods and methodologies that enframe how we conceive of the instrument. By applying technology such as a game controller, a sensor of some sort, a network protocol, an FFT or deep learning library, we typically incorporate into our instruments much of the ergodynamics and ideas embedded in those technical elements. Considering the speed of development and the competition in the respective areas of high tech, whose products we integrate in our new instruments, we can question whether it is realistic to expect our instruments to stabilise, or concretisise in the Simondonian terms (Simondon 2017). The technology moves faster than musical practices and what we are getting are snapshots of technics applied in musical composition and performance, technics whose materialities will be quickly replaced with new ones, but whose ergomimetic structures continue and become re-implemented in later technical objects. For us, researching in the domain of musical performance, it is therefore the *ergomimetic gesture* that becomes concretisised, not the technological object. The next two chapters explore the epistemic structure embedded in new instruments, how musical movement and technological objects relate, and further study what a digital organology might entail.

3

Epistemic Tools

The introduction and first chapter of this book studied the millennia-old rift between theory and practice in Western culture, carved into stone by the early Greek philosophers. Stiegler launches his *Technics and Time* by declaring: "At the beginning of its history philosophy separates *tekhnē* from *epistēmē*, a distinction that had not yet been made in Homeric times" (Stiegler 1998: 1). One of the important effects of Stiegler's work is the questioning of this enduring dichotomy between *techne* and *episteme*. Previous sections of this book verified this by pointing at the lowly status of musical instruments compared with the voice in the Middle Ages, perhaps with the exception of the harp and the organ. Given that musical instruments were the most sophisticated technologies available at the time, embedded with theory and ingenious craft, this is indeed peculiar and has to be explained with reference to the philosophy and theology of the period. We have seen how St Cecilia, the saint of musicians, was depicted with broken instruments, flutes and lutes, in the soil by her feet, whilst she gazes up into the sky at angels singing from books, sometimes playing the harp. The message is clear: music should be inspired by divine voices, in language, not abstract sounds. The flute, *aulos*, distorts the performer's face, transforming their organ of rationality, the mouth, into one of breath rather than words – *psyche* versus *logos*. Stringed instruments, such as harps, were acceptable to the Greek theorists as their tuning represented the harmony of the world, mathematics and music theory; the player of the harp simply executed a theoretical piece of music through playing, but with the mouth free to recite poems and chants.

Instruments for thinking

The role of musical instruments changed with the scientific revolution beginning in the mid-sixteenth century. Scientific instruments, such as weights, rulers, lenses, clocks, and other lab equipment, became important tools for thinking, and musical instruments were seen as an integral part in this study of the world. With these instruments at hand, new thoughts and experiments were possible. In the seventeenth century, Francis Bacon, often called the father of empiricism and modern science, wrote in his *Organum Novum*:

> Neither the naked hand nor the understanding left to itself can effect much. It is by instruments and helps that the work is done, which are as much wanted for the

understanding as for the hand. And as the instruments of the hand either give motion or guide it, so the instruments of the mind supply either suggestions for the understanding or cautions. (Bacon [1905] 2011: 259)

Just as a shovel is a useful instrument for digging a hole, the scale and the lens help us to think about and understand the world. Bacon advocated a new method of conducting science by seeking empirical evidence, and using instruments for that purpose. Instrumentation was becoming a critical extension of the human intellect, as well as its body: Galileo's telescope, the *organum*, was one such prosthetic extension of the eye, and so was the microscope, but other epistemic tools included instruments of calculation, geometry, time, astronomy, logic, and so on. The change can be represented by the move from the scholastic use of Aristotle's book of logic, the *Organon*, which was seen as the key instrument of reasoning, to developments of physical instruments that provided empirical data that might not only corroborate scientific hypothesis, but also provide new data that had to be explained *ex post facto*. These instruments were sometimes called "philosophical instruments," and the natural philosophers talked about "torturing" nature to reveal her secrets by means of these new instruments.

In their work on *Instruments and the Imagination* Thomas Hankins and Robert Silverman enunciate how scientific instruments can serve as the foundation of scientific thought, as "[i]nstruments have a life of their own. They do not merely follow theory; often they determine theory, because instruments determine what is possible, and what is possible determines to a large extent what can be thought" (Hankins and Silverman 1995: 5). Instruments thus serve as extensions of our bodies and amplifiers of our senses. They frame the boundaries of our perception and thinking, whilst mediating reality at the same time. For Don Ihde, instruments provide distinct phenomenological modalities in the way we relate to the world: "embodiment relations" extend our bodies or amplify reality; "hermeneutic relations" provide us with data, like a spectrograph or a heat map, that we have to interpret; and "alterity relations" where the technology becomes something we engage with for its own sake, for example in the use of virtual reality headsets (Ihde 1990). Musical instruments can exemplify all three modes. Furthermore, as Baird points out, instruments are impregnated with knowledge and can serve as demonstrators, models, or exemplars of how our conceptions of reality work (Baird 2004). This is how, for example, Newton's prism was key to the development of the theory of colour. Pascal's arithmetic machine (a mechanical calculator) from 1642 is also a good example of such a new instrument of the mind. Indeed, for Bacon, the implication of the new logic (as opposed to the mere perception of the senses) is that the "mind itself be from the very outset not left to take its own course, but guided at every step; and the business be done as if by machinery" (Bacon [1905] 2011: 256).

Although troubadours and other folk musicians would travel across Europe with their secular music, music theory in medieval times would largely be the domain of the Church. The empirical methods that came with the scientific revolution resulted in an emphasis on practice and experimentation, which, in music theory, shifted the focus to the domain of sound, musical instruments, and their function. Thanks to the new scientific instruments, which included musical instruments, the science of acoustics established itself across the continent: Bacon presented the new "Acoustique Art" in his *Advancement of Learning* (1605) where musical instruments become tools for scientific investigation. Marin Mersenne conducted experiments on string vibration and pitch in the 1620s and 1630s (Gouk 1999: 157) and Gaspar Scott's second book from *Magia naturalis* (1657–9) was called "Acoustica". Not since the early Greeks had acoustics been studied in this way and music been conceived of as a domain of empirical investigation. Furthermore, Bacon was eager to mend what he saw as an unfortunate division between the theoretical understanding of sound and the practice of music, blaming tradition for some of the problems: "Musick in the Practice, hath been well pursued; And in good Variety; but in the Theory, and especially in the yielding of the Causes of the Practick very weakly; being reduced into certain Mystical subtleties, and not much Truth. We shall therefore, after our manner, joyn the *Contemplative* and *Active Part* together" (Bacon 1670: §101). For historian of science Jean-François Gauvin, the drastic change in the new science of the seventeenth century was the role instruments played in research. With the proliferation of philosophical instruments, the scientist's laboratory became a habitus in which knowledge was embodied in epistemic artefacts, ranging from books to microscopes: "Virtue was not only found in the scientia of geometry, but also in the instrument and its operator" (Gauvin 2011: 317). However, the biggest change was in the new conception of the instrument, the organon, moving from Aristotle's conception of an instrument as a means to an end (*terminus ad quem*) to a more speculative conception of the instrument itself as a starting point for a journey into the unknown (*terminus a quo*) (Gauvin 2011: 333).

Figure 3.1 The Physical Laboratory of the Academie des Sciences. An ink and wash drawing by Sebastian le Clerc, from circa 1700. © Science Museum / Science & Society Picture Library.

It is difficult to stress enough the impact of the printing press in the sixteenth century for the field of music. Printing created a market for composers, publishers, printers, lay and professional performers. Not only did movable type printing of musical scores increase the popularity of music and thus musical instruments, but books on instruments and the theory of acoustics were also printed. The seventeenth century introduced the study of the engineering of musical instruments, for example in Praetorius's *Theatrum Instrumentorum* from 1620, which is often seen as the key work on musical instruments, and Mersenne's *Harmonie Universelle* from 1636, which contained an encyclopaedic wealth of knowledge on musical matters, unsurpassed at the time, and a book that is still the main source of information about instruments of this period.

Musicians in the sixteenth and seventeenth centuries were not very concerned with original personal musical expression. They were more interested in the functional role of music, ancient Greek music, and the exploration of the moral value of the art. This included Jewish and early Christian traditions. The music of the past was often played in opposition to what was seen as corrupt contemporary practices in music, but "[p]aradoxically, this appeal to ancient authority was often a means of legitimating entirely new practices, while at the same time discrediting established ones" (Gouk 1999: 115). As we will study later (in Chapter 14), the combinatorial arts were seen as methods imitating God's work-processes in creating the world, but for Baroque musicians it was important to see the laws of the world's creation in their music. God had created the world, where the fundamental notion was number, harmony, and ratios, and mathematics was the science of understanding God's creation. As Eberhard Knobloch describes it, "The world was a universal realization of God's music. Or the other way around: number was God's instrument for creating the world" (Knobloch 2005: 334).

Theoretical instruments

In early modern thought we find the notion of the instrument as a container of theory, and the musician engages with that system of thought when performing on it. This view changed with more personal compositional practices in the Classical and Romantic periods, but I will argue later that our conceptions and implementations of current music technologies are reminiscent of early modern thought, where the instruments themselves become the vehicles for scientific investigation and musical exploration. The notion that theory is increasing in musical instruments can largely be defended by referring to the number of written structures that underpin digital musical instruments. Behind our simple instrument are layers of microchips, busses, memory; operating systems, code libraries, classes; protocols, standards; and all kinds of auxiliary hardware that supports all of the above. The degree and role of representational thought thus differ in acoustic and digital instruments. In acoustic instruments most of the time spent with the instrument does not involve symbolic thought, at least not in kinaesthetic or proprioceptive terms. In digital technologies, the majority of time is spent designing, composing, and developing techniques through symbolic instructions, code, and involve electronic parts with clear functions, manuals, compilers, and APIs. The thoughts are therefore more symbolic (e.g., "I'll map this bend sensor to a resonant low-pass filter ranging from 100 to 12,000Hz"). When practising a physical digital instrument, we can, of course, get into the non-symbolic learning of the technology, and there are many cases of such practice (consider Waisvisz as an example), but more common is the engagement with the technology at the symbolic level of language, numerical control, and representational interfaces.

Instrumental agency

Technology equally conceals and reveals. We have already seen how technology can never be neutral. Technologies organise, select, focus the task, and shape the environment (or the material of the work) through various transformations. This is what Ihde calls amplification-reduction transformations (Ihde 1979: 56), and he takes the example of a dentist's probe that becomes a prosthesis of the hand, amplifying any unevenness in the tooth, whilst reducing information such as wetness or warmth. By selecting a technology, one is choosing a lens, a background theory, a specific worldview, and thus the interpretative scope of potential outcomes. This selection involves what kind of perceptual resolution we want to engage with the world and what type of embodiment and theoretical inscriptions. It is a necessary property of all technological artefacts to amplify certain possibilities of experience while at the same time reducing that very experience. Throughout the history of computing, and to a large extent today, computers do not serve as extensions of our body (embodiment relations) but rather as extensions of our mind (hermeneutic relations). When "mind" is used in this context, we are observing that the computational technologies are creating a language, a system with which to think. It changes many of our common habits of interpreting (or being in) the world. The aim is not to introduce yet another dichotomy between the body and the mind, but rather to point out the change in activities that are needed to perform the tasks at hand.

This nature of categorisation and abstraction of human knowledge and actions (both "know-that" and "know-how") is the essence of computer software, a fact that is very apparent in the realm of musical software tools. Software is, as Geoffrey Bowker and Susan Leigh Star point out, a "contingent" and "messy" classification that has diverse implications and, in our case, aesthetic, cultural, and cultural-political effects. As sociologist Bruno Latour (1987) so elegantly demonstrates, objects establish themselves as black boxes through repeated use and in the process their origins disappear. The object is naturalised through heavy use, a fact prompting Bowker and Star to observe that the "more naturalized the object becomes, the more unquestioning the relationship of the community to it; the more invisible the contingent and historical circumstances of its birth, the more it sinks into the community's routinely forgotten memory" (Bowker and Star 2000: 299). Musical patterns and musical styles become "blackboxed" in software. Most people do not know why the standards, implementations, patterns, or solutions in musical software are there (such as halftone equal tempered tuning, temporal quantisation, or note-based event model, to name but a few). It is often clear how these elements limit musical expression, but sometimes the suggestive nature and constraints of the software are concealed by the rhetoric, the streamlined functionality, and the slick interface design of the software.

The software instrument maker is embedded in a knotty infrastructure of operating systems, programming languages, protocols, standards, and interface limitations. This is a factor that involves social and technical elements well described by Bowker and Star: "Systems of classification (and of standardization) form a juncture of social organization, moral order, and layers of technical integration. Each subsystem inherits, increasingly as it scales up, the inertia of the installed base of systems that have come before" (Bowker and Star 2000: 33). This intricate, reticulated, technological structure is wide-reaching and applies techniques of heterogeneous origins. Digital musical instruments are a good example of Deleuze and Guattari's notion of the assemblage (1987) in how things often self-organise or are the result of a bottom-up emergence. Furthermore, technological artefacts incorporate politics, as Langton Winner (1980) pointed out, but they also have aesthetics, morals (Latour 1992), and style. The past few decades in the philosophy of technology have included studies of technological agency, and how our instruments affect the development of our ideas (Pickering 1995). Our technological systems are increasingly gaining autonomy as we like to delegate to the machine tasks that we find tedious or effortful, but in music technologies, this manifests through *play*, through playing the instrument, but also the playful interaction that can emerge between a human and machine if the latter is given agency. In music, in particular through conversational interfaces (Johnston et al. 2008), agency can become a highly appreciated feature.

The analysis of the deterministic role of technology is not new. Consider the Marxian analysis of the machine where technology is predicated as frozen labour, a concept that is increasingly interpreted into the context of our digital machines, software: "Thus, values, opinions, and rhetoric are frozen into codes, electronic thresholds, and computer applications. Extending Marx, then, we can say that in many ways software is frozen organizational and policy discourse" (Bowker and Star 2000: 135). Since computer music is inevitably mechanical, it would be interesting to conduct a Marxian analysis of music software: the engine that drives the music is not a human, but a machine. The human is merely an operator of the machine that performs routine tasks, repeating and automating laborious and boring rote movements. The digital music software, where music is inscribed, organised, and played back from a mixture of data structures and algorithms, is essentially an engine. The underlying core of the system is organised in blocks of repetitious routines. As such, it is ideal for repetitions, perfect timing, standard dynamic of notes, etc. What the digital system does not easily lend itself to are tempo changes, variable dynamics, metre changes, flexible durations, (accelerando, crescendo, etc.) – in other words, what the human being excels at when performing music. These temporal and dynamic changes in the human typically derive from emotional engagement with the music performed, a fact that is strongly emphasised in classical music education. We therefore operate with distinct ideals: the machine (time perfect, repetitious, logical, perfect, normative) and the human (flexible, fallible, instinctive, emotional, individual).

Epistemic tools

The awareness of the relationship between tool use and aesthetics has long been acknowledged in the crafts (Pye 1968; McCullough 1996). The practised hand using highly designed tools results in a unique aesthetic that is historically rooted within a tradition of workmanship (Wilson 1999). The tradition of violin making since Stradivarius is a good example. The same applies to digital tools although their material-symbolic nature might be harder to scrutinise than their physical counterparts. Computer systems are abstractions, interfaces or black boxes to designed activities. For the system designer it is therefore important to work with the users of the software (as we find in participatory design) and analyse the context in which it will be used (the focus of ethnomethodology). They should also be aware of the fact that the system itself will change the practice with its introduction in the workflow, as analysed by technomethodology (Button and Dourish 1996). It is precisely this abstract nature of system design that constitutes the epistemic nature of digital instruments. The tool is designed, its affordances and constraints are outlined, and the user's actions are predicted and delineated into the interface and interaction design of the tool. The designer decides what is revealed and what is concealed when the system is used. When computers are seen as mediation, the role of the system designer is to direct the channel of energy from the physical interface to the work (or the terminus) through the symbolic engine of the epistemic tool.

Baird demonstrates how technology can precede science, and afford scientific discoveries through its physical structure and functionality. The point here is not what precedes what, but rather that the instrument is an expression in itself, an externalisation of knowledge in a form that is not symbolic but material: "Knowledge can be expressed in many ways. Theories express knowledge through the descriptive and argumentative functions of language. Instruments express knowledge both through the representational possibilities that materials offer and through the instrumental functions they deploy" (Baird 2004: 131). Baird provides an analysis of three different ways in which technological artefacts can perform epistemological work, but he states that this list is not exhaustive. Instruments can embody knowledge as a material mode of representation (model knowledge); as a mode of effective action (working knowledge); and as a material mode of effective action that synthesises representation and action (encapsulated knowledge) (Baird 2004: 17). Technological artefacts are models of thought, performing epistemological work alongside scientific theories, and often preceding them. An example is the development of the water wheel, where the practical building of a model was more successful than theoretical calculations. Another example is of James Watson and Francis Crick discovering the double helix (DNA) through manipulation of physical models of atoms. It was the manual engagement with the physical model, experimenting with structural

Figure 3.2 A representation of Pythagoras exploring ratios of string length, weight, and size. From *Theorica musice Franchini Gafuri laudensis*. 1492. © Bibliothèque nationale de France.

combinations of the model that led to the discovery. The third example could be the new *Sketching in Hardware* conference series, where the doctrine is that hands-on knowledge and manipulation of physical stuff and hardware yields design results that are different from the design done in 3D packages on computers. Due to the representational nature of computer software, practitioners who engage with it at the level of code have always understood how its theoretical basis serves as a model, laying out principles and methodologies of practice (see Buxton 1977: 59).

In a Heideggerian manner, Baird shows that instruments are not simply "instantiation of ideas." Materials and ideas are both necessary in science and materials do not behave like ideas. This is the point of Baird's instrumental epistemology: "[A]n artifact bears knowledge when it successfully accomplishes a function" (Baird 2004:

122). Theoretical scientists are "concept smiths," whereas people who build instruments are "function smiths" (Baird 2004: 123). Function smiths develop, replace, expand, and connect new instrumental functions from existing affordances. The functions of a measuring instrument are semiotic; they are built according to our theoretical understanding of the domain to be measured. As such, measuring instruments do not extract information from the world: they generate new data signals that have to be hermeneutically translated and prepared into a suitable form, and only then can the data be understood as information about the world.

These threads in the philosophy of technology can be defined as an epistemological material turn, or what Karen Barad calls onto-epistemology (Barad 2003, 2007). If the problem of Western philosophy has been the separation of *techne* and *episteme*, this dichotomy is now being questioned, following a better understanding in cognitive science and the role of active perception (ranging from the enactivism of Varela et al. (1991) to the predictive perception of Clark (2016)). This new material epistemology opens up the possibility of analysing the acoustic, the electronic, and the digital beyond our alien phenomenology, as we did in a previous chapter.

Material affordances

Traditionally the luthiers of acoustic instruments acquire their skills through the embodied practice of making the instruments, as an apprentice under the supervision of a master. Many of the finer parts of these learned skills, including the intuition of material properties and sound physics, are in the form of tacit knowledge (Polanyi 1966), that is, embodied and non-theoretical knowledge derived from discovery, exploration, and refinement. The design of an acoustic instrument involves iterations of designing, building, and testing. A network of communication between the instrument builder and the performer extends to the composer and players of other instruments. The instrument makers work with what physical materials afford in terms of acoustics; a change in curvature, in width, or in material, results in a change in timbre. The interface is necessarily conditioned by the properties of the materials used, and various solutions are therefore introduced to adapt to the anatomy of the human body with the use of levers, keys, and other mechanisms. Acoustic instruments are firmly and equally grounded and conditioned by the human body vis-à-vis ergonomics and the capacity of learning – an instrument that is too complex will not succeed. Unlike electronic or digital instruments, acoustic instruments do not come with a manual, they *are* manual.

Electronic instruments are made of different materials. The material properties of electricity, magnetic waves, tubes, oscillators, capacitors, inductors, and transistors are what the inventors have at hand to explore. Similarly to the evolution of an acoustic instrument, the creative process is one of designing, building, and testing through a cycle of iterations, actively working with the materials to push their boundaries and see how users respond to the invention. The difference is that the electronic materials come with instructions and schematic diagrams that describe their behaviour. There is an increased logic of calculation, science, and engineering. The acoustics of Fourier and Helmholtz are applied in the design of oscillators and filters. However, some of the characteristic sound in electronic instruments depends on chaotic or entropic properties of the materials used, and, as Trevor Pinch and Frank Trocco illustrate so well, often the results of this experimentation took the inventors by surprise.[1] The user interfaces in electronic musical instruments can be built in any shape and form. Regarding ergonomics, we are still constrained by physical mapping, so when the instrument has been wired up, its fundamental functionality is not easily changed.

Digital instrument makers exist in a different world altogether.[2] The computer becomes the "workshop," and observing the digital luthiers at work might not suggest any associations with musical activities. Code as a material is not intrinsically sonic; it does not vibrate; it is merely a set of instructions turned into binary information that is then converted to an analogue electronic current in the computer's sound card. The materials of the digital instrument are many: a computer, a monitor, a sound card, an amplifier, speakers, and tangible user interfaces. Behind this material surface lie other materials: audio programming languages, digital signal processing, operating systems, mapping mechanisms between gestures and sound engines, and so on. From the perspective of Latour's actor-network theory, the networks enrolled in the production of digital instruments are practically infinite. There is an impenetrable increase in complexity, which means that the inventors have to constantly rely on black boxes. Furthermore, the materials used in the digital instrument originate from technoscientific knowledge. There are relatively few physical material properties at play (although, of course, at the machine level we find matter) compared to the amount of code that constitutes its internal (and symbolic) machinery. The inventors have knowledge about digital signal processing, sound physics, audio synthesis, gesture recognition, human–machine interaction, and the culture of musical performance.

Conclusion

Considering the above, it is therefore possible to define epistemic tools as instruments whose origin and use is primarily symbolic. Yet, as we explore in later chapters, their application-function is at the level of signal, not symbols. The concept of epistemic tools is not intended to define exclusively computer-based tools; it includes all tools that serve as props for symbolic offloading during our cognitive process. The chessboard with its black and white pieces is also an epistemic tool (although in this case one might want to swap "tool" with "props").[3] However, it also emphasises the way technology changes the cognitive structure of the user and explores the importance of the highly integrated and complex network of thought systems that constitute any technological object. Epistemic tools are cognitive extensions of the human, and their machinery is first and foremost symbolic operations and less physical operations.

This chapter has traced how the early moderns began to apply instruments in their quest for knowledge, developing what we define as the scientific method. Instruments extended our senses and ideas of the world, and through them, another world emerged.

Baird's material epistemology demonstrates this, and Ihde's post-phenomenological analysis helps us to understand the different modalities of our instruments. The argument here is that acoustic instruments are clearly impregnated with knowledge, but that their electronic and digital counterparts are imbued with an increased level of epistemic dependencies. Clearly acoustic instruments demonstrate acoustic theory, often music theory (e.g., with discrete pitch, as in the flute or the harp). Before movable type print, the acoustic instrument was the key material source of knowledge of its own use, and it was in the practice of playing that this knowledge was passed on orally. With print, and the subsequent developments in Classical and Romantic music, the instruments begin to increasingly serve the function of media where they channel the composer's ideas through the performer's play. Modernist music then coincides with electricity. The instruments become epistemic, composed, often directly fusing the instrument with the composition, as exemplified in the work of Gordon Mumma, David Tudor, or Erkki Kurenniemi; where the instrument constitutes the piece, for example in the work of Eliane Radigue or Morton Subotnick; or where a specific technique becomes the theory and aesthetics of a new piece, as with Stockhausen or Xenakis. The challenge of designing and performing with digital instruments is to understand their material properties and symbolic writing, their internal guts in the form of code, yet functioning as tools whose end arena is at the level of the signal. The digital system is material in its body, electric in function, and symbolic in control. Live coding is a good example here: tactile keyboards are used to write symbolic code that generates an electric signal. How do we analyse such instruments? I propose we need an organological approach for understanding them, a critical analytics of musical instruments, and I call this study *musical organics* (Magnusson 2017).

4

Digital Organology

An instrument is a theory of music. Since the beginning of documented history, musical instruments have been thought of as containers of music, demonstrators of natural laws, an embodiment of theory, ideology, and often aesthetics. The previous chapter discussed the heightened symbolic and theoretical level of design in digital instruments. While acoustic instruments are defined by their materiality, developed from a bottom-up exploration of the acoustic properties of the materials, digital instruments apply technical elements that are designed to conjoin through hardware and software protocols. The skills move from the hands to the brain, from exploring materiality and manual practice to reading manuals and experiment with control. The acoustic instrument's sound is given by nature herself: we had complex instruments in the form of lutes, flutes, and organs millennia before we ever understood sound. The physics of wood, strings, and vibrating membranes were there to be discovered and not invented, if such a distinction is entertained. These technologies developed slowly and reached a state of concretisation (Simondon 2017) through iterations of design and fitness in the context of other technologies.

The body of the digital instrument is largely theoretical and developed from a top-down methodology. In order to make the instrument, we need to know precisely what we require from the programming language, relevant protocols, the DSP theory, the synthesis theory, generative algorithms, and music theory, and we need to have a good understanding of human–computer interaction. Knowledge in all these areas has to be applied for the instrument even to work – never mind creating a good instrument. This is distinctively different from acoustic instruments where building a basic instrument can be relatively easy, for example out of a cigar box. While the *design* of a digital instrument is a process requiring knowledge of sound, signal processing, music theory, and interface design, the *use* of the instrument happens through habitation, through working with it, discovering its functions, and progressively building an understanding of its affordances and constraints (Magnusson 2010) – its ergodynamics. This is what philosopher of science Andrew Pickering calls the "performative idiom" (Pickering 1995: 144), and it applies equally to scientific discoveries and explorations in laboratories. As users of digital technologies, we do not need to know as much about synthesis or music theory, as the musical black box is intended to be operated through its simple and user-friendly interface. Ergonomically, the interaction happens primarily through a symbolic channel, which gradually teaches the user to operate with technical terms (such as "dBs," "filtering," or "compression"), but knowledge builds up with the habituation to the model. The predefined quality of the digital instrument means that its functionality can be exhaustively described in a user manual – all is supposed to be explicit; nothing is covert, hidden, or discoverable. Where the digital instrument exhibits any chaotic or non-linear behaviour, it has often been due to a failure in design, a bug in the code, or loose wiring in the hardware. Recently this rational functionality is beginning to be seen as a problem in the design of digital musical instruments, and instead we seek the imperfection, surprise, and mystery of acoustic systems. Researchers have therefore begun to investigate how the embedding of non-linear behaviour can benefit the user experience of playing instruments (Mudd et al. 2015), and with the use of machine learning at the core of our instruments, we might discover behaviours that no one designed – behaviours which the instruments learn from being played, and which slowly become part of their body.

Performers of acoustic instruments develop an intuitive understanding of their instruments, the music theory, and the tradition in which they operate. They internalise styles, techniques, and finger patterns. The interaction with the instrument is embodied and learning the instrument means negotiating a space for a dialogue with it. The instrument is not a mere channel for ideas, but a designed thing in the world that contains a strong agency, primarily in its musical affordances, but also in terms of character, quirkiness, or individuation (the object's history and change through time as an effect of manoeuvre and use). The instrumentalist "plays the instrument in," meaning that its features adapt to the musician. A luthier who takes an instrument they made for a maintenance check will be able to read off the instrument the musical character of its owner. As opposed to the generic explicitness of the digital instrument, the acoustic instrument contains boundless scope for exploration as its material character contains myriad ways for instrumental entropy, or chaotic non-linear behaviour that cannot be mapped and often differs even in the same type (brand and model) of instrument (including enlarged audio generating chains such as instrument + effects + amplifier). This uniqueness of the instrument, its haecceity or individuation as an effect of time is appealing; the ergodynamic nature of the instrument is there to be explored: what kind of sonic textures can be written with it, and what kind events or patterns played. Good performers enjoy exploring not only the unique affordances of instruments, but their limits as well, and will often play at the edge of these limits where the sound breaks or the instrument displays a non-linear behaviour that yields unpredictable sonic results. Examples include Miles Davis, Evan Parker, Jimi Hendrix, Angharad Davies, Hildur Guðnadóttir, Axel Dörner, or John Butcher. The same could be said about musicians exploring the boundaries of their voice, with Diamanda Galas, Nina Simone, Joan Le Barbara, Trevor Wishart, Blixa Bargeld, Stephanie Pan, Audrey Chen, and Phil Minton coming to mind.

Acoustic instruments evolve slowly, with iterations that might take months, or even years, before the instrument maker gets the necessary feedback needed to develop a technique further. On a larger scale – over centuries – we follow the long history of the viola da gamba evolving into the cello, and here we have a different feedback structure, from users to luthiers as passed down through generations. Conversely, evolution in digital instruments follows the innovation and speed of general information technology and high-tech culture. Faster computers, new interfaces, new protocols, new machine learning libraries – all these are applied in our instruments, supporting the old adage that whenever a new technology appears, someone will apply it for musical purposes and take it further. The digital has engendered a sense of novelty, curiosity, and originality in terms of performance, sound, and music. The musical results are strongly dependent on the instrument, and often *are* the instrument. Part of the excitement in the domain of new musical instruments in the twenty-first century can be attributed to this fact as the musical creativity goes beyond the sound itself and includes the system through which it is performed. A downside of this situation, however, is that the novelty and digital features of the instruments create a sense of discontinuity with tradition, alienation, and lack of understanding by the audience as to what the instrument or performer is actually doing.

Evolutionary organology

The evolutionary development of acoustic and digital instruments are therefore different in kind. To understand the organological conditions of each, we might look closer into the evolutionary aspects of their materiality. Biologist Niles Eldredge, a co-author of the theory of punctuated equilibria (Eldredge and Gould 1972), is an ardent collector of historical cornets. In 2002, curious about the vast variety he observed in the design of cornets, Eldredge began collecting instruments and arranging them in phylogenetic or taxonomic relationships of shape, style, manufacturer, and age. Eldredge's biological theory of punctuated equilibria suggests that evolution does not happen through gradual transformation of whole biological lineages, but rather through quick morphological ruptures and discontinuities followed by longer periods of stability. Eldredge observes a similar evolution in the phylogenesis of musical instruments: "I knew that there were periods of stasis in cornet development, and also periods of radiation and innovation" (Walker 2003: 38). By running the design of selected cornets (seventeen characteristics or anatomical elements were identified, such as the bell-position and keys) through phylogenetic software, Eldredge retrieved results clearly indicating that musical instruments (and in general any designed artefacts) do not follow the same evolutionary development as biological systems. Artefacts are subjected to different and more non-linear laws than living beings, as designers can borrow features from other species of instruments, go back in time and re-implement features that have disappeared, or introduce features of different material modality, such as the electronics added to the halldorophone. The phylogenetic evolutionary lines thus differ in biological and cultural evolution and this is why plans, sketches, patents, drawings, schematic diagrams, photographs, and manuals are important in the study of cultural and technical artefacts.[1]

Biological science, including the phylogenetic theory of Eldredge, lacks the vocabulary or the discourse to analyse cultural artefacts as evolutionary phenomena. Dawkins's proposal of memetics (Dawkins 1976) is a prominent contender, but for some reason this field has not developed as other sciences have, perhaps because it lacks the empirical basis, or Popper's falsifiability, for it to evolve further. In any case, within philosophy, anthropology, and science and technology studies, there are myriads of theories regarding the evolution of technology. Simondon's theories of technicity and transindividuation (Simondon 2017) and Stiegler's epiphylogenesis (Stiegler 1998) are relevant, as they underscore how the technological potential or tendency precedes the implemented technologies. The system of technics in its totality serves as an environment in which ideas emerge as through the system's own logic. For example, Stiegler writes that "the logic of invention is not that of the inventor. One must speak of techno-logic, of a logic literally driving technics itself" (Stiegler 1998: 35). He borrows this idea from Bertrand Gille's "loose determinism," or the notion that although one cannot anticipate technological evolution a priori, innovation is never random, even if it appears so. Technological progress is driven by a specific logic innate to technology itself: "The system's dynamic offers the possibility of invention, and this is what is essential to the concept of technical system: the choice of possibilities in which invention consists is made in a particular space and particular time according to the play of these constraints, which are submitted in turn to external ones" (Stiegler 1998: 35). In other words, the technical infrastructure that evolves offers or suggests inventions (in the classical understanding of the term), instances of objects, or solutions that derive from the engulfing techno-social structure. For us – musicians – the embodied knowledge of play is part of that structure. Since the kinetic *action* already exists, we are offered the option to transform it into an ergophor and apply it in a new technology.

An example of how technological systems influence artistic ideas might be how the science of cybernetics became an omnipresent ideology in the 1960s, equally influencing economics, sociology, politics, arts and music. Gordon Pask pioneered this artistic research with interactive cybernetic installations in the 1950s, Pauline Oliveros created her tape echo cybernetic system in the early 1960s (Oliveros 1984), and Brian Eno and Robert Fripp popularised the technique with the development of the *frippertronics* in the 1970s. The environmental conditions of these inventions are those of economy, science, technology, society, and culture. The potential of invention is already there in the technological system and infrastructure: just as it is already certain that we will be sitting in a driverless car on our way to a lab to modify our genes. Philosopher of science Gaston Bachelard has pointed out that instruments are "materialized theories," admittedly discussing scientific instruments (Bachelard 1984), but my aim with this book is to demonstrate that the distinction between scientific and musical instruments is quite unnecessary to boot: it is simply that the study of music as an expressive art form, as opposed to a phenomenon that epitomises human cohesion, communication, ergonomics, and psychology has been more prominent in our culture.[2]

Musical organics

Evolutionary organology is easier when analysing acoustic instruments, compared with digital, since the digital is to a much higher degree an assemblage of technologies of diffuse origins. The attempt to analyse digital musical instruments through phylogenetic criteria is harder work due to the invisibility of the underlying and blackboxed mechanisms (code libraries, protected protocols, obscure standards, and closed hardware). We, therefore, have to acknowledge this complexity of the digital stratum and the impenetrability of the history of the digital instrument and rather look at its nature through dynamic taxonomical and ergodynamic orders. Simon Waters recently said in a keynote talk: "The challenge presented by hybrid physical/virtual instruments is not a minor typological one. It's a fundamental ontological shift, which exposes weaknesses in the typologies and methodologies which underpin organology, and the normatising nature of its organizational zeal" (Waters 2017). We therefore ask: how could we go about organising the organological classification of digital instruments? As Kartomi (1990) points out, there are no limitations to the number or types of analytical taxonomies that can be designed for musical instruments. The phylogenetic analysis of musical instruments provides us with one alternative history and classification. We can build a taxonomy of musical instruments from it, but that is a taxonomy based on genetics and not their various functions, such as the sonic, or ergonomic qualities. Hence, although it is tempting to understand musical instruments from the evolutionary perspective, there are other classification models that will take precedence, in particular that of interface materialities, which resembles the most popular existing classification systems.[3]

In current compositional practices in which the musical instrument becomes a system, or a piece on its own, and all available materials – whether it is horsehair or a gyroscope sensor – are equally applied in the new instrumental assemblage, it emerges that tree-like classifications break down. There is a demand for establishing organisational principles for these new digital instruments: inventors want to learn from each other (McPherson et al. 2016; Paine 2010), performers benefit from a stronger contextualisation of their musical practice, musicologists need a terminology to analyse and reference developments in the field, and composers wish to understand the instrumentation principles of these new technologies. The critical analytics of musical instruments can be helpful to all of the above. Organologists have presented a plethora of useful approaches to classifying and sorting musical instruments, useful for comprehensive musicological knowledge, for structuring information about instruments, or for organising a museum's instrument collection. The organisational principles are many and they differ amongst the world's cultures, often focusing on the material substrata of instruments and their vibrational function. Originating in the classical Indian *Nāṭyaśāstra* system, adopted by Mahillon in 1880, and redesigned by Hornbostel and Sachs in 1914, we now operate with the following classificatory divisions of instruments: idiophones, membranophones, chordophones, and aerophones. In 1940, Sachs added the electrophone category as a response to new musical materialities, such as oscillators, filters, pickups, and amplifiers (Sachs [1940] 2006). The electrophone category has proven insufficient for today's context, and various improvements have been proposed (Weisser and Quanten 2011). However, the problem is extremely complex as the field of new electronic instruments has dramatically increased in size, activities, and technological solutions since the 1940s, both in the analogue and digital domains.

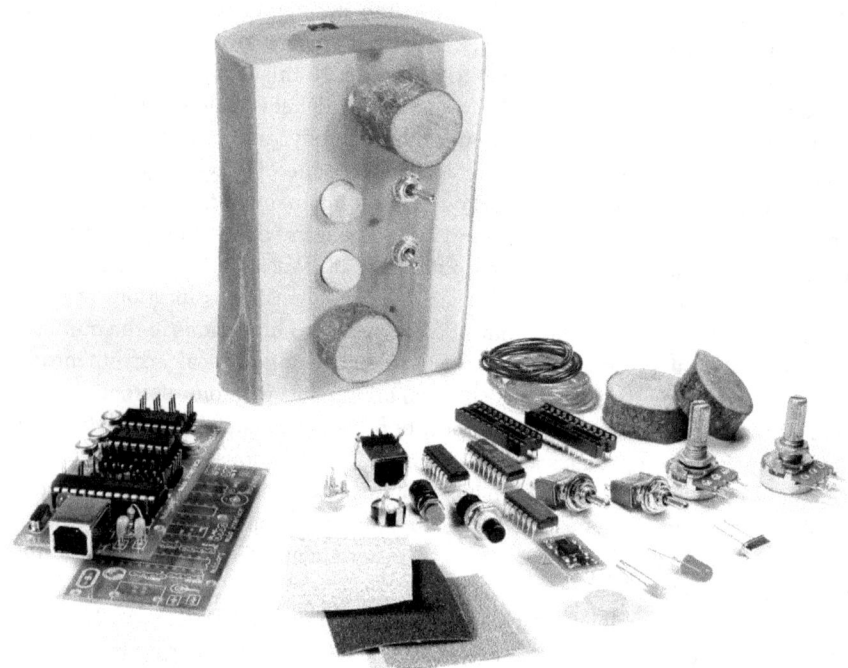

Figure 4.1 An Owl Project mLog. A gestural interface built with accelerometers, potentiometers, and buttons. Here with its organological guts displayed. © The Owl Project.

The problem we are faced with can be illustrated by the following question: what constitutes the organisational principles of digital instruments? We are surely not only interested in their physical materiality, as the plastic, rubber, or metal they are made of do not produce the sound they emit. Instead, we hone into the type of sound-producing algorithms, performer gestures, perception-modalities of sensors, mapping strategies, ensemble or inter-instrumental potential, to name but a few. The frameworks we apply to understand instrumental qualities from these principles form a study I call *musical organics*, or the critical analytics of musical instruments. I have argued that a comprehensive classification system is needed that supports such multi-perspective analytics, and this is now made possible with machine learning and dynamic multimedia representations (Magnusson 2017). The difficulty in attempting to continue the classificatory strategies of the nineteenth and early twentieth centuries for the new material reality of digital musical technologies is evident. For this reason, much of the work attempting to add digital instruments to existing organological classifications is never completed, or the authors acknowledge that the proposed attempt merely represents the first steps towards a new or improved organology.

Digital organology, the title of this chapter, can equally be understood as an approach to studying instruments applying digital humanities research methods (as Eldredge does with his cornets) and the study of digital instruments, where organology is the science of musical instruments. Either way, there is no study of digital instruments without reference to the past. It is also important to keep in mind that the organology we are performing here is not so distinct from Stiegler's "general organology" which he divides into three tiers: "the thinking of grammatization calls for a general organology, that is, a theory of the articulation of *bodily organs* (brain, hand, eyes, touch, tongue, genital organs, viscera, neuro-vegetative system, etc.), *artificial organs* (tools, instruments and technical supports of grammatization) and *social organs* (human groupings ... and social systems in general)" (Stiegler 2010b: 34). Although serving as the director of IRCAM between 2002 and 2006, Stiegler's work does not focus on musical instruments; his organology is wider (thus the "general organology"). However, I suggest that any instrumental organology needs to be that general too: we need to include the physiological, the technical, and the social in our understanding of musical technologies. The epistemic nature of our digital instruments has amplified the grammatisation of music and its production, it has changed our bodily relations to our instruments, and employed new sections to the totality of techno-social structure. A musical organics organology therefore relates to the technologies of the body, of our prosthetic tools and instruments, and what kind of social functions these are inscribed into.

New instruments, new ideas, new instruments

Musical instruments are considered a special type of material object because they both mediate and resist at the same time, they channel ideas at the same time as constituting the possibility of those ideas.[4] The contemporary awareness of the nature of media, the fractured self, the interconnectedness of things, or object-oriented philosophies, place instruments in a primary position as boundary objects through which we negotiate activities. Musicians have always been deeply attached to their instruments, but we are experiencing a new form of questioning by designers, manufacturers, and composers regarding what a musical instrument means. The so-called "material turn" in the humanities also relates to findings in cognitive science regarding embodiment (Ihde 1990), enactivism (Varela et al. 1991), and extended cognition (Clark and Chalmers 1998), which in turn can be coupled with social activist movements such as DIY, DIWO (Do It With Others), open source, modular synth building, home brewing, knitting, carpentry, and other activities that seem to provide hands-on quality time away from screens and networked media.

In the craving for the real, acoustic instruments become the epitome of material objects with which we build a personal history and connection. The digital has pointed a different lens at instrument making, similar to the way in which photography redefined the fine arts, or how artificial intelligence has created a mirror of ourselves in cognitive science. If relational philosophies framed the musical work as an assemblage of a composer, performer, audience, score, instrument, architectural space, instrument maker, manufacturer, media, and so on, it is only natural that composers and performers began to analyse and rethink the roles of each these elements. A good example is how Helmut Lachenmann introduced his "musique concrète instrumentale," or concrete instrumental music, in the early 1970s. This approach emphasises the concrete nature of the instruments, re-thinks their potential as sound sources, and congruously presents a musical notation that prescribes performance actions and extended techniques. Here notation is about what to do with the instrument, and not about its desired pitch: "This means a music in which sound events are chosen and organised in such a way that the nature of their origin is considered at least as important as the resulting acoustic properties themselves" (Lachenmann in Craenen 2014: 84).

The focus on the instrument itself as a source of embodied activity – a focus on relationship, skill, and presence – has yielded a design approach where the instrument becomes a material source for sonic investigations, rendering it more as an end in itself than a medium for expressing neutral musical ideas. Lachenmann's composition requires that performers become conscious of their own body and their relationship with the instrument's body, an awakening of sorts where it is not only "a concrete body's ability to touch, but the exploratory touch of a spirited body. Thus touching the instrument gains a double function: it is simultaneously an instrumental action with the intention to produce sound and the proprioceptive experience of touching" (Craenen 2014: 91). Lachenmann's method of composition shifted the focus onto the materiality of traditional instruments, their quality of wood, strings, and horsehair, at a time when instruments were increasingly made out of plastic with mapped control elements and inbuilt screens: bodies made of information rather than natural materials.

While many composers and performers of acoustic instruments engage in the study of borders, edges, chaos, and non-linearity as creative resources, there are other composers who feel that traditional instruments have been exhausted of expressive potential. For composer Giovanni Verrando, the traditional acoustic sound paradigm had become impractical and inadequate, so he embarked upon becoming a luthier himself, gaining a sufficient amount of skill and traction for him to be able to set up a dedicated course of study in Lugano, Switzerland, where composers study the relationship between composition and luthiership. Here students learn the craft of building new musical instruments that inspire sonic imagination, an imagination that builds on the twentieth-century focus on timbre, yet emphasises the twenty-first-century focus on material objects, instruments, and new notations. Verrando characterises the twentieth century as a period of rethinking the symphonic orchestra, freeing it, concentrating on the sound it affords, as opposed to simpler pitch range compositions, and concurring with Pierre Boulez's critique of the orchestra, which "still carries the imprint of the nineteenth century, which was itself a legacy from court tradition" (Verrando 2014: iii). For Verrando, this demonstrates how our compositional theories and technical evolution do not evolve in a linear fashion, but rather in more complex and manifold ways. The move is one from quantitative to qualitative compositional approaches, or from thinking with symbols to operating at the level of the signal:

It was not until the twentieth century, however, that timbre would be definitively established as a structural parameter. More precisely, it was in the second half of that century that, thanks to information technology, electronic music and the work of spectral composers, we came to understand that the concept of timbre could no longer be conceptually and materially separated from those of pitch, dynamics and temporal flux. (Verrando 2014: 36)

New compositional ideas are often inspired by new technological potential (such as spectralism or the focus on timbre deriving from electronic and digital technologies), and the current ideological framework in contemporary music, ranging from clubs to conservatories, have hatched an environment where musicians become technologists. Composition involves system design or instrumental building. For instance, Marianthi Papalexandri-Alexandri's work exemplifies an approach that transcends past distinctions between an instrument and a piece. Papalexandri makes her own instruments and installations out of wood, strings, loudspeakers, and electronics, often performing the pieces herself. Combining acoustic and digital technologies, her work feels contemporary and strongly post-recording, as the power of her pieces, with heavy visual elements of instruments and their performance, go beyond the sonic and become multimedia. Sound or video documentation does not do the pieces the justice they demand.

This relationship between material technologies of sound and our aesthetic musical ideas is complex and multifaceted. Technology embodies theory and this affects the music, as much as our musical ideas are written into our software and hardware. Digital tools have democratised music, and we now have a situation – contrary to the electronic music studios of the mid-twentieth century with privileged access – where advanced music technologies can be in the hands of anyone. The composer at IRCAM and a school kid in a developing part of the world might use the same tools in their music (CSound was installed on the One Laptop Per Child computers that were distributed all over the world) and the live coder in Mexico might use the same language as an astronaut in the International Space Station (the Sonic Pi live coding language is pre-installed on the Raspberry Pi computer that went into space). For decades, electroacoustic composers and electronica musicians have shared the same tools, resulting in music that is often similarly sounding, but with aesthetic ideas and social context that are often separated by light-years.

Conclusion

The previous sections have shown us that musical instruments, whatever materials they consist of, are vehicles for musical and scientific thinking. Let us not forget Pythagoras's *kanōn* ("measuring rod"), an instrument he used for studying the harmonic rationality of the world (Rehding 2016). In the West, we talk about musical canons which are the musical works considered central to a tradition (and which we use as a standard against which to "measure" other works), but in the Arab world the word "canon" continued to be used for a string instrument and today we find the popular Arabic zither-type instrument called the *qanun* (قانون), typically tuned in a quarter-tone equal tempered tuning (with twenty-four notes to the octave). Interestingly, the qanun is a great example of an epistemic tool that defines our musical world and vision, one that, owing to its unfamiliar tuning, opens up doors to musical thinking profoundly different from that found in the Western tradition.

Against the logic that sees musical instruments as a direct channel for our musical thoughts, seamless media that materialise our imagination, the critical analytics of musical instruments reveals them as epistemic objects with agency; as technologies that are not supposed to merely channel but to also to converse, resist, challenge, surprise, and reject their performer. This is where much of the exciting work in new music technology design currently resides. Instead of being a dead medium of musical means, the digital instruments becomes the object of focus, an end in itself, akin to, and perhaps surpassing, its acoustic predecessors. The compositional and performative act becomes that of exploring the ergodynamic nature of the instrument, to "torture" the music out of the object, much as scientific instruments were used to torture nature to reveal her secrets in the seventeenth century.

This chapter has looked at instruments as objects with which we think musically – both as epistemic tools and ergodynamic artefacts to explore – and for twenty-first-century musicians, they become material forms into which we write our musical theories through complex grammatisation processes. In the next part of this book, we will look into a different type of sonic writing, that of musical notation, as this also forms an integral part of our digital music technologies.

Part II

Symbolic Inscriptions

5

Writing Music

How do we write music? What is musical notation? Why do we write music? These are simple questions with complex answers that relate to culture, media, syntax and semantics, communication, practice, education, and performer skills. La Monte Young was so impressed by David Tudor as a performer, by his predilection for breaking out of instrumental habits in his improvisational and interpretative practice, that he considered Tudor to be "an instrumentation" on his own.[1]

> Bring a bale of hay and a bucket of water onto the stage for the piano to eat and drink. The performer may then feed the piano or leave it to eat by itself. If the former, the piece is over after the piano has been fed. If the latter, it is over after the piano eats or decides not to. (Young in Schaeffer 1996: 27)

Young's *Piano Piece for David Tudor #1*, quoted in full here above, exemplifies the fractured mosaic of notational practices in contemporary music that we need to understand in order to grasp how musical notation is transduced when represented in digital media. Young's piece is clearly a program: it describes the objects, actors, variables, parameters, functions, and the conditions required to set up the process of the piece. It is an instruction, like a cookie recipe, or a map that is to be followed to reach a destination. The piece is left open to the accidentals and possibilities of performance: either the piano eats the hay or not. Young, who clearly has a solid understanding of instrumental agency, makes the piece rest upon the instrument's agency, and the performer serves the instrument, like a stable boy feeding and grooming a horse.

Music precedes and was a constitutive force in humanity's evolution: in the introduction we explored Tomlinson's theories of music as a pattern-making activity that formed our ancestors' perception of time, rhythm, collaboration, memory, and social cohesion. In early hominid cultures, repetitive actions that applied primitive tools for work (such as cracking nuts or cutting meat) are forms of work patterns with accompanying sound that would eventually become a musical activity as an end in itself; they are the result of a process that separates the symbolic from the functional. These types of accounts open up a questioning as to what came first, musical instruments or language, as the premise here is that the very possibility of spoken language depends on developed vocal cords. However, the voice does not have an external referent for tunings or scales. With instruments we get prosthetic technologies that extend beyond the body, forming a tertiary memory (Stiegler 1998) where music is inscribed into material objects. The early sound-generating objects were our instruments, but they were also the ergodynamic vehicles that framed the social potential and further continuity of musical practices.

In the grand scope of musical evolution, as told by authors such as Tomlinson (2015), Cross (2001), Mithen (2005), or Patel (2008), the practice of musical notation is very recent, and the attribution of music to a specific composer has existed only for a fraction of that time. Musical notation becomes possible with the advent of writing technologies; the oldest known musical notation is the Hurrian clay tablet from around 1400 BCE, found in the Syrian city of Ugarit (Nettl 2015: 268). That music is written in cuneiform notation, containing a hymn to the goddess Nikkal. The transcription shows it is for voice and an accompanying instrument, a lyre or a harp, and an accompanying tablet contains information on how to tune the instrument. There are also instructions regarding how the music is to be read, thus both program and data on the same medium, a practice we are very accustomed to with digital computers. In the 1970s, researchers at the UC Berkeley music department, under the lead of Professor Richard Crocker, released what they called the "world's oldest music" on a vinyl album. The music has been recorded and performed many times since, for example by Ensemble De Organographia.[2] Whatever success this ancient music is experiencing now, musical notation did not become part of any prosperous musical practice until three millennia later.

On writing

In *Phaedrus* 275a, Socrates warns of writing that if "men learn writing, it will implant forgetfulness in their souls," stating that this is a technology not for memory, but for reminding. We may smile at Plato's sceptical logocentrism (Derrida 1997), but we know he was right, although this is somewhat paradoxical, considering that his own works were written down. Plato pointed out what the advent of writing would do to our mnemonic practices: no one can remember Homer's *Iliad* anymore, nor would anyone embark upon memorising Joyce's *Ulysses*, and we know that melody and metre were important mnemotechnical devices for memorising such long poems. However, with the advent of writing, theoretical treatises on music emerged, preserving the music theory of the ancient Greeks such that it largely continued throughout the Middle Ages. In this early music theory, the notation of musical works was not given much importance as music was primarily transmitted orally and by imitation, through a lineage of musicians that preserved the musical tradition. As is customary in oral cultures, tradition has priority over innovation, as we find exemplified in Plato's *Republic*, where it is written about music that the

> overseers of our state must cleave and be watchful against its insensible corruption. They must throughout be watchful against innovations in music and gymnastics counter to the established order, and to the best of their power guard against them,

fearing when anyone says that that song is most regarded among men "which hovers newest on the singer's lips," lest haply it be supposed that the poet means not new songs but a new way of song and is commending this. But we must not praise that sort of thing nor conceive it to be the poet's meaning. For a change to a new type of music is something to beware of as a hazard of all our fortunes. For the modes of music are never disturbed without unsettling of the most fundamental political and social conventions, as Damon affirms and as I am convinced. (*Republic* 424b)

As we have seen in earlier chapters, there was plenty of music theory in ancient Greek writings. These were mostly theoretical studies of harmony, ratios, tunings, scales, etc. There are some existing musical notations from the time of the ancient Greeks, in the form of fragments, on tablets and papyri (Pöhlmann 2001), but these works are few (sixty-one pieces, extending from the fifth century BCE to the third or fourth century CE), compared with the abundance of writings on music theory. In terms of information theory, and compression of data in an age of scarce media formats, this makes sense. Before the printing press and the mass distribution of musical scores, what would be more valuable to write down, the rules for making music, or an instance of music resulting from those rules? Only through cheap paper and streamlined printing and dissemination practices, resulting from the invention of the Gutenberg press in the fifteenth century, do we begin to see a focus on the musical work itself, celebrating the composition and its author. As an example of this division between theory and practice, Aristoxenus, a pupil of Aristotle and a key music theorist of Greek antiquity, excluded musical composition from music theory on the grounds that the former involved mere skill (*techne*) and not theory or science (*episteme*). For him, notation contributed nothing to the understanding of the art:

> Now some find the goal of the science called Harmonic in the notation of melodies, declaring this to be the ultimate limit of the apprehension of any given melody. Others again find it in the knowledge of clarinets [*aulos*] and in the ability to tell the manner of production of, and the agencies employed in, any piece rendered on the clarinet. (Aristoxenus in Barker 2007: 194)

For Aristoxenus, such views are "conclusive evidence of an utter misconception" (Barker 2007: 194). Aristoxenus also pointed out two extremes in the music theory of the time: on the one end of the spectrum were the Pythagoreans, who favoured reason and mathematics, and viewed perception only as mere sparks for the intellect to rationalise musical theory; on the other end were the more empirical *harmonikoi* (harmonicists) who based their musical evidence on perception alone, and the *organikoi* (instrumentalists) who found musical coherence in their instruments, also preferring perception and instrumental structure above theory (Barker 2007: 439). This division between musicological rationalists and empiricists would continue in the following centuries, although the emphasis and weight was always on theory as opposed to practice, arguably derived from the schism instituted between *episteme* and *techne*.

With Aristoxenus declaring the notation of musical pieces to be of no use for scientific enquiry, most ancient and medieval theoreticians on music ignored musical notation. Around the first century CE, Greek music theory in the form of a scientific and philosophical tradition was beginning to fade, instrumental skills were deteriorating, and theorists were paying little, if any, attention to the notational symbols in earlier writings (where alphabetic letters signified pitch). Whatever existed in the form of musical notation was largely forgotten, although music theory and harmony continued to be studied. In terms of Greek writers on musical notation, Alypius (fourth century CE) was the most systematic and coherent author on notation. He was a musicologist and author of the *Introduction to Music* (*Eisagoge mousike*), which includes a complete tabulation of Greek musical notation. His work never became well known, perhaps due to the general apathy for musical notation amongst Greek and medieval theorists. It was not until Girolamo Mei and Vincenzo Galilei, the father of Galileo, began studying it in the sixteenth century that it began to have its revival in music theory.

Throughout antiquity and the medieval period, musical composition in the form of notation was seen as a domain of *techne*, as memory aids for practitioners. These works were not copied, nor were they discussed or analysed by music theorists; notation was a technology for training instrumental technique and performance practice, as opposed to music theory. Music theory (*mousike*), on the other hand, had a far more important role than theorising the structure of performed music. Music theory explained how the world is construed, it theorised the rationality and harmony of existence through logic that could be tested by resolving to actual instruments like the harp. Unlike theory, the practice of music was a skill (again, *techne* as opposed to *episteme*), and something that should not be studied in detail. Aristotle says, in Book VII of his *Politics*, that "since we reject professional education in the instruments and in performance ... we must therefore give some consideration to harmoniai and rhythms ..." (Strunk 1998a: 32), which is clearly an imperative statement that frames the study of music in favour of theory. We can speculate about this lack of interest in notation, and about the focus on education and practice within the context of how an alternative myth portrays the Muses. They are traditionally said to be the nine daughters of Zeus and Mnemosyne (personification of memory – see Chapter 1), but in some myths there are only three muses, called *Melete* ("practice"), *Mneme* ("memory"), and *Aoide* ("song"). Thus we find, in ancient mythology and thought, a notion of music as something that relates to memory and practice, not to be instantiated through writing but through playing, although the *science* of music, of ratios, harmony, and rhythm, became been well established following the work of Pythagoras.

On the possibility of writing sound

In the fourth century CE, St Augustine of Hippo presented his interpretation of Platonic cosmology in *De Ordine*, where he stated that: "Take now music, geometry, the motion of the heavens, number theory. Order is so overpowering in these, that anyone seeking its source will either find it there, or will be led to it through them without error" (Augustine 2007 §2:5). The science of music was given a status next to the numerical sciences in the quadrivial arts, which were part of the liberal arts defined by Cicero in the first century BCE. The liberal arts were seen as a preparation for the mastery of oratory, and included geometry, music, grammar, poetry, natural science, and moral and political philosophy. For St Augustine, music had a special role in that it combined the abstract and the empirical, the mind and the senses: "Now number is a mental construct and, as such, ever present in the mind and understood as immortal. Sound, on the other hand, is temporary and fleeting, but can be memorized" (Augustine 2007 §2:14). St Augustine here expresses the enduring duality derived from the Greek understanding of music in which the science of music is about harmonic relationships, a representation of the world, but the performance of music is a skill, a mimetic form, which suggests but cannot embody truth.

A few centuries after Cicero, Boethius (480–524 CE) formally structured the liberal arts into the trivium (grammar, dialectics, and rhetoric) and the quadrivium (arithmetic, music, geometry, and astronomy).[3] His *De institutione musica*, together with Cassiodorus's *Fundamentals* and St Augustine's *De musica*, were key works that brought Greek music theory into the medieval period. These texts define music as an intellectual discipline, a *musica scientia*, and they were "probably not intended to be read by musicians or composers" (Palisca and Bent 2001). This distinction between theory and practice, where the valuable knowledge of music is primarily seen to remain in the theoretical domain, is clearly expressed by Cassiodorus: "Music is indeed the knowledge of proper measurement [*Musica quippe est scientia bene modulandi*]" (Strunk 1998b: 34). It is the discipline that deals with numbers expressed in sound, and emphasis on the theoretical as the key focus in the study of music is also evident when Boethius writes: "Nor do I affirm that all those who handle an instrument lack knowledge [*scientia*], but I say that not all of them have it" (Jeserich 2013: 59). Indeed, in *De institutione musica* he divides philosophy into theoretical (speculative) and practical (active). Boethius's work became the standard text for music in European universities, as it was well known to the eleventh- and twelfth-century scholars who founded the curriculum of the new European educational establishments.

The writings of Isidore of Seville in the early seventh century confirm the central position music held across the sciences in medieval times: "Thus without music no discipline can be perfect, for there is nothing without it. The very universe, it is said, is held together by a certain harmony of sounds, and the heavens themselves are made to revolve by the modulation of harmony" (Strunk 1998b: 40). Isidore summarises St Augustine in *De ordine*, stating that music is the art of measurement, in particular of song and pitch, and the "sound of these, since it is a matter of impression upon the senses, flows by into the past and is left imprinted upon the memory. Hence it was fabled by the poets that the Muses were the daughters of Jove [*Zeus*] and Memory [*Mnemosyne*]. Unless sounds are remembered by man, they perish, for they cannot be written down" (Strunk 1998b: 39). This famous statement demonstrates how inconceivable musical notation was in the seventh century. Sounds cannot be written, Isidore says, pointing to the important role of oral transmission, to the practice of *imitatio* where practical music is seen as a form of mnemonic art, one which typically concurrently supports and is supported by lyrical poetry. This relates to the role of the voice, which was seen by authors such as Cassiodorus as the perfect instrument, created by God, in contradistinction to human instruments, which were only imperfect simulations thereof.

After Boethius and Cassiodorus, very little new original work of music theory was produced, including Isidore's writings, which are mainly summaries of existing knowledge. One of the reasons for the oblivion of earlier music theory is that by the sixth century, the ancient Greek language was largely forgotten in the West, although Arabic and Byzantine scholars maintained and developed this body of knowledge. However, it was not until in the eleventh and twelfth centuries that this treasure of knowledge preserved and progressed in Arabic scholarship begins to be translated into Latin, a period which prepares the ground for the European Renaissance (Mathiesen 1999: 609), and in which new theoretical knowledge on ancient music begins to emerge. Examples include ninth-century musicologist Al-Kindi, the tenth-century philosopher and musicologist Al-Farabi, and the twelfth-century Ibn Rushd (Averroes), all of whom were translated into Latin. These writers wrote extensively about the music theory of the ancient Greeks, from Pythagoras to Aristoxenus.

Symbols notating music

In many ways, Isidore is right: it is impossible to write sound. If we are thinking of reproducing music, the act of interpretation, audience, and architecture will always be part of the performance, rendering no two performances the same. And, paradoxically, machine recording and reproduction can never be perfect either, as the choice of microphones will colour the recording, the uniqueness of the time and space of recording will affect the performance, and the audio equipment used to listen to the music will further condition the music. Even the music as "written" in our own memory can never be reproduced for these very reasons.

The notion that music could not be written down began to be questioned in the ninth century. Neumes began to be used to indicate articulations or inflections of notes, not the pitch or the duration of the note. The neumes are a form of ekphonetic notation, or symbols added to texts, a practice that became common in early Christian, Jewish, and Arabic texts. Neumes were used as memory aids for monks in learning to sing the psalms. The term *neume* is Latin for a hand gesture or a sign, and the lines above the alphabetic writing imitated such hand movements. The neumes were dots and curly lines that were written above the text and included symbols for going up or down (*virga* – "/" or *punctum* – "."), indicating melodic movement, but only vaguely signifying intervals, and certainly not notes. This type of notation is first and foremost an aid to memory, a descriptive system that supports musical performance, in this case singing. The neumes were not used for instrumental music. As some authors have pointed out (e.g., Treitler 1992), neumes should not be seen as imperfect forerunners of staff notation. The music would still have to be memorised, transmitted orally, and with quite some effort too, according to contemporary accounts. This origin of notation in sung words as opposed to instrumental practice meant that notation developed in a particular way – for example, we do not find much focus on timbre, amplitude, or duration of the notes, as those parameters would be defined by the syllables of the words sung, as well as their semiotic context.

Hucbald of St Amand, in the late ninth century, developed a way to indicate pitch more precisely by combining neumes with alphabetic notation, inspired by ancient Greek notation. The ninth century saw the birth of music treatises, such as *Musica enchiriadis* (often attributed to Hucbald), making use of Dasia signs, which operated with writing syllables on horizontal lines and with symbols signifying pitch changes. Daseian notation demonstrates that the technology to write exact musical pitches, as opposed to the relative and indefinite neumes, was available already in the tenth

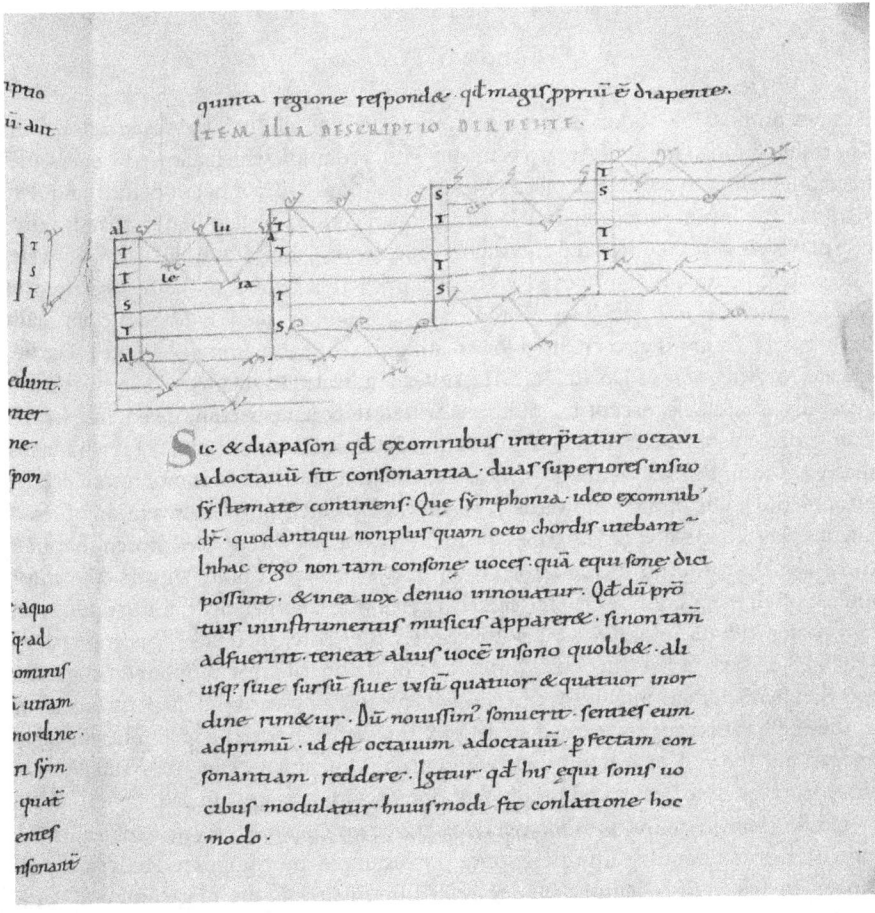

Figure 5.1 An example of Daseian notation in *Musica enchiriadis*, from the ninth century. © Biblioteca Apostolica Vaticana.

century, but the general acceptance of music as an "art of memory" meant that notation, as we conceive of it today, was not of much interest.[4] Behind these early musical notation systems lay established song-models from the period. The melodies were often simple, seen from later standards, but this allowed performers to improvise from the melodies, to place their own mark on the skeleton provided by the composer. All kinds of musical techniques were applied to the melodies, such as polyphony, heterophony, and other ornamentations. In polyphonic plainchant singing, the principal voice (*vox principalis*) would be accompanied by an "organal voice" (*vox organalis*), in notation systems often called *organa*. A similar model of improvisation from a known musical scale is also central to Arabic and Indian music, with their respective systems of the *maqam* and the *raga*. Music theory and procedures for music making were set up, often quite sophisticated, but exact and linear prescriptions of how to perform absolute pieces were not seen as important aspects of musical life.

Guido of Arezzo

It is not until 1030 CE that we get a precise description of pitch in Western notation. The Italian monk Guido of Arezzo came up with a solution which allowed singers and instrumentalists to perform a piece of music that they had not heard before, and this signified the invention of musical writing proper. Guido's system had four staff lines, two of which were coloured red and yellow, representing the notes F and C respectively. The system included the use of keys (or clefs), where a symbol at the beginning of a score, written on the staff lines, would represent the key, such as F (bass clef) or G (treble clef). Guido is also credited for creating the solfège, naming the notes (Ut, Re, Mi, etc. known today as Do, Re, Mi). The solfège indicates the degrees of a key, and are generally modulable, except in traditions where they have become fixed and where "Do" represents the pitch of C. A highly rational project, it left out some of the nuances that were found in the neumes, such as articulation, local rhythm, microtonality, or portamento (slide of pitch, upwards or downwards). For Guido, the invention of fixed pitches was a necessity; he describes a situation of chaos within the Church that had come about as a result of the imprecision of the neumes, calling singers "the most foolish of all men" because "[o]ne scarcely agrees with another, neither the pupil with his master, nor the pupil with his colleague." Therefore, to relieve this monastery squabble he says, "I have decided, with God's help, to write this antiphoner in such a way that hereafter any intelligent and studious person may learn the chant by means of it; after he has thoroughly learned a part of it through a master, he will unhesitatingly understand the rest of it by himself without one" (Strunk 1998b: 102). This system allowed for pitch to be notated for the first time, as Guido states himself.

Guido's staff notation proved useful, and for two centuries it was applied in the monasteries for notating liturgical songs, for example, by composer Hildegard von Bingen in the twelfth century. Hildegard is often listed as the first composer with attributable works, as her name was on the scores, but most music was anonymous in this period. Her compositions were monophonic, descriptions of pitch only, as there was no note length, tempo, or rhythm available in Guido's system of notation. This problem of time was addressed by the Notre Dame school of polyphony of twelfth-century Paris, where Leoninus and Perotinus were instrumental in developing a system of notation that would enable two or more voices to sing different melodies in parallel, and was often called modal notation (modal for "measure," "size," or "length"). Leoninus compiled and composed a work around year 1200, called *Magnus liber organi* or *Great Book of Organum*, in which both pitch and rhythm were incorporated into the musical score. This work was greatly improved by Perotinus, who composed four-voice polyphonic pieces. The new rhythmic inventions of musical notation were not relevant in monophonic music, but suddenly become crucial when notating for two or more voices. This early type of polyphonic music was based on the plainchant, but when expanded into separate voices, the term *organon* became applied for this musical technique. The modal notation developed into mensural notation which afforded the description of rhythmic durations and polyphony, a system that was heavily used until the seventeenth century.

While these emergent notational practices were primarily used for the voice, an interest grew in instrumental notation, and new development took place in the fourteenth and fifteenth centuries with the tablature system. Melodic or single voice instruments could be notated for using the same notation as the voice, but for polyphonic instruments, like the keyboard, the harp, the lute, or the guitar, tablatures proved useful for the notation of simultaneous voices played on one instrument. This new system also coincided with (and we can speculate whether it influenced or was influenced by) a change in playing technique. In the fifteenth century, lutenists and players of other stringed instruments began to use their fingers as opposed to a plectrum, and this resulted in increased polyphony. Tablatures are an idiomatic language for writing instrumental music, for example where each staff line of the score represents a string on the instrument. When melodies are written in tablature form, it is called "intabulation." This system is still used today, typically in guitar instruction or in new digital instruments. The power of this notation is in its functional manner: the location on the instrument is specified, not the pitch, so the layers of the sign are different; in Peircean terms, Guidonian notation is symbolic, while tablature is indexical (see, for example, Peirce 1868).

Mnemonics and notation

As the story of notation is told, we might consider polyphony appearing as a result of notation, through the externalising and spatially arranging temporal sonic information. However, some authors have argued (e.g., Navarro et al. 2009) that the large cathedrals in which the plainchant was sung had such a long reverberation time (up to ten seconds) that notes would overlap and thus artificially creating polyphony via the technology of reverberant stone-wall architecture. New notes sung should ideally be in harmony with the previous notes as they would have to coexist for a few seconds in the reverberant space. These historical accounts explaining the origins of polyphony are, of course, following a rather narrow European musical history, as polyphonic music can be found in sub-Saharan Africa (e.g., with the Aka Pygmies) and Oceania. However, this music of faraway lands was not available to medieval Europeans, so the question of notating it never came up. Chain-singing was also a common musical technique in the European Middle Ages, and this is effectively a polyphonic technique, although not notated polyphonically.

Guido's staff notation effected a spatial spread of notes on the X and Y axes, allowing visual representation and thus an atemporal understanding of the music, as data spatially laid out in front of the musician. This was not an original intention of the system, but with such a tool, memory becomes spatialised, visualised, externalised, offering thought processes that were previously hard to conceive of. This spatial or synchronic character of notated music meant that performing it becomes a form of re-temporalisation, of reinstating its diachronic temporal nature. The effect of externalised musical thinking through the instrument (organum) of notation quickly materialised: after polyphony, we begin to see counterpoint (*contrapunctum*, a term that relates to the neume notational practice), the musical techniques of the fugue (mirroring, transposing, inverting, retrograding, expanding, delaying, etc.), and polyphonic instrumentation for more than one instrument; all techniques that benefit from notation as a technology of composition and arrangement.[5]

In her book *Medieval Music and the Art of Memory*, Anna Maria Busse Berger (2005) develops the theory that before notation, composers would memorise complex rules of composition, and generatively apply them in musical situations with singers or instrumentalists, or both. In the typically formulaic oral cultures, the student is given a specific rule for every problem. Berger describes how, in "the discant and counterpoint treatises, consonance tables, interval progressions, and note-against-note progressions are systematically committed to memory by endless and tedious 'rules' and examples, which were clearly considered the central part of learning counterpoint" (Berger 2005: 156). However, with literacy the rules would become fewer and more general, as compositions could be noted down and disseminated, and practitioners could refer to the score as written evidence: the external object to consult when in doubt, rather than having to consult the memorised rule set of note-against-note progressions (Berger 2005: 157). This most likely relates to how printing enables cheap dissemination of musical works, so they do not have to be compressed as a generative theory: their instances can be written down (see the related discussion about genotypes and phenotypes in Chapter 8).

Berger's theory demonstrates that with writing, orality does not disappear: the two modes coexist. She emphasises how composers would follow strict rules, and compose in the mind applying the strategies of *ars memorativa*, the medieval system of mnemotechnics. This was done through various devices, such as generative rule tables, classification of musical materials, hand gestures, scales and didactic verses, and other instruments of memory involving graphs of trees, maps, circles, ladders, and more. It is indeed peculiar that the inventors and users of neumes worked for decades, writing musical symbols above liturgical texts, without ever coming up with a system that defines absolute pitch relations. Berger reminds us, in this context, that composers and performers could not be separated for "both relied on the same practice" (Berger 2005: 128), stressing that the use of notation up to the fourteenth century was primarily an aid to memory in recalling music that had already been memorised by the people who sang it.

Conclusion

This section presented how our scholarly ancestors viewed the domain of music primarily as a theoretical discipline, though with practical dimensions of actual performance. Performed music was an expression of music theory, a phenomenon which has shifted in modern times, where music theory becomes an expression of composed music. The written theory of music was seen as a key part of the liberal arts, a sonic expression of mathematics, harmonics, and rhythm or metre. Musical practice, however, existed primarily as oral culture. It had a theoretical, literate underpinning, but individuated musical works were transmitted orally, learned by heart and taught by imitation and participation. Notation emerged as a memory aid within ecclesial practice, and although the technology was applied by practitioners of secular music, they did not have the same infrastructure of copying and disseminating texts as their clerical counterparts, and so most of that music is now lost. As studied in the first part of this book, the instruments themselves were also conceived of as vehicles of music and musical thought. For medieval practitioners, the primary function of music, as recommended by Plato, was to preserve music as cultural memory, as something that represented the harmony of the world, and innovation was therefore frowned upon. The next section studies how the advent of movable type printing introduces the writing down of tradition, of extensive musical archiving, thus freeing cultural memory for the possibility of new works.

6

Printing Music

Before musical notation, music could only be preserved through repetition, by continually re-inscribing it in people's memory through performance. The temporal and ethereal nature of music means that it is never there, present, in front of us; it simply materialises as a succession of fleeting moments in time, as an effect of vibrating objects. With notated music, an external reference point is established via script that represents and preserves the music, offering new methods of composing, performing, and disseminating music. It becomes possible to preserve and archive old music, and consequently another effect emerged: since tradition can be stored on this external medium (as opposed to memory), the act of composing new musical pieces does not constitute a threat to the tradition. In the *Republic*, Plato often applies the same arguments to music and gymnastics, namely that innovation should be prevented at all costs. This type of thinking seems alien to us today, but we might perhaps understand his point against musical innovation better if we were to imagine a situation in our current sports, where composers (or game designers) continually redefine the rules of the game. Imagine if football had a repertoire of games written by specialist game-designers, with new games (rule sets) regularly composed. The players would then have to study the new rules, interpret them, practice, and perform. For us, the spectators, the excitement of going to a football match would partly be to experience the new rules. And imagine all this without any broadcasting or recording media! We can argue about the usefulness of this analogy, but it illustrates, to some degree, what Plato had in mind.

Written music offers transmission through text, the possibility of referencing and quoting other works, and it becomes a shared external cultural memory. It spatialises time by arranging temporal forms into visual patterns that can be manipulated in a very different way than in repeated recitals of music through song or instrumental performance. Spatially laid out, written music offers itself for considered analysis. Notation is a prime example of what Stiegler calls "grammatization," or the "analytical process by which temporal and perceptual flows of all kinds are rendered discrete and reproducible through being spatialized" (Ross in Stiegler 2018: 20). This property of the symbolic grammatisation of music also opens up for experimentation of polyphony and other structural operations. Furthermore, by organising music in the synchronic domain of the score, musical notation brought with it the possibility of a composer as a profession, since in earlier practice composition was a dispersed activity among performers and theoreticians.

With the increased use of musical notation, a focus was directed at the medium itself, as typically happens when a new medium appears. This resulted in gradually more complex compositions which began to explore the formal properties of the new media format. Sometimes, this manifested in a focus on the art of notation itself, where musical manuscripts became works of eloquent design and sophisticated expression. This can be observed in fourteenth-century work, where Senleches's score for "La harpe de melodie" is designed in the shape of the harp, whilst Cordier's love song "Belle, bonne, sage" is notated in the shape of a heart. That piece is from the Chantilly manuscript which also contained "Tout par compas," a circular notation resulting in a practically infinite canon. The style of which these pieces were symptomatic was called *ars subtilior*, and it emphasised complex rhythmic and notational practices.[1] Due to the experimental format, these pieces were sometimes called *Augenmusik* (music for the eyes). Experiments with the format of the musical score were common in the early days of music, for example, canons written as a square for four singers standing on each side of the score, or Ghiselin Danckerts's 1535 *Ave Maris Stella*, a non-linear rule-based notational piece, where the melodic parts were scored onto the squares of a chess board, and the music would evolve according to the chess play (Westgeest 1986).

The advent of notation brought considerable changes to the established practices and technologies of music, but this transformation was not immediate. Views of the nature of notated music, its role in science and society, had yet to be experimented with and debated. For example, during the fifteenth century, the idea of a composed notated work as an artistic product was quite alien. This is exemplified by Leonardo Da Vinci, who, eager to include painting within the liberal arts (quadrivium), argued that music is subordinate to painting due to its immateriality. With an argument that echoes that of Isidore of Seville, Da Vinci described music as being ethereal in nature: "painting excels and surpasses music, because it does not perish as soon as it has been created, which is music's misfortune" (Vergo 2005: 142). Although musical notation existed in Da Vinci's time, his statement demonstrates an understanding of music as a theoretical discipline that is reflected through practice, but not one of individual written works. This can be illustrated by the fact that among the diverse topics studied and developed in Da Vinci's extensive notebooks (about 6,000 pages of notes and drawings), including instruments, music theory, and acoustics, there are no musical compositions. And yet, Da Vinci was a fine performer of the lyre, an instrument he built himself, often performing at the Sforza court, where he improvised his music with such virtuosity that he was considered by his contemporary Vasari to be "the best improviser of verses of his time" (Isaacson 2017: 119).

These experimental designs of musical notation in the fourteenth century became less prominent by the middle of the fifteenth century, when composers began shifting their focus to the harmonic content of music, or polyphony, which was something the technology of notation made easier to do, indeed suggesting it by the means of its spatial nature. Musical notation was becoming concretisised, a standard, and it diversified the functions of music across society. The mastery of musical instruments became more common, which in turn advanced the trade of instrument making. What

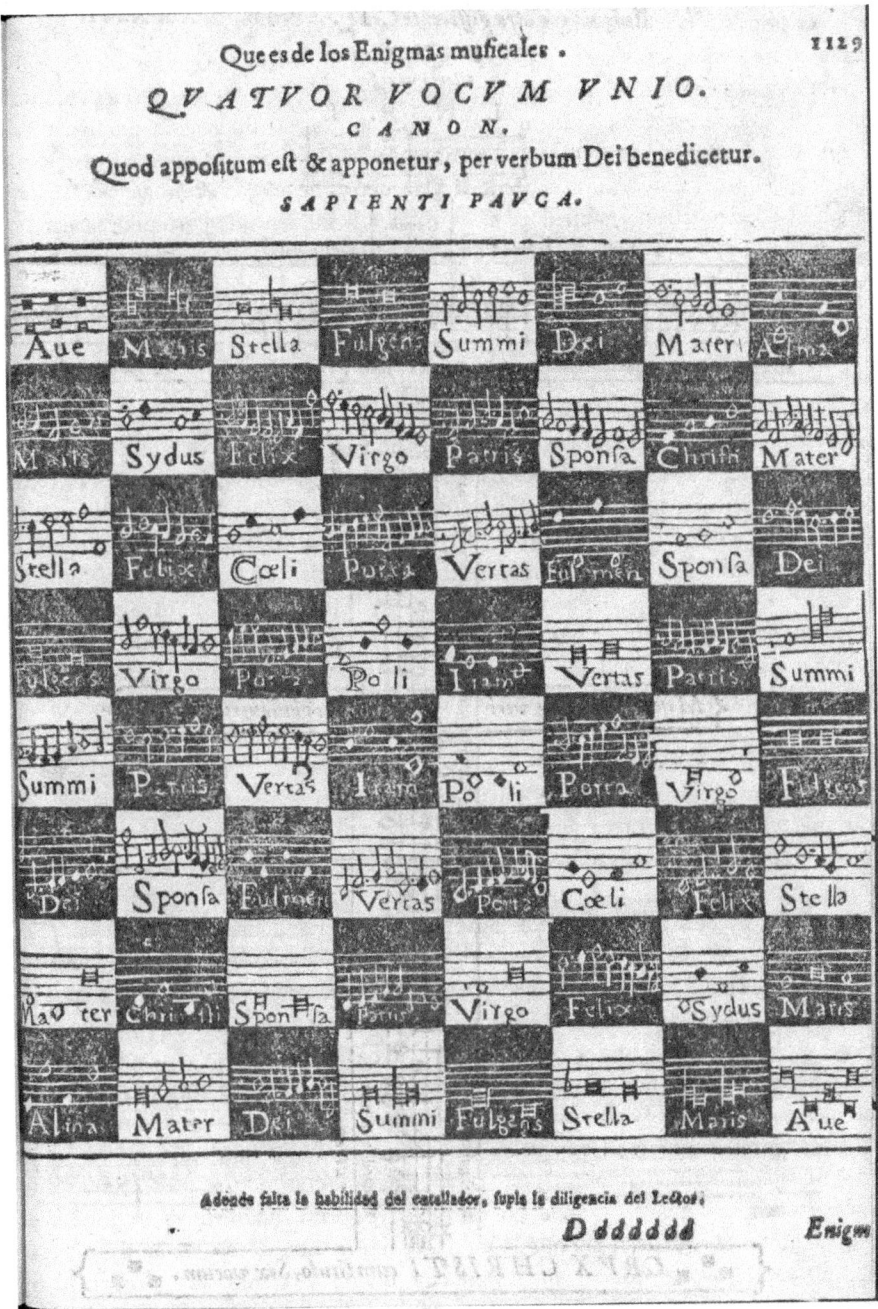

Figure 6.1 Ghiselin Danckerts's open score of *Ave Maris Stella* from Pietro Cerone's *El melopeo y maestro: Treatise on the Theory and Practice of Music*. Published in 1613. © Biblioteca Nacional de España.

was previously a bricoleur practice of instrument making had now become a proper trade, a profession with accumulating and refined knowledge. Artisans increasingly explored mathematical relationships in art, patterns, and decorations. Ornamentation and skilled woodwork in musical instruments, as well as interest in mathematical forms, often inspired by architecture, resulted in developments of instrument shapes, sound hole forms, and tuning techniques. Marquetry, the art of decorative wood inlay, emphasised geometric ornamentation, which greatly influenced instrument design.

The printing press

Development in movable type printing and the Gutenberg press heralded a new age of musical media. The first volume of printed music was the *Mainz Psalter* (or *Codex psalmorum*) from 1457. The book was the second volume printed using movable type, after the Bible, and the first one in colour. However, interestingly the pages did not contain any musical notes; the staff lines were embedded in the text, but readers were supposed to write down or compose the music themselves. People creating musical apps today will note how contemporary this idea seems. The first printed volume of notated work is the *Constance gradual,* published in Germany in 1473. From that moment, various religious texts appeared that included musical notation. Woodcut printing was typically used, but around the new century, Ottaviano Petrucci (1466–1539) developed a musical printing technique using sharp metallic type. In 1501, Petrucci published a volume of music, and soon after thousands of collections with madrigals, masses, motets, chansons, frottolas, and other musical forms began to be printed using the new technique. Finally, in the 1520s, Pierre Attaignant developed movable type technology specifically designed for music, which enabled note-by-note layout on the five-line staff to be printed in a single impression. This increase in speed and improvement in musical engraving strengthened the music industry, with various European cities becoming centres of music printing. Attaignant published over 1,500 chancions by different composers, as well as tablature scores for lutes and keyboard. At this time, anthologies of instrumental music began to appear and secular music was finally printed and disseminated, to much enjoyment in European households. This relates to how, in the first part of this book, we discussed how instrument building became a prominent industry during this period, coinciding with and dependent on the increased availability of printed music, and we demonstrated the interconnected relationship between media formats, music technologies, and cultural practices.

While hand-written musical notation did have value for the preservation and communication of music, the advent of movable type print and the mass production of musical scores changed the economic model of music. Composers began to conceive of their music as a product which would be distributed, sold, and performed by professional musicians all around the continent in dedicated music halls.[2] As a result of notation, the act of musical performance gains a new dimension, in that in addition to repeating known music, it involves performing new and unknown work, and the concepts of novelty and originality are emphasised. However, this was a very different

Figure 6.2 Movable type notational symbols could be arranged onto a grid and thousands of prints made by the press. © Rosendo Reyna of Music Printing History Website.

notion of originality and ownership to the one that emerged with recorded music; composers were still happy to reference, borrow, copy, or rewrite other composers' work, without it being seen as a breach of intellectual property.

Printers would organise commercial book fairs and these promoted composers at an increasing rate and reach, resulting in an acceleration in the diffusion of musical styles and techniques. By the middle of the sixteenth century, the proliferation of musical scores, in tandem with a new profession of performing for and teaching the aristocracy music, established performance as an integral part of education, in addition to the already established scholarly status of music as part of the quadrivium. With relatively inexpensive books of printed music, new works also reached the bourgeoisie who took up performing instruments as a form of home entertainment. During the late stages of the Renaissance, the Church's influence over musical practice was dwindling and instrumental music was becoming more ingrained in the general culture. This can be seen in the fine art of the time, where instrumental performance was clearly a key element of most societal functions (Ausoni 2009; Philips 1997). We must be careful when reading written histories of musical practices, as pre-Gutenberg literacy is necessarily elitist, and written accounts often ignored the exciting world of troubadours, travellers, bohèmes, gipsies, and other groups who were part of European culture, performing in taverns, festivities, town squares, and during work. The visual art of the time is often better at depicting this undocumented reality than written accounts. Good examples are the thirteenth-century painting of a moor and a Christian playing together or the fourteenth-century illustrations of seedy gatherings in the French collection *Roman de Fauvel* (1310–14).³

Figure 6.3 A page from *Roman de Fauvel*, a fourteenth-century French allegorical poem with early examples of musical notation embedded. From 1316. © Bibliothèque Nationale de Paris.

Polyphonic vocal music became more complex as an effect of notation in the fifteenth century, and this called for instrumental accompaniment. During the Renaissance, an increased democratisation of knowledge and the influence of Greek poetic themes resulted in heightened interest in secular music. Although instrumental music had existed throughout the Middle Ages, new instruments such as the lute, the guitar, bowed instruments, organs, and keyboard string instruments, often arriving via Arabic cultural influence, began to be more prominent in public and private life (Zecher 2007). Zecher points out that with increased instrument ownership across diverse demographic layers and through the printing of instrumental music, a process of democratisation began to take place, in which instrumental music was reaching the public through notated scores:

> Between 1500 and 1600 only one volume of instrumental music was printed for every twenty volumes of vocal music, and nearly all of this repertory is for the lute, cittern, and guitar ... with regard to the books of instrumental music that do exist, there is evidence that more *copies* of these were printed and sold than of vocal books, which indicates that although the instrumental repertory was smaller, it was reaching a wider public than just the court, which was the primary consumer of printed vocal music. (Zecher 2007: 12)

Before 1500, very few instrumental music pieces were written for specific instruments. We find general notation of music that could be played by "voices or viols," as was often written on the title pages of sixteenth-century music books. In her book *The Imaginary Museum of Musical Works* (Goehr 2007), philosopher and musicologist Lydia Goehr argues that the reason we do not find much instrumental music until the middle of the eighteenth century is that since ancient Greek music theory, music was considered a mimetic art, following the tradition of representing the perfect model of the natural world, the beautiful, and the good. The aesthetic principle of *mimesis* was primarily achieved in music through the voice, where stories, objects, relationships, movements, and meaning could be imitated through text and verbal intonation. For the Greeks, instrumental music could not be considered mimetic, thus failing as an art form (Goehr 2007: 143).

Musica poetica

Gioseffo Zarlino (1517–90) was one of the key theorists who brought music theory down from the abstract domain and connected it with practice. In his work, the *Institutioni harmonische*, Zarlino stated that the two parts of musical knowledge, *theorica* and *practica*, were inseparable, unlike the views commonly held by his contemporaries: "music considered in its ultimate perfection contains these two aspects so closely joined that one cannot be separated from the other" (Palisca 2006: 38). We are witnessing here a joining of theory and practice, a predecessor of the scientific revolution, which emerges partly from a new use of scientific instruments.

What is materialising at this point in time is a belief in the role of instruments as tools to think with, of universal scientific methods becoming formalised, as in the work of Descartes who attempted to provide a *methode* over and above local *habitus* logic (see Gauvin 2006). We find the same focus in the work of Mersenne, who applied scientific methods in studying the nature of the string, discovering the formula for pitch (width, length, and tension of the string) and harmonics (how it vibrates in the ratios of integer multiplications of its base frequency, such that a string tuned to 100 Hz would have a second harmonic of 200 Hz, a third of 300 Hz, and so on).

With the work of scholars like Zarlino and Mersenne, the science of music moved from abstract music theory into more empirical domains where the material qualities of musical instruments were studied. The instruments themselves became an obligatory part of other lab equipment such as weights, rulers, prisms, and other scientific instruments. Musical instruments were used as "epistemic tools" to study the nature of sound and how we hear it. At the same time, composers began composing more freely, paying less attention to music theory as an overarching structure and understanding it more as a set of tools available for the development of musical progression. Monteverdi, for example, would later declare that he was not qualified to study *musica theoretica* – people like Zarlino could do that (Palisca 2006: 165) – but instead he studied composition, discovering compositional techniques to influence affect, for example in devising a new genre of madrigal, the *madrigale guerriero* (Palisca 2006: 173).

In the sixteenth century, with the advent of printed music, a new form of musical practice began forming, one that was seen as distinct from musicology (*musica theoretica*) and performance (*musica practica*): this was *musica poetica*, thus defined by Nicolaus Listenius in his 1537 work *Musica*:

> Poetica is that which strives neither for knowledge of things nor for mere practice, but leaves behind some work after the labor. For example, when someone writes a musical song, the goal of this action is the consummated and completed work. For it consists in making or fabricating something, that is, a kind of labor that leaves behind itself, even after the artist dies, a perfect and completed work. Therefore the musical poet is someone engaged in the occupation of leaving something behind. (Listenius in Palisca 2006: 49)

Lydia Goehr argues that "[m]usica poetica seems to have been introduced to give emphasis to the idea that an understanding of making music – compositional principles – was as important as performing or theorizing about it" (Goehr 2007: 118). Readers familiar with Heidegger's *The Question Concerning Technology* will be reminded of how he elucidates the Greek concept of *poiesis*, as signifying the creation out of nothing, *ex nihilo*, which is different from *techne*, or skilled craft practice (Heidegger 1977). What emerges here is the birth of original musical thinking, a specialisation of the musician's style that distinguishes him or her from other musicians, and yet is distinctly different from what we find in twentieth-century concepts of the musical work.

Music as (re)search

Isidore of Seville writes in his *Etymologies* that the term music, deriving from the Muses, has its etymological origins in μάσαι, which means to "seek" (Barney 2005: 95). Isidore is here echoing Plato's etymological dialogue, *Cratylus*, in which the Muses are said to have derived their name from their search for truth. In the sixteenth century, musicians increasingly began talking about the "discovery" of music, or the process of finding new combinations of musical materials as possible outcomes from the established rules.[4] The Italian word *ricercare* was also used for this process of searching, seeking out, and finding musical material from the system of rules. There is a clear etymological relation to the word "research," and there is a development from the ricercare to the *étude*, or the instrumental study of a "learned form" (Stein 1979: 207). Musicians were those who "found" musical tunes in the combinatorial possibilities of musical theory, and, of course, the French origins of the word troubadour are indeed related to searching and finding, or *trouvère* (Stevens et al. 2001).

This philosophy of musical creation as search or investigation was not an isolated case: music, as one of the liberal arts, was seen to follow the classical methods of creation. Cicero defines the canons in the rhetorical arts as being five-fold: *invention* (determination of a topic), *disposition* (arrangement of materials), *elocutio* (style and ideas into sentences), *memoria* (memorisation), and *pronuntiatio* (the delivery). Music applied similar methods: in seventeenth-century Italy, a parallel strand to the ricercare developed where improvisations turned into notated *inventions*, a musical form later further modified and defined by Johann Sebastian Bach. Invention is a term borrowed from rhetoric, and serves as a metaphor for "the idea behind a piece, a musical subject whose discovery precedes full-scale composition" (Dreyfus 1996: 2). It is the subject matter of a piece as well as the mechanism for discovering good ideas therewithin. Dreyfus argues that central to the compositional process during Bach's time was the notion of discovering an invention. The question of invention and its role in the scientific revolution in early modernity is explored in depth by Rebecca Cypess, who asks: "In what ways could music composed for and made with instruments have reflected and contributed to the culture of invention, exploration, and discovery that characterized the early modern era?" (Cypess 2016: 8). Later chapters will develop an argument of how this ontology of music relates to contemporary notions of music, ones we might identify as post-recording, where musical systems (in the form of instruments, software, apps, installations, generative scores, instructions, cybernetics, etc.) are invented that spawn instances or versions deriving from the original invention. Each instance is the result of a certain exploration or search.[5]

The musical work

Baroque composers would write so-called "skeletal" melodies and expect the performer to improvise, extemporise, and ornament the music by adding trills, mordents, slides, turns, etc., and the concrete execution of which would depend on the musical style of the period and place. From the perspective of grammatisation, or extended mind theory, the music is written in layers – onto paper, into instruments and performers' bodies – as well being the musical culture itself. Music is here seen as manifestation of a larger cultural assemblage, and this type of ornamentation or performer addition to the skeleton of the music is characteristic of the aesthetic of the period. Improvisation and ornamentation in Baroque music-making demonstrates how the divide between a composer and a performer was less distinct than it later became, and practically every performance of a musical piece involved a degree of cocreation by the instrumentalist (Cypess 2016: 25). Performers were typically well versed in music theory and composition, as well as performance; the modern division into the distinct practices of musicology, composition, and performance are professional divisions that emerge later. Baroque music would not necessarily be written for any particular type of an instrument: it was a skeletal structure that could be expressed through all kinds of instruments. Printed music brought new performance opportunities, new musical instruments improved playability, and these resulted in a professional class who knew their instruments extremely well. Slowly, composers began to write idiomatic pieces for instruments that were being increasingly standardised, honing the sound of the instruments and the bespoke skills of their performers.

After the Baroque and Classical periods, with Romanticism, the idea of music was less that of a system reflecting God's creation, but more as personal expression that originated in the individual, as explored by the philosopher Immanuel Kant: "creative imagination is the true source of genius and the basis of originality" (Kant in Dreyfus 1996: 2). There is a shift in emphasis towards the composer and the performer in the nineteenth century, whilst previously music had been more tied to, and in large part constituted by, its extra-musical *function*. In her book, Goehr tells the story of the musical work-concept – the *opus* or *ouvre* – and how it emerged by the onset of the nineteenth century. In particular, she examines how the concept of the musical score changed from being an imprecise blueprint for musical interpretation, where the performer would add ornamentation and improvise over certain passages, to a compositional practice where the score became more nuanced and precise. This was happening in tandem with increased standardisations in notated music, which gave composers more control over their work, but consequently diminished the creative role of the performer. For Goehr there is a clear distinction between what we might call "open works" (Eco 1989) and fully notated works. She writes that such "a contrast emerged fully around 1800, just at the point when notation became sufficiently well specified to enable a rigid distinction to be drawn between composing through performance and prior to performance" (Goehr 2007: 188). Instead of relying on solid theory and instrument knowledge where performance was practically a real-time composition, the musical work became an ontological entity that existed apart from its numerous performances and is expressed in the score. The work-concept forms part of an ideology that became possible after the advent of mass production print technology and increased skill in instrument making. With standardised instruments, improved technology, and a common repertoire of playing, composers could now assume the acoustic timbre of the instrument and the skill levels of instrumentalists. This development defined new musical norms, cultural patterns, and expectations related to music. With the work-concept new social roles became established, intrinsically weaved into complex power relations: the composer, the conductor, the virtuoso/soloist, the orchestra, the audience, publisher, event organiser, etc.

Although there were sixteenth-century precursors that contributed to the understanding of the musical work, such as we read in Listenius's quote above, the concept only fully emerges with the philosophy of German Romanticism, which articulates and develops the concepts of autonomy, authorship, form, product, and interpretation. It connects with the late eighteenth-century establishment of musical orchestras that began to perform music for its own sake, not as part of other social functions, such as religious, dance, aristocratic social life, work music, etc. Before 1800, composers typically conducted their own work, but around this time, works began to be written with such great precision that they could travel independently of the composer, consequently calling for the role of the orchestra conductor. Thus arrives the concept of *Werktreue*, a faithfulness to the original, as if the work embodies a transcendental reality, or represents a Platonic idea. The concept is clearly one of fidelity and authenticity; the performer and the conductor have to be true to the intentions of the composer. For Goehr, this change in the ontology of the musical work symbolises a reversal in the role of musical notation: instead of supporting the musical performance by serving as an ancillary mnemotechnic, the written score became the primary manifestation of the music. Concurrently we witness a development where the primary function of music ceases to be that of imitation (mimesis), thus freeing instrumental music from the limitations of being the accompaniment of singing, and opening it up for new dimensions in composition. And with refined musical notation and established performance practices the ontological status of music is utterly transformed, rendering performances of the work as mere instances of a work that, ontologically, can never equal its true being and manifestation in the musical score.

Conclusion

Previous chapters have investigated how the technological process of grammatisation has spawned new functions in innumerable areas of musical practice: there are now hundreds of professional roles involved in an "industry" that a few centuries ago might have involved very few people. Standardisation has resulted in cultural expectations and norms that are rigidly written into our social practices.

This chapter has attempted to demonstrate how the printed score is a fundamental element in defining the ontology of the musical work as we understand it today; the score is a file format that shapes society around it, through established musical practices. Musical notation is not defined by theory; it can express theory but the craft (*techne*) of the composer involves organising musical event in time through a notational language, a bit like the poet or the dramaturge. The musical score changed the way musicians work by means of its new mnemotechnics. The score is clearly an extension of the mind, and with notation music can be architecturally composed on the surface of the paper. This allows us to move parts around, explore multiple streams of patterns (e.g., polyphony), sketch, erase, overwrite. Another effect of the musical score is the changed composer–performer relationship. The composer can send the performer a score, get feedback, and then alter and improve the piece. The score offers a communication channel between the respective know-hows of the composer and the performer, two roles that became increasingly divided and defined in the nineteenth century, when many composers would not be able to perform their own works. But eventually, when the piece is finished, the score becomes a gift to the world, an object that offers study, interpretation, engagement, and ultimately the domain of the interpreter. It is more than a file format for musical distribution, it becomes a world into which the performer enters, more akin to a computer game than an MP3 file. Composer Yolande Harris expresses this well when she says that "[b]eyond theorizing the score in terms of notation, much can be learned from reconceptualizing the score as *relationship*" (Harris 2013: 195). The next chapter explores how machine notation changes the concept of musical composition, including the work-concept, in music notated for humans and for machines in equal measure.

7

New Languages

Musical notation is no one singular thing: it is multi-purpose, interdisciplinary, culturally conditioned, and technologically dependent. We have discussed the origins of the score as a method for communicating and preserving music, with the corollary that it becomes a commodity that can be transmitted over space and time. When the musical score is seen as a technology for the preservation of music, the criteria for standardisation and repeatability are important. However, with the invention of phonography, of sound recording or the machine writing of music (the topic of Part III), we witness an emancipation of musical notation that can be compared with how photography liberated painting (Costello 2018: 13). The argument is that when we have machine technologies for drawing light or writing sound, the function of representation can be delegated to the machine, and we humans begin to dig deeper and further in the expression of our music. In the fine arts, this resulted in impressionism, abstract painting, expressionism, and surrealism. For music, the new technology of sound recording was now capable of accomplishing many of the established functions of notation, which subsequently changed both the ontological and epistemological status of the musical work. The musical score became a different material object, where we can detect a shift in its function from prescriptive to catalytic aspects over the course of the twentieth century. Questions arise such as: how does the score encourage creative thinking and performer engagement? How does it yield interesting interpretations of the work, yet maintain its identity? What is its place and role in contemporary musical practice?

Descriptive and prescriptive notation

The twentieth century brought new musical ideas and media technologies, and to understand how this evolution feeds into the design and culture of new digital instruments, we can benefit from exploring how the role of the musical score has been theorised. Ethnomusicologist Bruno Nettl follows Charles Seeger in making a distinction between the *descriptive scores*, which "seem mostly to have or have had an archival or preservative role, perhaps serving as mnemonic devices for performers rather than as aids to composers in controlling and manipulating their structural building blocks" (Nettl 2015: 30), and *prescriptive scores*, which prescribe clearly the performance of the music, the exact pitches, duration, and to some degree, dynamics and tempi. For Seeger, one is "prescriptive and subjective, the other descriptive and objective" (Seeger 1958: 185). This is a question of directionality: whether scribes are notating what they *want to* hear or what they *have* heard. Descriptive scores aim to be objective and give as much detail as possible for analysis, understanding and possibly reproduction, but prescriptive scores are dedicatedly written for an interpreter (typically a human, but also a machine, or more obscurely, animals, plants, or other natural phenomena, such as wind or tides). Most often the assumption is that the performer is embedded in the cultural context of the particular musical practice, which allows the composer to omit all kinds of information that are tacit features of the performance context.

We could engage in creating a complex typology of musical scores (e.g., descriptive, prescriptive, representative, action scores, event scores, code scores, study scores, listening scores, secondary scores, etc.), but these categories will often overlap and most fall under the general rubrics of the descriptive and prescriptive dichotomy, which I argue is sufficient for our discussion. Descriptive scores are about representing what is *heard* for documentation, communication, analysis and study, by both humans and machines. Under the concept of descriptive scores belong forms of traditional notation, study scores, graphic visualisations, textual descriptions, waveforms, spectrograms, machine parsed feature detection, information display, and more. Any attempt to write the sonic features of sound, the topic of Part III of this book, should fall under descriptive scores. Before the advent of sound recording, descriptive notation was clearly the key method for documenting music, and in the twentieth century descriptive scores have gained an increased role in musicological study, teaching, and other musical discourse. They are also the nuts and bolts of machine-listening technologies in artificial intelligence.

Prescriptive scores, on the other hand, are *directives* to performers. These can be mappings from visual elements to pitch or other musical parameters, which we are familiar with from the centuries-old conventions originating with the work Guido of Arezzo. For Seeger, early musical notation was largely descriptive, but became increasingly prescriptive in the eighteenth century, with advances in symbolic syntax and the establishment of the notational tradition. The technological concretisation of the musical score thus enables the possibility of the musical work, as theorised by Lydia Goehr. With increased standardisation by composers, educational bodies, and performers, the potential for objectivity in musical notation increases, yet it can never be complete, as humans will always interpret any written text differently. Goehr writes about the advent of computer music musical notation, that the "kinds of programme or algorithm [that] have been produced in recent years ... serve to reinforce the emphasis on notation albeit now somewhat more broadly viewed.... It even increases the means by which one could translate inadequate scores into adequate ones. We could simply produce for all music computerized scores containing neither ambiguity nor imprecision" (Goehr 2007: 33).[1] This is a wonderful observation, as it clearly articulates the struggle mid-twentieth-century composers were having regarding the author- and performer-function in new musical work. This relates to the expressive power of machines as programmable devices for sound, greatly surpassing humans in precise interpretations of instructions, which then resulted in a revised view of the role of the human performer. It also reverberates a certain discourse of earlier machine music in the form of musical automata and player pianos, a musical trend that began in the seventeenth century with musical clockworks and mechanical toys.

Notating for (human) machines

Thomas Patteson's book, *Instruments for New Music*, tells a fascinating story of mechanical and electronic music in the early twentieth-century music of the Weimar Republic. The machinic in music was then called "New Objectivity," an aesthetic and methodological attitude that celebrated the objective rendering of machine performances, over and above the personal, interpretative, imperfect human performances. The new machines liberated music from the limitations of human technique, and offered perfection in notation. Oskar Schlemmer, an interdisciplinary artist affiliated with the Bauhaus school, exclaimed: "Not the misery of mechanisation, but the joy of precision!" (Schlemmer in Patteson 2016: 47). The fascination with the machine as music technology can be traced back to early musical automata, for example ancient Greek water organs, or Arabic scholars ranging from Banū Mūsā in the ninth century to al-Jazarī in the twelfth century. These trends strongly galvanise in the seventeenth century with the renewed focus on new instruments in both science and music for empirical investigation. Beethoven wrote in 1826 about the role of the metronome in his music, describing it as a great addition to musical practice. He replaced the customary tempi markings (e.g., *adante* or *andiago*) with precise metronome values (in bpm – beats per minute), and stated that the performance of orchestras had greatly improved thanks to the new time-keeping technology (see Goehr 2007: 225). With more sophisticated machines, many composers began to write for musical machines, such as player pianos. Handel wrote *18 Pieces for a Musical Clock*, Mozart wrote *Fantasie K.608*, "an organ piece for a clock," and Beethoven frequently collaborated with a well-known automata inventor, Johann Nepomuk Mälzel, also known for marketing the metronome. In a 1928 interview, Arthur Honegger describes this development as one in which machines might perform tasks that humans find difficult:

> Mechanical music permits the establishment of the master-interpretation. The future is with the completely mechanical orchestra, which will offer first the advantage of being no longer limited by the human possibilities of extend and duration ... I believe in the future of the mechanical in the domain of music ... which alone [is] capable of the problems created by growing demands of human interpreters. (Godlovich 1998: 155)

New music technologies in the form of notational standards, machines, automata, and sound recording all contributed to the changing conceptions of the role of the human subject in music making in the first half of the twentieth century. With common techniques of musical notation becoming increasingly complex in response to changes in musical practices, some commentators began noting problems with this set-up, and in

1954 Thurston Dart expresses that composers "like Stravinsky and Schoenberg leave the interpreter no freedom whatever; every nuance of dynamic, tempo, phrasing, rhythm and expression is rigidly prescribed, and the performer is reduced to the status of a gramophone record" (Dart in Cole 1974: 127). Ideas of *Werktreue* are reinforced and there are demands such as "complete compliance with the score" (Goodman 1968: 186), where, from an ontological perspective, a bad performance without mistakes is considered to be an "instance of the work," whereas a more expressive performance with one mistake is not. With musical notation becoming an increasingly technical and rigid system, together with the advent of machines capable of rich musical performance and executions of sounds by means of exact notation, an artistic response was due, and an ideological movement that responded to the rigidity or "tyranny" of the musical score was born.

Thus, around the mid-century composers began to rethink the role of human performers, conceiving of them less as "machine substitutes," in the words of Stockhausen, establishing a more open and participatory relationship between composer and performer. Again, Yolande Harris's notion of the musical score as a formation of a relationship rings very true for this understanding of the score. Stockhausen points out that in earlier approaches "there no longer remained any room for 'free decision,' for interpretation in the best sense of the word" (Stockhausen 2004: 378). He subsequently explains how it fell on the shoulders of the musicians working with machine technologies for music – creating electronic synthesizers, recording with tape, programming computers – to transform views on what human interpreters could and should be doing when interpreting musical work. They began to write compositions where "the performer is granted fields for free, spontaneous decisions, to which machines are not amenable" (Stockhausen 2004: 378). The advent of the machinic thus brings forth two distinctive responses. Firstly, we encounter the use and imitation of the machine, at times where the machinic becomes part of the aesthetic, with examples like Russolo, Nancarrow, Kraftwerk, electronica, or beatboxing serving as good examples. Secondly, there is a contrasting development that can be seen as an antithesis of the machine, where the machine is seen as humanity's "other."[2] Various composers began to work from this standpoint, often writing scores with algorithmic rules for human interpreters to follow. Examples of such compositions include Pousseur's *Scambi* (1957), Cage's *Fontana Mix* (1958), Stockhausen's *Plus-Minus* (1963), and Cardew's *Treatise* (1963–7). At the same time ethnomusicologists were increasingly introducing recordings from across the world, and Indian, Arabic, and African music, as well as jazz and all kinds of folk, became a strong influence on the practices of notated music. These globalist developments were complex and happening in all areas of culture and society, and concepts of tradition and musical canon began to be questioned and transformed.

Open notations

The altered role of the musical score and increasing performer involvement in the creation of the work, prompted Umberto Eco to theorise about the "open work" (Eco 1989). For Eco, the open work is where a new level of autonomy is given to the individual performer, not merely in terms of interpretation, but also in terms of creative judgement in how to develop the piece. This kind of work breaks with the tradition of the precise score, more resembling the "components of a construction kit" (Eco 1989: 4). The interpreter becomes a co-creator, a composer, different in nature to but still reminiscent of early music performances in which extemporisation and ornamentation were essential elements in musical performance. It is clear that the openness under discussion is not a passive interpretation of the piece, but rather an active engagement in which the structural elements of the work are manipulated by the performer or the listener. The open work is therefore a system that enables the interpreter to actively engage with the score itself, reinterpret it, and appropriate it to the context in which it is performed. The new scores were prompts, junctures of composer and performer creativity, or locations that enabled new and unforeseen musical results. New notational languages were integral to the new ideology emerging in the arts and humanities, sometimes affiliated with the "death of the author" (Barthes 1977; Foucault 1984), and it is no coincidence that the Introduction of Deleuze and Guattari's *A Thousand Plateaus*, on the Rhizome, starts with a picture of Bussotti's Piano Piece for David Tudor.

Under prescriptive notation we would therefore place *action notation*, where the score consists of instructions to the performers as to how they should play their instrument or move their bodies, what Lachenmann calls "concrete instrumental music" (see Craenen 2014). Other examples of prescriptive scores include *verbal notation* (see Lely and Saunders 2012), *event scores* (Kotz 2007), *generative software* and *live coding* (Cox 2012), *instruction pieces* (Ono 1964), *realisation scores* (Dack 2005), *soundpainting* (Thompson 2006), *S-notation* for DJs (Sonnenfeld and Hansen 2016), and *animated notation* (Hope 2017). Tablatures and cheironomics (hand signing) are the oldest form of prescriptive notation, emerging at the same time as staff notation, but whereas staff notation describes pitch, typically for the voice, tablatures represent the instrument, and depict the string or key to be struck: the action, not the pitch. This difference can be explained by the fact that the voice has no external reference, so the notation had to be written for the sonic end result, not the means of producing it. That said, common Western musical notation has developed into a very expressive syntactic system for melody, harmony, rhythm, and much more. Pieces written in tablature could, for example, be tuned differently but have the same notation.

The above forms are all examples of prescriptive scores, supporting a general view of musical communication where the composer gives instructions to an interpretive actor, in order to perform a piece according to a generally well-defined context. The piece is the work, but now this work does not need to be identical or repeatable, which is one of the key changes in the aesthetics of twentieth-century composition. Although musical notation since the eighteenth century has largely been about symbol-to-pitch relationship (as opposed to symbol-to-action, as we find in tablatures), there are examples of scores in the form of action notation, for instance in some of Bach's cello pieces where the instrument's tuning is altered but the instrument is played as if tuned normally. This practice is called *scordatura* and can be characterised as a tablaturisation of the traditional score, as it changes the signified of the notational symbols from signifying pitch to performer action. This becomes gestural notation, as in the action score, where the marked note signifies the finger location on the neck as opposed to the pitch. A similar break in signification is also found in Cage's *Sonatas and Interludes*, where a note in the score does not signify pitch, but rather a key to press on the keyboard. Since the piano has been prepared with screws and bolts between the strings, the notation is not about the sound or the pitch, but the mechanism by which hammers will hit the strings. Some argue that notation has never been about pitch at all, but the action to perform on the instrument itself. This idea should be considered in the context that Guido's musical notation was originally devised for vocal music, where it is harder to separate pitch and the state of the vocal cords, and this is most likely where the symbol–pitch (as opposed to symbol–action) relationship derives from.

Graphic scores

The new notational experiments of the 1950s added syntax for indeterminacy, algorithms, instrument manipulation, and other techniques to the composer's compositional palette. Since language can be seen to behave like a virus (Burroughs 1962), these ideas spread fast: just as the tangible knob on the synthesizer, or a slider in the software calls for its use, the new graphic language in composition enticed composers and created affordances for new compositional ideas. Although traditional notation is graphic as well, with all its shapes and forms, those are clearly defined symbols that signify a meaning determined by the users (composers and performers). After Feldman, Brown, Wolff, Cage, Patterson, Stockhausen, Boulez and others explored non-linearity and shared authorship through new notations, composers around the world such as Pauline Oliveros, Toru Takemitsu, Ramon Sender, Robert Moran, Martin Davorin-Jagodić, Milan Adamčiak, Roman Haubenstock-Ramati, Llorenç Barber, Daria Semegen, and Sorrel Hays began developing and exploring further graphical signs as new language elements of musical notation. The manner in which the audiovisual sonic writing systems of Oskar Fischinger or Daphne Oram influenced these developments warrants a study on its own, as the conceptual leap of writing for machines versus humans was sizable. While these developments were partly of ideological origins, with new artistic grounds being ploughed, they were also technological. With lithography, music engraving had become easier than in previous print techniques, and further diversity and ease of use with offset printing, phototypesetting, and xerography facilitated the mass-production and hand-out of scores to larger ensembles. Cornelius Cardew wrote that "if a composer wants a string orchestra to sound like a shower of sparks, he can interrupt his five-line staves and scatter a host of dots in the relevant spaces, give a rough estimate of the proportion of plucked notes to harmonics, and let the players get on with it" (Cardew in Walters 1997: 28). Scattering "a host of dots" on the score was not possible with previous music engraving techniques, but the new print technologies made mass production of copies possible (e.g., photocopying dozens of copies of a graphical score for an orchestra). This is a good example of how new media technologies enable new compositional thoughts: the photocopier facilitates certain compositional practices, just as movable type or MIDI allow for others.

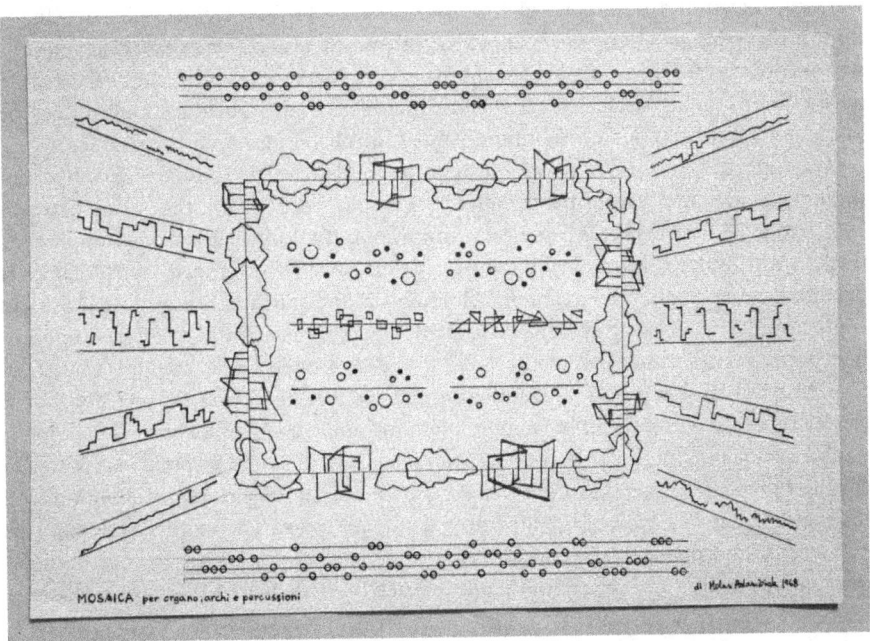

Figure 7.1 *Mosaica*, a graphical score piece by Milan Adamčiak from 1968. A piece that exemplifies the art of the score as well as the musical results. © Michal Murin.

The practice of adding non-traditional graphical elements to musical scores was inspired by the visual arts, and in the American and European scenes at the time there were strong communication channels between musicians and other artists, whether in the fine arts, sculpture, architecture, or dance. Composers often had extramusical backgrounds, such as Cage in architecture, Xenakis in engineering, and Cardew in design, and the interdisciplinary nature of new music involved people who made music, but might not consider themselves to be musicians, such as Yoko Ono or Allison Knowles, or indeed people for whom such distinctions were irrelevant, such as Ben Patterson. For many, the graphic score was a blueprint for musical actions, a piece to be interpreted by instrumentalists or singers, but also an art in its own right. In a manner resembling fourteenth-century *ars subtilior* score art, discussed in the previous chapter, the focus shifts back onto the musical score as an aesthetic object in itself. This can be evidenced, for example, by Cage's *Notations* book (Cage 1969), Smith's *Scribing Sound* exhibition in 1984, or more recently the *Notations 21* compilation (Sauer 2009). Indeed, Bill Hellermann began to use the term "score art" in the mid-1970s to signify the artistic nature of the score and its openness for interpretation. The *Scribing Sound* curator Sylvia Smith writes: "To standardize notation is to standardize patterns of thought and the parameters of creativity. Our present abundance of notations is as it should be. It makes our differences more clear" (Smith 1984). This is a good specimen of how new notations were conceived and developed in the 1960s and 1970s. The technologies developed during this time are heavily implemented in our *current screen-based digital media*, where code has become a form of scripting and the screen has become the canvas for new compositional strategies. The ergodynamics of the open symbolism of the (animated) graphical sign offers innumerable fruitful compositional approaches.

We might ask why and how graphic notation came to be so prominent in composers' thinking in the latter half of the twentieth century, in particular in the period between 1950 and 1970. The answer is complex, as innovations typically happen on multiple time tracks with innumerable contributing parameters. Innovation in print technologies is one obvious explanation, as are interdisciplinary art and collaborations with science and engineering. The breaking up of the stagnated hierarchies and politics of the composer-interpreter relationship also emerges with the idea of open works. There was also the Eastern spiritual dimension of abandoning the ego and observing the interdependent arising of all things; that graphic notation became an instrument for such expression is doubtless no coincidence. In addition, spectral analysis of sound inspired composers to consider that any parameter can be named and notated as long as we pay attention to it: for example, the insertion of a mute into a trumpet's bell can be written in notation as *con sordino* and indicates the composer's striving after a more nuanced control of the instrument's timbre and even of the quality of each individual note. Wagner, for example, introduced a specific notation for muted notes, with a + sign above the note (Read 1979: 359). Composers increasingly applied different alternative and extended notations equally for nuanced control and open interpretations (Dimpker 2013).

Multimedia notation

The practice of multimedia and intermedia art boomed in the 1950s and 1960s, echoing the 1920s and 1930s experiments in visual music (Fischinger, Pfenninger, Richter, Ruttman), with Harry Smith, Norman McLaren, John and James Whitney, and Steina and Woody Vasulka creating work that broke down disciplinary boundaries. Their work was neither music, nor visual art or film: it was the combination of all of these. Visual music was becoming an art form in some ways reminiscent of nineteenth-century ideas on the *Gesamtkunstwerk*, but its origins were different, rooted as they were in modernist conceptual art and strongly focused on formal investigation into new mediums, such as abstract film, video art, sound installations, happenings, cybernetics, electronic sound, and video. Furthermore, in the 1960s and 1970s, the philosophy of Fluxus (Friedman and Smith 2006) would encourage many composers to rethink their compositional practices, in particular to rethink the roles of the artist and the viewer, listener, and performer. Although the intermedia Fluxus artists are best known for their visual art and happenings, its impact on music is indubitable: Ben Patterson created influential scores in the early 1960s (Patterson 2012), Toshi Ishiyanagi's pieces were often exhibited as visual art on their own, and Yoko Ono published her book *Grapefruit* in 1964, which contained instruction pieces, many of which are musical scores written for the reader to complete. These musical pieces were humorous, astute, surprising, and often culminated in some spectacular event. George Brecht, for example, created a work called *Incidental Music – Five Piano Pieces* in 1961, where the performer is instructed to build a tower out of wooden blocks on top of the strings of a grand piano. The aim is to build a tower as high as possible, and it is its eventual collapse that results in the main sonic event of the piece, although the whole performance (including the silent construction of the tower) is, after Cage's 4'33" (from 1952), clearly a musical piece.

The notion of *play* is important in these works, fusing as they do the concept of playing music and playing a game (see Moseley 2016). And yet this is not a frivolous concept of the game, as David J. Getsy points out in his book on the role of games and play in twentieth-century art. Indeed, Marcel Duchamp, the godfather of modern art (including music), took his game-playing very seriously (Getsy 2011), in particular chess – a practice that became a musical piece with John Cage in 1968. Duchamp wrote various musical pieces, and in this context his 1913 *Erratum Musical* is of interest, but in this piece the performers follow instructions to generate the score before the piece is played, originally with Duchamp and his two sisters pulling notes out of a hat. Art understood in the context of games emphasises rules and the generativity that results from them, often practically infinite combinations of colours, movements, or notes. If the rules are too strict, there is no game to be played: a good game involves clear rules and the potential for creative exploration within those rules – suffice to mention here Go, chess, football, or poker – but we might also include practices that are not typically considered games, such as dance and martial arts. The notions of play and gaming in art are notions brought to the fore by Fluxus, and we benefit from them across the arts, including video game design (Pearce 2006). Cage, who was a strong influence on Fluxus artists, saw the score as an environment with which performers can express themselves, in many ways akin to how computer games are environments for exploration and sometimes expression. Cage expressed this with the metaphor of a camera: "the composer resembles the maker of a camera who allows someone else to take the picture" (Cage 1939: 11), an attitude towards expression that composer Henry Cowell amusingly described as "one of comparative indifference, since he believes the concept to be more interesting than the result of any single performance" (Cowell [1952] 1981: 135).

Rules and process

The themes of performance, execution, instruction following, were called into question in the 1960s. Composers began to use the notion of *process* for more than a simple descriptor of their compositional methods: they saw it as a fundamental aesthetic element in their music. As Steve Reich writes in 1968, "Though I may have the pleasure of discovering musical processes and composing the musical material to run through them, once the process is set up and loaded it runs by itself" (Reich 2002: 34). The story of how Reich came up with his compositional method in 1965 is well known: he was listening to two tape loops playing together, but one at a faster speed than the other, resulting in an out-of-phase development of the sounds, yielding surprising results that could not be foreseen with notated music. The technique formed the compositional approach of well-known pieces such as *It's Gonna Rain* and *Come Out*, from 1965 and 1966 respectively. Brian Eno has given a similar account of himself noticing how electronic equipment running unsupervised at a low volume was actually an interesting musical process in itself.[3] Tape here manifests as a particular media technology that becomes an aesthetic actor in how music is composed and performed. And not just music: we are reminded of Beckett's *Krapp's Last Tape* discussed in the Preface of this book or Burroughs and Gysin's tape cut-ups. Later, Reich would stop using tape in his work, but continue composing phase pieces, incorporating techniques discovered with the use of tape into his notated music for human performers. This is not the first or the last example of how a new technology brings forth a technique that, as an ergophor, is translated into practices in another domain, yet still retaining a reference to its origins.

Above we read Stockhausen's explanation of why graphic notation and open works became popular amongst composers in the 1960s, but this shift to more open and improvisational music also coincides with the increased popularity of released recordings of the world's musics, in particular Indian, Arabic, and African. These musical traditions have a drastically different relationship with what we are calling here the "work-concept," being fundamentally based on improvisational or loose musical structures. The sudden availability of this unfamiliar music, which was often of undecided length, procedural and process-orientated, created shockwaves amongst musicians who rejoiced in how rhythms and melodic forms emerge as a result of playing together, and in a context that is unique to the time and place, never to be repeated again.

Improvisation and the aleatoric

The concept of "improvisation" does not travel fully defined between musical cultures and languages; what a Western musicologist might consider a prime example of improvisation might not be defined as such in the language of another musical tradition. The concept itself might not be an interesting problematisation for that particular musical culture. The explorations of open works took place before the ethnomusicological releases and global travel of the 1960s, but around the period of the 1940s–1960s the ideas of non-linearity were omnipresent in musical circles. The influence of jazz music on notational practices, ontological conceptions of the musical work, and the role of the performer as a co-creator in the music is extremely complex and cannot be overlooked. Jazz has been a major force in Western music, but it was largely neglected by the composers of the European tradition until the 1960s. It is a remarkable story that has been told by composer and musicologist George Lewis, who argues that "Eurological" composers in America largely ignored improvisation as a compositional strategy in their work, as it was the domain of jazz. Lewis marshals the views of Carl Dahlhaus, as the epitome of the Eurologic notion of the musical work:

a composition is, first, an individually complete structure in itself ("ein in sich geschlossenes, individuelles Gebilde"). Second, this structure must be fully worked-out ("ausgearbeitet"). Third and fourth, it is fixed in written form ("schriftlich fixiert") in order to be performed ("um aufgeführt zu werden"). Finally, what is worked-out and notated must constitute the essential part of the aesthetic object that is constituted in the consciousness of the listener. (Dahlhaus in Lewis 1996: 96)

Lewis contrasts this with an "Afrologic," which is based on improvisation, a different conception of the musical piece and the role of the performer. He is writing in the context of American musicology, but there is a pertinent question whether the notion of the "Afrologic" would not also apply for Javanese, Indian, Arabic, Irish folk, Flamenco, and other such musics.

Lewis's potent criticism of Cage's role as an "aesthetic arbiter" whose views present "whiteness as a normalizing position from which others are judged" (Lewis 1996: 104) is complex and beyond the scope of this book, but it is clear that Cage was never interested in improvisation: he was more preoccupied with indeterminacy as a *compositional* process, not a performance method. For Cage, performers should stick to his instructions; the aleatoric and the indeterminate were techniques he designed as part of a piece, and the music might be different every time it is played for that reason, but this was not because the performer was improvising. Here is Lewis's point, then: the structure of improvisation found in jazz was rejected by the Eurological composers who maintained the idea of the Romantic genius – the author. Indeed, in spite of arguments of shared compositional decisions, non-linearity, generativity, site-specificity, collaboration, and other ideas that attack the notion of the composer as genius, the work-concept kept living a good life in twentieth-century music. This is perhaps best illustrated by how many times Cage's *4'33"* piece has been performed around the world, the number of scores that have been printed by Cage's publisher C.F. Peters, and how many copies have been sold of the piece on albums and services such as iTunes.[4] If there is one element of the Eurologic that has proven difficult to eradicate, it is the work-concept itself.

The cost of freedom

There are many reasons for why graphical scores and other forms of open works, so prominent between the 1950s and 1970s, became less popular over time. From the perspective of the performer, there is the educational aspect: people have not been trained in reading graphic or other non-conventional scores in the conservatories. For them, it can be a source of annoyance to be presented with such a thing, unless they are genuinely interested in these new modes of expression. Contemporary composer Miguelángel Clerc describes how, "In my experience most of the performers with whom I collaborated wanted to have a traditional visual relation with the score" (Clerc 2013: 155). Furthermore, in an IRCAM symposium organised as part of this Sonic

Writing research project, cellist Frances-Marie Uitti noted that works expressed in new notations are also an economic problem for musicians, as it usually takes much longer to practice them. This can be for many reasons, but it involves the performer having to learn a new notational language, understand the compositional system applied, in addition to practising the actual piece itself.

Another reason for the decline of the graphic score is that composers often complained about musicians' use of clichés when performing open works. Boulez raised concerns regarding how the audience perceives such performer freedom: "With improvisations, because they are purely affective phenomena, there is not the slightest scope for anyone to join in. Improvisation is a personal psychodrama, and is regarded as such" (Boulez 1976: 65). Jonathan Harvey, whose book on Stockhausen (Harvey 1975) was rather critical of the latter's "open work" period after the 1960s, wrote: "As soon as there is more than one person improvising, enormous simplicities or chaoses arise. There are either boringly obvious climaxes and lulls or there is a veneer of complexity which sounds all too obviously the handiwork of chance" (Harvey 1975: 123). Even Lachenmann decided to make a revision of the score for his 1969 work *Pression* in 2010, possibly due to concerns with unsatisfying interpretations, an approach also taken by Lucian Berio when he rewrote his flute sequenza in 1992 using more standard notation. Berio explains: "I wanted the player to wear the music as a dress, not as a straightjacket. But as a result, even good performers were taking liberties that didn't make any sense, taking the spatial notation almost as a pretext for improvisation" (Berio in Orning 2013: 101). From the perspective of twenty-first-century music, one wonders if this reaction to the improvisatory is due to authors wanting to redeem the work-concept and their compositional authority, as, and we will explore this in Part IV, contemporary composers are not finding this problematic to the same degree.

Nevertheless, the new notation practices of the 1960s and 1970s drastically challenged the notion of the musical work and its linear and deterministic ontology. Nelson Goodman's theories of musical notation and the requirement of an exact reproduction of the written in performance, resulted in him arguing that some of Cage's music could not be defined as musical work (Goodman in Goehr 2007: 32). Goodman's thesis lays down the key criteria of what constitutes a musical work on the reproducibility and fidelity of its notated signs. In the notation system itself there should be no ambiguous semantic elements, and an object (sound) should be clearly assignable to the symbol. In performance, failing to play the right note renders the performance faulty and therefore not really expressing the work. In contrast to such reactionary theories, we might consider the strictly notated work of Iannis Xenakis or Brian Ferneyhough, some of which is physically impossible to perform owing to its complexity, and it is this impossibility of performance that makes the work indeterminate. Goodman's form of analytic argumentation appears to be diametrically opposed to the musical developments of the time, and many would argue that the musical work has a much wider sociological reference than the perception of its existence merely as a score: who is the author, how is the work produced, where is it performed, what is its name and descriptive contexts, how is it notated, sold and received? Goodman's work-concept elevates the status of the musical score, and thus strongly depends on it, but the coming chapters will explore how new musical practices go beyond the score and coincidentally also the work-concept itself.

Conclusion

There are various trends that transcend the work-concept in modern music. Some forms of music do this through a critical questioning of the concept; other musics simply do not care about it. One critical approach is represented by the renewed interest in pre-Romantic music, typically played with period instruments. As Goehr so elegantly demonstrates, the tradition of early music performance is able to yield a reconsideration of our rigid ideas of the musical score: "More than any other movement currently existing within the European tradition of classical music, the early music movement is perfectly positioned to present itself not only as a 'different way of thinking about music', but also as an alternative to a performance practice governed by the work-concept" (Goehr 2007: 284). The other trend is found in the experimental musical practices of computer music.[5] This is not about experimental music as a genre (which, somewhat paradoxically, it has become in journalistic parlance), but musical activities that are truly opening up new fields of knowledge through research and venturing into the unknown, asking questions equally in language, through the building of objects, or by setting up performance contexts where the nature of music is fundamentally questioned.

Current research into new musical instruments is developing musical artefacts or ergodynamics that offer the investigation and innovation of new musical aesthetics. I am also thinking of new notational systems, through screens, code, physical objects, where the musical performance emerges from new modes of inscriptions, often undecided or emerging from collaborative context. Finally, with machines listening to and creating music, perhaps the real musical experiments of our times are happening at Google (e.g., in the Magenta project), Sony, Spotify, Native Instruments, and academic research labs across the world. It is clear that the research groups developing these technologies are not very concerned with the author-function or the work-concept; they are simply dealing with different issues.[6]

Many of the practices in new music question established institutional frameworks from different angles. However, a centuries-old tradition does not die so easily: our musical infrastructure is based on the work-concept, with publications, releases, performances, funding bodies, media appearances, and more, all presuming the work as an entity clearly definable and identifiable. Policy, education, media, copyright, and other economic structures maintain, and indeed depend on, the work-concept and will continue to do so for some time. Against such entrenched practices, we find that new music is busy exploring and developing the ideas and language of improvisation, live coding, NIME instruments, computational creativity, generative systems, sound art, apps and games, and other forms where algorithms, artificial intelligence, or interactivity become parts of the compositional elements that define the music. This development will be explored in Part IV of this book.

8

Machine Notation

Musical instruments are extensions of our bodies; they become part of our bodily organs, as etymology teaches us. Machines can be instruments too, but they have traditionally been separate from our bodies: their power derives from elsewhere, as if by magic (and we note that *machine* and *magic* are etymologically related words, relating to the Indo-European root *magh- (as in "might")). Computer code is a form of musical notation. Any Turing-complete programming language is capable of expressing musical instructions, and there are domain-specific programming languages for creating and controlling virtual instruments or robots playing acoustic instruments. With the advent of the digital computer we thought we had gained the ideal environment for writing more sophisticated instructions of algorithmic expression, but it quickly became apparent that new technologies were quite imperfect for musical expression. Bernardini and Vidolin state that "live electroacoustic music currently possesses notational conventions and practices that can be compared at best to tablatures of the Middle Ages" (Bernardini and Vidolin 2005). Contrary to analogue equipment, digital computers are flexible in their internal structure, allowing for the definitions, encoding, and processing of musical data of complexity that goes well beyond human capability. But while the computer is an excellent instrument for the composition and execution of algorithmic and generative music, a workshop for designing "any sound you can imagine,"[1] one of the main recurrent problems is that of the interface, both in terms of performance and the notational languages available.

Algorithmic music

Throughout history, music theory has described sets of generative rules that are applied in the realisation of music through notation or performance. In addition to being an integral part of the quadrivium and natural philosophy, medieval music theory was also the generative support for live performance, in which the performer would "execute" the rules of music. The evolution of music technologies (from the instrument, through musical automata and electronic machines, to the digital computer) is a history of increased theoretical encapsulation in musical instruments. We are now accustomed to software generating chord progressions for us, improvising with our playing, generating music in specified genres, and so on. Computational creativity in the arts using digital computers has a history extending back to the 1950s, but it is really in the twenty-first century, in particular through machine listening and deep learning, that the use of the computer for this purpose becomes realistic and achievable. People are often sceptical of such new developments and interesting questions are asked about creativity, the creative mind, and the concept of art. Many discussions about computational creativity become extremely complex and difficult, especially if conducted in an interdisciplinary context, as the three key concepts in that discussion – "creativity," "the creative mind," and "art" – do not have a clear definition, and any proposed definition will always be open to dispute.

Computational creativity is by no means a twentieth-century phenomenon. Indeed, much of what can be done in the machine can be done by hand too. Computational methods of composing, or algorithmic music, can be traced back into antiquity, but for our purposes (i.e., focusing on notation and formal creativity) we once again encounter the eleventh-century figure of Guido of Arezzo, one of the key protagonists of music and musicology, and the inventor of staff notation. Now we come across Guido in the context of his definition of algorithmic compositional rules that translate the vowels of sung words into musical pitches. For this reason, Guido is often considered to be not only the inventor of notation, but also algorithmic music.[2] Previous chapters have demonstrated that while this might be a simplistic historical account, Guido's work has proven important, influential, and quite ingenious. As an example of such innovative forging of connections, musicologist Helmut Kirchmeyer, writing in 1960 on the new music of open compositions, looks back to Guido's *Micrologus* for such compositional strategies. He writes: "The year 1026 provides the historian with the first instance of the problem of the rationalistic control of the creative spirit" (Kirchmeyer 1962: 11). What Kirchmeyer calls "rationalistic music" would today probably be called generative music, algorithmic music, or described as computational creativity. His aim is to show how the music of the 1960s, and what might be considered a break with the idea of the work-concept, derives from a much older tradition:

> Over the past few years a number of modern compositions, whose structure derives from principles similar to construction-kit combinatorics, have found their way into the concert halls. There has been a tendency in some quarters to deny these works any profound meaning or artistic value, but the fact is that few if any other compositional phenomena of recent years (we are speaking only of such as have acquired a certain prominence) can boast such a complete and unbroken genealogy – both historical and ideological – as these attempts to attain to artistic production on the basis of number-speculation. (Kirchmeyer 1962: 11)

Guido's rules translate vowels into musical pitches, and for Guido "everything that can be spoken can also be sung" (Kirchmeyer 1962: 11). Kirchmeyer illustrates how, in the eleventh-century "extremely rationalistic" and holistic mindset, the idea that musical melodies would naturally accompany text was taken for granted and Guido's rule-based system of vowel-to-pitch algorithm was unquestioned. This should be understood in the context of the period's world-view, in which people believed that a divine being had created the world, and in which physical laws, language, and logic all formed a perfect structure that could be reflected in poetry and music. This approach to musical composition was not a one-off: a century later an English music theorist, Johannes de Affligemensis (c. 1100), came up with a similar system, notably claiming that the music produced would never be better than the system composed, and we should thus put much care into designing our systems. This is a realisation also known to contemporary generative music system designers, as "you can never get more out of automation than has already been put into it" (Kirchmeyer 1962: 16).

This historical musicology of algorithmic music of the mid-twentieth century coincided with and inspired studies of non-linearity, rule-following, and emergence in new compositions. For example, Herbert Eimert, composer and main engineer at the WDR electronic music studio in Cologne, observed in 1957 how post-serialist composers became interested in pre-notated musical practices, as if they were discovering a connection that was ruptured with the entrenchment of the work-concept. However, this does not mean going back in time, as "[o]ur illuminating reference to historical precedent does not imply that we in any way seek justification in medieval theory for electronic music, which is characterised by a meeting of acoustical and compositional developments which are particular for our epoch" (Eimert 1958: 7). For Eimert, the new musical sounds of the electronic machines were a new "foreign musical language," one that had to be developed, learned, and spoken by composers and listeners alike.

Combinatorics and creativity

The generative rules of Guido and Johannes described a direct mapping between vowel and pitch. With the work of Catalan mystic Ramon Llull (1232–1316), who has often been considered a central force in the foundation of combinatoric creativity, we begin to see thinking in more abstract and layered rule sets. A fundamental idea in early combinatorics was that the world is entirely rational, a place where the small mirrors the large, and where language directly maps to the world. Llull was an encyclopaedic scholar who believed that all knowledge could be subjected to combinatorial methods, often reaching mystical dimensions. A student of Arabic, Llull was influenced by the use of *zairja*, a remarkable device used by Arab astrologers to generate ideas and gain answers from questions entered, by algorithmic means. The zairja is an excellent example of an epistemic tool, a mechanical structure that embodies logical procedures. Llull is often described as the father of informatics (Bonner 2007) and he inspired Leibniz's work *De arte combinatoria* from 1666. But Llull's work should be seen in the context of the thirteenth century, where games of chance using dice and other such techniques were also popular methods to arrange, combine, and compose both lyrics and musical work. Most of the outputs of these systems were never written down, as it was the generative system that was of importance, not the individual renderings of its outputs.

The technique of mapping vowels to pitch was called "soggetto cavato" by Zarlino in his *Le istitutioni harmoniche* from 1558. An important musicological work, the description of this rule-set became the foundation for further numerical, combinatorial, and algorithmic approaches in musical composition, used by composers from Bach to John Cage. Similar theories to Zarlino's that are foundational for combinatorial musical systems and that foster automaticity in the composition of music are also found in Mersenne's *Harmonie universelle* (1636). The logic of combinatorial arithmetics was seen as aiding compositional discovery, exploration of possibilities and instances of possible musics. We should note that this is before the advent of the work-concept where the ontology shifts to the written piece: here, music is not even considered to be a written work, but rather the space of possibilities that enables the explorations of versions of versions; phenotypes, in other words, whose genotypes derive from other phenotypes. Mersenne's work was also influenced by Llull's combinatoric art, and both subscribed to the belief that God was the first combinatorist in creating the universe, so for humans to be creative we would also need to apply combinatorics. Mersenne's research was wide-reaching, and he conducted pioneering work in mathematics that he wanted to prove through music (Nolan 2000). Both in his *Harmonicorum libri* (1635) and *Harmonie universelle*, he describes various combinatorial rules for musical composition or performance.

Athanasius Kircher's work on combinatorics was also influential. For him, the verbs "compose" and "combine" were synonyms, and he developed a mechanical algebraic method for musical composition, something that was later described as "sounding algebra" (Knobloch 2013). Kircher's approach was characterised by analogical thinking and he applied Mersenne's theory of combinatorics in his substantial work on musical arts, which included rules for permutations of musical material over time. His encyclopaedic *Musurgia universalis*, from 1650, became one of the most important musical texts of the seventeenth century, and gives a compelling and cogent testament of how music and its theory was conceived of at the time. The book abounds with diagrams, tables, and illustrations that range from harmonics to algorithmic theory to organology. For example, Kircher provides a "new, true, certain, and comprehensible method of composing all types of melodies" (Kircher Book V). Book six, related to the first part of this current book, is called "Organic music, or, instrumental music." In the eighth book, Kircher presents his "miraculous musicology", and here the subtitle reads: "new musarithmetic art, by which anyone, no matter how unskilled in music, can attain a perfect knowledge of composing in a short time."[3] In book nine, we find a section on musical automata, or rather, how cylinders can be notated for the machinic performance of instruments (organs, bells, etc.). Kircher calls the technology *cylindrus phonotacticus*, and it is an early form of machine notation later developed into the punch card technology also used in the Jacquard weaving loom and mechanical pianolas.

Compose a system and run it

For philosopher Gottfried Leibniz, a master of the combinatorial arts and inventor of numerical machines, "[m]usic is a hidden arithmetic exercise of the soul, which does not know that it is counting," and in his *Precepts for Advancing the Sciences and Arts* (1680), he writes that the algorithmic rules for musical composition can be so well defined that laypeople can be instructed to compose music following certain rule sets, without having any knowledge of music at all (Leibniz [1680] 1951). A little later, in 1692, Swiss mathematician Jakob Bernoulli (1615–1705) gave a speech in which he presented his methods of combinatorics, from his work in progress, *Ars conjectandi* (art of conjecturing), posthumously published in 1713. Here, Bernoulli presented a mathematical method of combinations, which he then demonstrated with a hymn composed using those rules. He presented the method as extremely useful, as it would help musical practitioners of his day in composition, arranging, and generally putting things together (Knobloch 1979: 258). Combinatorics was widely studied in the seventeenth century, by scientists including Schooten, Leibniz, Wallis, and Preset, and the deepening of knowledge through scientific investigation resulted in the critique of the thirteenth-century Llullist approach to combinatorics and of Kircher's theories of music. These mathematicians often applied their combinatorics in music, as it was still considered a form of mathematical art at this time.

In the Classical music of the eighteenth century, melodic phrases became clearer and simpler, albeit more personal and innovative, as if the old mechanical theories were now left behind and perhaps used more in the form of games or interesting technique, for example with the *Musikalisches Würfelspiel* where dice are applied to generate a new piece. Bach, Haydn, and Mozart, to name but three, all applied generative systems of composition in their work (Loy 2006: 296). In this context, Haydn wrote of the "game of harmony, or an easy method for composing an infinite number of minuet-trios, without any knowledge of counterpoint," and Mozart discussed the "[i]nstructions for the composition of as many waltzes as one desires with two dice, without understanding anything about music or composition" (Niehaus 2009: 38). In addition to a much freer attitude to borrowing material and ideas from other composers than we have today, the strong compositional rules of Classical music also meant that composers working within that framework could produce more music over their lifetime than their Romantic and post-Romantic successors. "If Bartok and Schoenberg composed fewer works than Mozart, it was not necessarily because they were less gifted than he, but because the styles they employed required them to make many more conscious, time-consuming decisions" (Meyer 1989: 5).

Combinatorial and computational ideas of creativity were prominent in the nineteenth century and mechanical computers and other related devices began to feature in the imagination of music theorists. In 1821, Dietrich Nicholas Winkel created a "machine that could compose" called the *Componium* (Fowler 1967: 47). This was a mechanism consisting of two revolving barrels that take turns performing. This generates multiple variations of the music and is a good example of a generative music machine. Interestingly, the Componium was inspired by Johann Nepomuk Mälzel's *Panharmonicon*, a mechanical orchestra system first exhibited in Vienna in 1804. Mälzel is often listed as the inventor of the metronome, although reportedly it was Winkel himself who shared with him the idea of the device (Fowler 1967: 47). Mälzel was a frequent collaborator with Beethoven, who passionately agitated for the use of the metronome in music, a good example of how a new technology infiltrates and informs compositional aesthetics.

With advances in mechanical computers and logical machines, a large area of research inspired by the work of Leibniz, music was seen as an example of a data structure that such machines might parse. This was clear to Ada Lovelace, often identified as the world's first programmer, when working on Babbage's *Analytical Engine* in 1842. For Lovelace, the conceptual leap of translating from a number to a note was trivial, and she wrote that the engine might "compose and elaborate scientific pieces of music of any degree of complexity or extent" (Roads 1996: 822). The historical, and cognitive, relationship between music and numbers is so strong that this was, of course, also obvious to Lejaren Hiller and Leonard Isaacson, who wrote the *Illiac Suite* in 1956, using one of the first electronic computers available in educational institutions (the ILLIAC I), and to Max Mathews at Bell Labs, who created the MUSIC synthesis environment the following year, iterations of which composer Laurie Spiegel used in her early computer music work and now forms the design basis of the popular CSound music programming language.

Generative music

The selected historical threads delineated above help us to understand the ideological background of what we now call generative art and music. These historical examples of computational creativity in Western culture form the conditions for the theory and practice of contemporary music technologies. However, there are plenty of similar examples of such mathematical and logical systems in other systems of art, for example in Indian and Arabic cultures with musical algorithms and mechanical automata (see Hill 1974; Truitt 2015). The general assumption has been that if the rules of a task can be formalised, they can also be written into hardware such as pin cylinders and punch cards. Winkel's Componium is an excellent example of this generative theory materialised in a machine (Collins 2018). The modern use of "generative" for what might be earlier denoted as combinatorial methods is inspired by Noam Chomsky's generative grammar in his *Syntactic Structures* (Chomsky 1957), where he argues that the infinite number of sentences in a language are made possible by the application of a finite number of rules. These rules can change over time, but they are the generative mechanism that makes communication possible. Similarly, musical theory is a set of changing rules that produce a theoretically infinite number of musical works. The premise of computational creativity is that if the rules can be formalised, then a computer can execute them. That is a somewhat problematic statement, involving aesthetic, philosophical, and psychological questions that we will study further in Chapter 14. A student of the history of generative art, as articulated in contemporary discourse, might come to the conclusion that such art is typically the output of digital computers. This need not necessarily be so. Let us remind ourselves of Philip Galanter's definition of generative art:

> Generative art refers to any art practice where the artist cedes control to a system that operates with a degree of relative autonomy, and contributes to or results in a

completed work of art. Systems may include natural language instructions, biological or chemical processes, computer programs, machines, self-organizing materials, mathematical operations, and other procedural inventions. (Galanter 2009)

Although it is not explicitly stated in this definition that the system itself can also be considered to be the work of art, we should note that such a carefully crafted system of procedures made by an artist could well be evaluated and reviewed by critics with just as much detail as its rendered outputs. Indeed, in computational generative art, it often does not make sense to separate the process from the output, particularly if the procedural system can itself be distributed as the work of art. Mitchell Whitelaw discusses in his book, *Metacreation: Art and Artificial Life*, how artists become meta-creators by applying artificial life in their art, an application in which artists design the system that creates the art. For Whitelaw, emergence is one of the key characteristics of A-life art, where the artist has formed the rules, but virtual agents execute them often with such complexity that the results are unpredictable to the artist (Whitelaw 2004: 207). With computational media we are able to design more complex systems than a dice game or a shifting pin cylinder combination of melodies, but the underlying principle is the same: creating a set of rules, and initiating a process that executes them. Performed music originates from its theoretical, instrumental, and mechanical (when theory has been implemented as machinery) foundations, just as in the oral-formulaic compositions that we find in the Greek oral tradition (of which the *Iliad* and *Odyssey* are examples) and how it applied an improvisational technique based on a collection of memorised verses (Foley 1988). Gottfried Michael Koenig expresses this generative potential of invented systems in his "Composition Processes" essay, where he describes the composition as a system to be explored, and the advice is: "Given the rules, find the music" (Toop 2010: 96).

Musicology of code and systems

For the musicologist, the question becomes what it takes to understand such a piece? Can the musical composition present itself sufficiently through repeated listening or do critics and musicologists have to engage with the code in which the piece is written? Indeed, is it important that the source code of the piece is made available – a privilege that has traditionally been granted to critics and students of music in the form of the musical score? These are open questions that have been dealt with to some degree elsewhere (Collins 2008; Cox, McLean, and Ward 2000; Fuller 2006), but it is certainly a new experience for a musicologist to ask "How is this piece made?" or "What are the compositional ideas?" and not be able to study the score, the instrument, or the performance context. The new context is a situation in which the musicologist is required to look under the hood and read the algorithmic composition of the piece in the form of code.

Let us quickly explore alternative methods of investigation: a common way of describing generative works of art is through the use of metaphors. Since metaphors

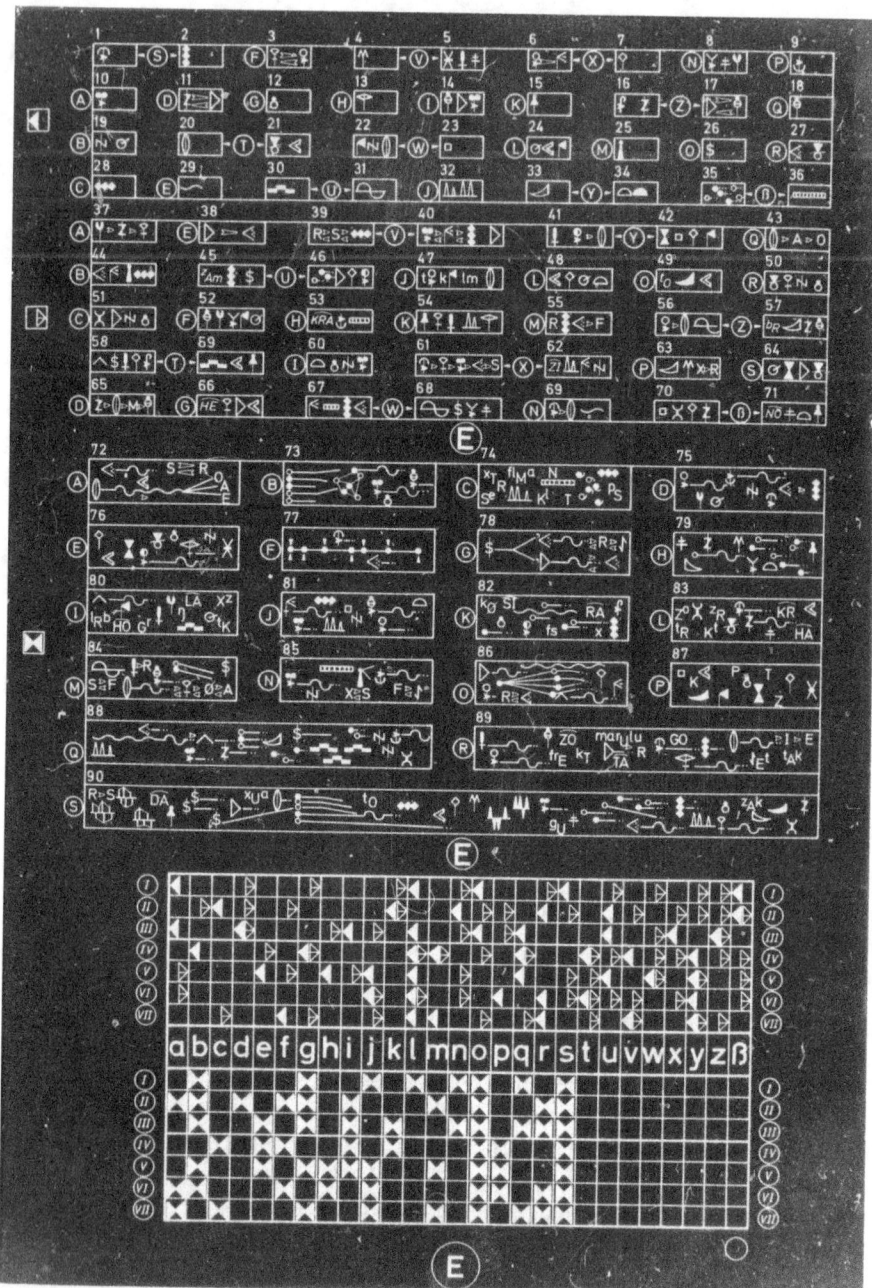

Figure 8.1 A generative score schemata of *Cybernetics*, a piece by Roland Kayn, released in three parts, from 1966 to 1969. © Ilse Kayn.

are often used to explain things of unfathomable complexity, we could ask: which metaphors can be applied to describe and explain generative music in a critical discourse? Defining generative art systems as a recipe that produces varied outcomes is sufficient, but does not explain much. Brian Eno has compared it with gardening, where the artist creates the conditions for vegetation to grow. Galanter has framed this in biological terms as genotypes and phenotypes, where the software is the genotype that engenders theoretically infinite musical instances as phenotypes (Galanter 2009). Another approach would be to see the problem in Deleuzian terms as virtuality (the software) that contains the potential to become fulfilled in the actual (the pieces) (Ayrey 2005). Yet another proposition might find explanatory power in Saussure's "le langue" and "parole" (Saussure 1986). But perhaps the most direct way of analysing, discussing, and understanding generative music is to talk about the work itself in its original form: the code. This might have been an unjustified requirement in the past, but with increased programming literacy and after decades of exposure to code and the act of coding as art forms – see, for example, the theory and practice of software art (www.runme.org) and live coding (www.toplap.org) – it has becomes easier and more natural to engage with generative music at the source level, which is the code itself, not through metaphorical refractions or the analysis of a large collection of rendered instances.

In an article from 1984, Marco Stroppa discusses the difficulty in analysing electronic music. The problem is that without the notated score we do not have the same possibilities of shared understanding of the musical material, for example in addition to the spatial analysis of being able to jump back and forth in the score, comparing elements and observing the progression of the pieces visually. Stroppa acknowledges the subjectivity of the musicologist: "Through analysis, we reveal the composer only after revealing ourselves, in an act of creation of something which is personal, controversial and alive" (Stroppa 1984: 177). For Stroppa, the responsibility lies with the composer, since neither the "operational data, with their cold, technical disposition, nor the composer's graphic representations may be considered as 'scores'" (Stroppa 1984: 179). This can only be solved by an effort on the part of the composers themselves. They need to provide the materials – research, descriptions, sources, explanations, sketches – that can help with understanding the work. Stroppa is not presenting the solution to the problem in his article: he is merely articulating it, and by doing so, he compares the disparate situations of notated music and electronic music.

Generative music and live coding thus call for a musicology of code. Code is but another system for musical notation that can be read by machines and humans alike, but ideally executed by machines. The musicologist would need to read the code and listen to variations of its output. Since reading scores is a common practice in musicology, this should be a natural extension of the musicologist's skill set in the modern age. However, the demand to see the code, in order to continue the musicological tradition of analysing the notated work, as text, also puts a pressure on composers to make the code accessible. Publishing source code could be seen as a requirement if the piece is to have the chance of receiving a fair, grounded, and informed review. This does not necessarily entail releasing all the code under open source licensing; all code-related elements can be copyrighted and protected, but an open source licence might result in further options in engaging with the work, such as remixing, recoding, releasing, and recording, thus spreading the work and increasing its impact.

Modular systems

With the resurgence in the interest of modular synthesizers in the twenty-first century, we observe yet another example of this desire to encapsulate musical theory into a physical object: to materialise music, formalise it, create a system in which we can explore musical paths through improvisation, *ricercare, invention, trouvère, étude*, or studies. Musicians build their own modules and design the combination of modules, often adhering to the so-called Euro-rack format (with affiliated standards of size, power, connections, etc.), where modules are fitted into a unique combination that represent the musical ideology of the owner. There is a distinct DIY approach in the modular synth culture, and people enjoy building their own modules from scratch, from pre-bought kits, or through developing designs shared online, exchanging expertise, and attending summits where people exhibit, discuss, and compare their creations. Modular synth-making culture exemplifies a postdigital practice (Berry and Dieter 2015) where the point is not to go back to pre-digital technologies – many of the modules contain digital chips anyway – but to transcend the general digital computer to arrive at a more distinctive, limited, designed, and characterful musical instrument. For a musician using modular synths, the design and construction of a unique module (such as an oscillator, a filter, or a timed gate unit) is clearly a creative act, but the creativity is not only in the technological domain: it is musical too, in that the module becomes the scope of possible musical creativity; it becomes the generative framework from which the music derives. The assemblage of modules in the modular synth becomes a compositional process in which the scope of the music is defined, and the musical personality of the creator becomes clearly identifiable. Compositions in new technologies transmit in ways beyond the musical score.

In a recent conversation with Suzanne Ciani (personal communication, 2017), she pointed out how her system descriptions, written in 1976, have begun circulating as a PDF document on the Internet, influencing people in the popular modular synth community. Indeed, this document, over four decades old, expresses in the first few sentences what we are beginning to see as the key character of electronic and digital instruments, namely that designing them entails the creation of a compositional system, and playing them is an ergodynamic exploration of that system, just as in Bach's inventions and versions of the same composition, or the results of Mozart's dice casts. Ciani writes in her report: "Following is an outline of a 'Basic Performance Patch' which I designed for a Buchla Series 200 instrument, and a brief description of some of the musical ideas that evolved as a result of working with this patch" (Ciani 1976: 2). Ciani designed a patch, which is a process of composition, and then she goes on to describe musical ideas that "evolved as a result of working with this patch"; in other words, the patch is about forming the research field and setting the boundaries of

Figure 8.2 A piece in the making. The picture is of Susanne Ciani's Buchla 200 modular synth during her "composition" of the piece, before a concert in Berkeley in 2017. Photo © Thor Magnusson.

thought and expression, but then it takes time and engagement to explore the ergodynamics of the patch, and get to know the system, perhaps similar to how a dancer begins to know their dancing partner. For Ciani, the report itself is an example of where her knowledge and system design is transmitted to new generations of practitioners. Her music (and we should see the concept of musical composition here as being beyond that of mere notes, as something akin to system design) continues in that report, just as Buchla's musical design manifested in her work. At the Buchla memorial event in San Francisco in April 2017, many of the composers and performers over that fabulous weekend described how working with a Buchla synthesizer felt like conversing and collaborating with Don Buchla himself. Ciani collaborated closely with Buchla for years, working in his lab and performing with the instruments, and is one of the most knowledgeable composers using the Buchla synthesizers.

Conclusion

We have seen designed systems and improvisations or versions derived from these, as if we were composing genotypes and executing phenotypes in our musical performances, or, to put it in computer science terminology: the work is the class and the performances are object instances of that class. For musicians up to the nineteenth century, this was quite a normal way of thinking. But, as Lydia Goehr demonstrates, all this changes with the work-concept, with a further drastic change in the advent of sound-recording technologies. Recording engenders an ontological shift in how we perceive music as a static fixed product, an idea that lasted for a whole century, but is now quickly changing in the second decade of the twenty-first century with its new technological practices in composing, disseminating, and enjoying music. Composing music with machines means engaging with different notational languages, such as formal rules, electronics, or code. Composition therefore becomes systems design which involves integrating an assemblage of heterogeneous elements – hardware, code, protocols, and standards – that, together, constitute the new music.

Part III

Signal Inscriptions

9

Inscribing Sound

Unless sounds are remembered, they perish, because they cannot be written down, wrote Isidore of Seville in the seventh century. Although Isidore was talking about music when he said "sound," he was not entirely correct in this distinction, even at the time of his writing, as he was using the letters of the alphabet to notate on paper the words he spoke (and during the medieval times people would read and write aloud). Unlike some alphabet systems, such as Chinese logograms or Egyptian hieroglyphs, the symbols of the Latin alphabet do signify sounds, and alphabetic writing is a type of sonic writing, of structuring the formation of sound in time. Like composers of later periods, Isidore implemented symbols to signify sound, but he was, of course, talking about music and not language. It was not until centuries later that *neumes* became a way for describing the movement of pitch, a type of notation we also find in Egyptian, Tibetan, Japanese, and other traditions. Following the neumes came Daseian and staff notations, and the history of ever-more sophisticated syntactical systems is well known. Our traditional forms of musical notation are indeed modes of writing music, but arguably St Isidore was right all along, in that these notational systems do not sufficiently represent the sounds themselves, and only serve as rough instructions on how to sing or play them. This part of the book explores the changes that follow the advent of mechanical writing of sound.

Notating what?

The attempt at designing notational systems to transcribe sound is fraught with difficulties, for such an endeavour is always a process of selective abstraction and classification. Any ethnomusicologist who has tried to use common Western music notation to transcribe music from cultures whose music is distinctively different will corroborate the number of problems that emerge. Notation systems determine which parameters are "valued," and these get abstracted out and assigned a symbol. Western systems often omit features such as microtonality (where the pitches are more fine-grained than in our common chromatic scale), different tuning systems (e.g., equal, just or mean temperament), diverse forms of musical metre, or perhaps timbre or gesture. It was clear to the pioneers of ethnomusicology, such as Kunst, Hood, Seeger, and Nettl, that this process of transcription is also a process of translation and interpretation

(Hood 1971; Nettl 2015). The transcribers will have to "fit" what they hear into the symbolic language available, which most likely is also the language in which they think, so it might be even harder for them to *hear* certain specifics in the music that are valued by the musicians of the foreign tradition. In short: they lack the ear to hear what is there.

In his "Writing Ghost Notes," Peter Winkler discusses the problems of transcribing popular music, arguing that it is impossible to transcribe what is of value in popular music using traditional notation. And yet he finds that there is something to be gained, namely that the transcriber engages with the music at a different level of inspection, discovering patterns and structures that are less obvious through listening only, and the music can then subsequently be shared with other listeners in a new form. For Winkler, to present a notated transcription as an objective representation of the music is an "act of arrogance," but if ethnomusicologists acknowledge the imperfection of their practice, the practice should rather be seen as an act of humility, of care, as the claim for objectivity is never upheld (Winkler 1997: 201).

Winkler describes his unsuccessful attempts to use automatic transcription in his studies. For him, technologies such as Steeger's Melograph (see below) are deeply problematic as they can only transcribe a single voice, and newer technologies are not sophisticated enough to notate complex audio signals. In Chapter 12 we will study how machine listening is improving in this domain, for example in areas such as instrument separation (or audio stream segregation) and pitch recognition. However, even if the audio can be identified as separate streams, this does not solve the problem of how to symbolically represent the subtle sonic parameters of the sound, as music is multidimensional not easily symbolised on the two-dimensional plane of paper or screen. Moreover, musical cultures are so diverse, and individuals within them too, that it is far from obvious *which* audio parameters should be represented. The nuances of vibrato, tremolo, timbral change, dynamics, microtones or micro-rhythms, methods of bow or plectrum application, embouchure, and more might be symbolically notated, but who defines and applies the symbols? Who understands them? The novel signs in any new notation system need to become a shared language, as it is the *use* of the language that constitutes its meaning, as Wittgenstein demonstrated (Wittgenstein 1968) in tandem with his rejection of the possibility of a private language.

Mechanical writing

The subjectivity of human listening begot the idea of dynamic mechanical writing of sound.[1] Scientists in the early nineteenth century, trained in the development and use of scientific instrumentation for measuring and capturing the world, and inspired by recent experiments in photography, began to investigate whether sound could be written down, similar to the possibility of writing light with photography. The result would be phonography, here defined as the mechanical writing of sound, equivalent to photography's mechanical writing of light. With laboratories containing scientific instruments (including musical instruments, as we have seen) to observe natural phenomena, the idea of studying and writing sound was not completely alien to scientists, but there were two problems: firstly, that the understanding of sound was very primitive, and secondly, that the frequency of sound, or the vibration of the objects generating it, is higher than can be observed with the eye (i.e., our eye cannot discern the pitch of a guitar string although we clearly observe its movement).

A comparison between writing light and sound might be due: in representational drawing, painting, or photography, humans or machines write onto canvas, paper, or screen what is present in front of the eye or the lens. The electromagnetic light waves are simply the medium between the object and the observer. However, with phonography, the aim is not to depict the instrument or the vibrating object, but the *effect* of their oscillation, the actual sound they emit, through invisible air, and this is at a rate far beyond what could be symbolically translated by a human, but that is achievable by a machine. Sounds are events, not things in the world (O'Callaghan 2007). Because of the fast motion of objects generating perceivable sound – resulting in the very fast movement of air molecules back and forth – it is impossible to represent sound itself with visual symbols isomorphic to its frequency rate. We might represent vibrational movement in the typical seismographic waveform format we are accustomed to in modern sound editors, but this is an analogue one-to-one visual-to-sound *signal* mapping, thus not *symbolic* in the sense of musical notation. Furthermore, we never see the sound itself, as if we zoom in to actually inspect its waveform, we cannot keep up with the speed of the sound, and if we zoom out to see the waveform from a temporal distance, we do not see the waveform anymore, just a high-level normalisation of it. For this reason, notation for humans is bound to operate a higher symbolic domain, such as the note to be played or the action to be taken. Machine notation, conversely, and until recently, has focused on writing the actual wave motion that is unavailable to human perception in its granularity.

The appeal of early machine writing of sound is that it writes the signal itself, not symbols representing it. This is an attractive idea, deemed impossible by Isidore, but with the scientific instruments that began to populate laboratories in the seventeenth century, we began to use instruments to extend our perception and measure the world in new ways. Thus, the desire to capture sound emerges again with the scientific revolution, as a result of developing technicity, but for us, in the twenty-first century, it would be a good exercise to imagine how our innovative ancestors might have speculated about writing sound. We know that instruments and other objects oscillate: if we touch them we feel their vibration, dust rises off a vibrating table, rings appear from a pebble thrown into a pond, and we can feel deep vibrations on our bodies. In our imaginary exercise we need to take into account that before the seventeenth century, the knowledge of air molecule behaviour was very limited, so the idea of particles bouncing into each other forming a longitudinal wave (back and forth, as opposed to up and down like ripples on a water surface) had not emerged. It was only through the work of Galileo, Mersenne, and Newton in the seventeenth century that we began to get a deeper understanding of sound. However, we observe things without understanding them, and oscillating bodies and the motion of matter on a vibrating surface had been explored by instrument makers for centuries; an example can be how fine-grained sand has been applied as an organological element in drums in many cultures. This practical knowledge of how the world functions, wanting for a scientific understanding, manifests clearly in many of the Enlightenment thinkers. A good example can be found in the work of Da Vinci, who, in the late fifteenth century, theorised about sound travelling as waves. He wrote in one of his notebooks: "I say then that when a table is struck in different places the dust that is upon it is reduced to various shapes of mounds and tiny hillocks. The dust descends from the hypotenuse of these hillocks, enters beneath their base and raises itself again around the axis of the point of the hillock" (MacCurdy 1938: 559). Galileo Galilei, the son of prominent organologist and music theorist Vincenzo Galilei, conducted similar experiments and in his 1632 work *Dialogue Concerning the Two Chief World Systems* he describes how the vibration of a brass plate would leave traces of the sound:

> As I was scraping a brass plate with a sharp iron chisel in order to remove some spots from it and was running the chisel rather rapidly over it, I once or twice, during many strokes, heard the plate emit a rather strong and clear whistling sound: on looking at the plate more carefully, I noticed a long row of fine streaks parallel and equidistant from one another. Scraping with the chisel over and over again, I noticed that it was only when the plate emitted this hissing noise that any marks were left upon it; when scraping was not accompanied by this sibilant note there was not the least trace of such marks. (Galileo 1914: 101)

It was in this context that Robert Hooke, the seventeenth-century natural philosopher and polymath, began to do formal experiments described in his *Musick Scripts* as an extension of previous research into harmonics (a longitudinal research programme that was launched with Pythagoras's theory of harmony). Hooke would use his "otoacousticons" made of glass or metal, and observe the internal motion of bodies through its vibration (Kassler and Oldroyd 1983). In 1671, Hooke performed an experiment at the Royal Society in which he observed the movement of flour in a vibrating glass, feeding different frequencies of sound into it, and thus changing the motion and patterns of the flour (Gouk 1980). In her research on the role of music in the scientific work of Hooke, Penelope Gouk writes that he was deeply inspired by Bacon's fictional "sound-houses,"[2] and indeed by his research into acoustics in general. A particular experiment of Bacon's, called "Of motion of bodies upon pressure," became the primary standpoint for Hooke's further impactful exploration in acoustics. Here, a relevant passage from Bacon's posthumous *Sylva Sylvarum* reads:

> Take a Glasse, and put Water into it, and wet your Finger, and draw it round about the Lippe of the Glasse, pressing it somewhat hard; And after you have drawn it some few times about, it will make the water frisk and sprinckle up, in a fine Dew. This Instance doth excellently Demonstrate the Force of Compression in a Solid Body. (Gouk 1980: 581)

Inscribing Sound 127

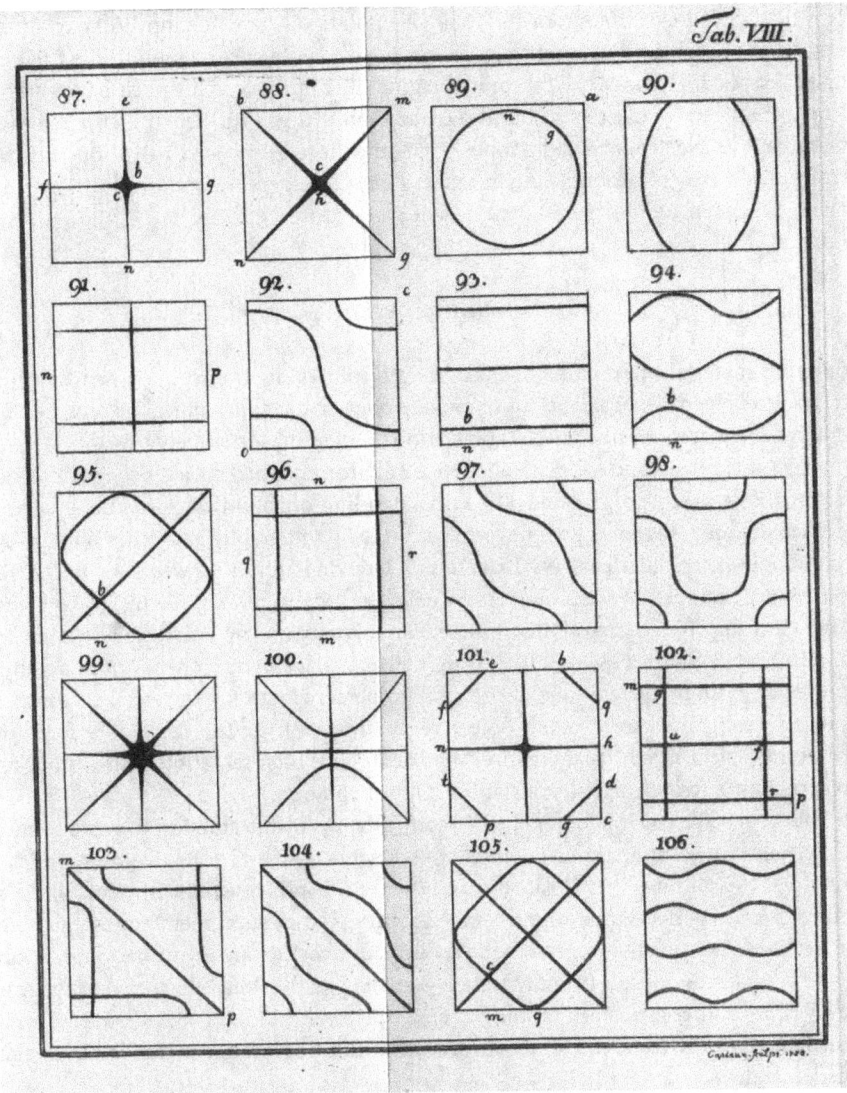

Figure 9.1 A diagram from Ernst Chladni's *Discoveries in the Theory of Sound*, a pioneering eighteenth-century work on acoustics. The figure illustrates what are now called Chladni figures.

A century later, Ernst Chladni (1756–1827), a German physicist and musician often described as the father of acoustics, began studying the vibration of plates when coupled to a sound source. Chladni used sand on his metal plates and played them with a violin bow such that the sand would form unique patterns depending upon the frequency of the plate.[3] These are now called "Chladni figures" and are the foundation for a field called *cymatics*. Adorno (1990: 59) points out how Johann Wilhelm Ritter referred to the Chladni sound figures as "script-like Ur-images of sound": the original images of sound, no less! We can only imagine the impression this must have had on people, finally to be able to see sound, a perception that was further strengthened with the advent of electronic oscilloscopes. Cymatics have been used in various artistic work since then – for example, Alvin Lucier's *The Queen of the South* from 1972 – and it has influenced computer-based models of audio representation.

Sonic pens

Vibrating materials, resonating objects! We are still trying to imagine how we would write sound given our observations of moving objects that emit sound. The nineteenth-century scientist was trained in the application of scientific instruments in laboratories, where sound, music, and the general science of harmonics had always been an integral part of the research programme. The understanding of sound as vibrations in air or another medium, together with new inventions in photography and other measuring technologies, inspired ideas of writing sound. Drawing vibrations was not a new idea, as there are examples of seismometers made in the second century CE by Zhang Heng at the Luoyang observatory of the Chinese Han Dynasty, and by Jean de Hautefeuille in 1703 in France (Ferrari 2006: 370). We are now at a place in our journey of imagining how to write sound, where we realise that if an object vibrates, it invites us to attach a drawing mechanism to it, or an object resonating with it. The result is a form of "vibrating-object-graph," that is a direct and analogous representation of the sounding object's state: a written signal, not symbolic representation.

This is how, in 1857, Edouard-Léon Scott came up with a solution to write sound. His elegant invention, the *phonautograph* ("sound-self-writer"), operates on the principles of a sliding crystal plate (one metre per second) covered with soot, where a trumpet horn collects the sound into vibrating tympanum membranes that are connected to a stylus that scratches the soot off, drawing an analogous representation of the actual sound. Scott proudly stated: "I make prints, positive or negative, of this new *writing*, rather crude prints still, but easily perfectible" (Scott [1878] 2010: 9). The questions that drove Scott's research are beautifully expressed in his letter to the Académie des sciences in 1857:

> Is there a possibility of reaching, in the case of sound, a result analogous to that attained at present for light by photographic processes? Can it be hoped that the day is near when the musical phrase, escaped from the singer's lips, will be written by itself and as if without the musician's knowledge on a docile paper and leave an imperishable trace of those fugitive melodies which the memory no longer finds when it seeks them? Will one be able, between two men brought together in a silent room, to cause to intervene an automatic stenographer that preserves the discussion in its minutest details while adapting to the speed of the conversation? Will one be able to preserve for the future generation some features of the diction

of one of those eminent actors, those grand artists who die without leaving behind them the faintest trace of their genius? Will the improvisation of the writer, when it emerges in the middle of the night, be able to be recovered the next day with its freedom, this complete independence from the pen, an instrument so slow to represent a thought always cooled in its struggle with written expression? (Scott [1878] 2010: 5)

What we might find remarkable today is that Scott had no intention of playing back what he recorded, to reproduce the written sound. It is as if the idea did not occur to him, and this could be set in the context of photography, for example, where light is written down, but never called back into existence other than by isomorphically emitting the light of what the picture is of. Furthermore, from the perspective of remediation, the phonautograph relates to the musical score, and its function is to be read and interpreted by a human. Finally, and most obviously, this conceptual block can also be explained by the fact that if your medium is scratched lamp-soot on a crystal plate, it is not easy to imagine how that could ever be reproduced. Digital technologies have enabled optical reproduction of Scott's recordings, and a quick online search will present renderings of Scott's *Au Clair de la Lune*, many of which were "educed" (or transduced from visual imagery to sound) by David Giovannoni (Feaster 2012). Scott here describes the application domains he can see for his invention:

> I propose to apply my process to the construction of a divider instrument; to that of a mathematical tuner for all instruments, of a stenographer for the voice and of instruments; to the study of the conditions of sonority of various commercial substances and alloys; and to produce industrial designs for embroideries, filigrees, jewelry, shades, illustration of books of an entirely new kind. (Scott [1878] 2010: 14)

Physicist and acoustician Rudolph Koenig worked with Scott on improving the phonautograph. Koenig was a scientific instrument maker, and in 1868 he built a flame manometer where a membrane controlled the amount of gas a light burner received and the fluctuating flame would draw onto a rotating cylinder. These manometric sketches were then transferred over to a more permanent paper format (Feaster 2012: 99). But the seeds for optical sound recording, where light is translated to sound waves – as we find later in the optical film soundtrack – have been planted in the landscape of sonic media. It should be mentioned that Scott's invention also inspired the first *electric* seismograph, which was invented in 1875 by P. Filippo Cecchi, who used sliding smoked paper and a movable stylus to draw earthquakes.

If Chladni was the father of acoustics – and he did indeed argue strongly for it to be an independent science – his heir was Hermann von Helmholtz. In 1863, Helmholtz published his work *On the Sensations of Tone*, a highly influential work, equally for musicologists, acousticians, and other scientists, and in that work he displays pictures of Chladni forms, as an explanation of his own acoustic theory (Helmholtz 1875: 65). Helmholtz is known for his invention of acoustic resonators, but those are hollow metal containers that resonate at certain frequencies depending upon the cavity size and shape of the resonator (similar to if you were to play loud white noise and put a wine glass around your ear, you would hear the resonating frequency of the glass).

The phonograph

The founding influence of Helmholtz on acoustic research is indisputable, and his work spawned a revival in the attempt to understand and represent sound. Alexander Graham Bell was one of the people Helmholtz affected. As a young man, Bell reportedly began the research that led to the invention of the telephone by giving Helmholtz's work a "hysterical reading" (Bell did not understand German): while looking at the pictures, Bell incorrectly thought Helmholtz had invented the telephone. He then later tried to replicate what he thought had been Helmholtz's invention, with some success (Ronell 1989: 287). However, as Jonathan Sterne describes in his *The Audible Past*, it was Scott's phonautograph that influenced Bell and Clarence Blake in 1874 to attempt to graphically write sound as an aid to elocutionists and the deaf. Both Bell's mother and wife were deaf and his father and grandfather were elocutionists, so this research had a deep personal relevance. The idea was that deaf people could learn how to speak using Bell's father's system of *visible speech* (phonetic symbols that depict the position of the speech organs when producing sound – a gestural notation, or tablature of sorts (Heselwood 2008)). Bell's new machine was supposed to provide visual feedback of sound, which would help the learning process of formulating the right sound from the graphic symbols of the visible speech system (Sterne 2003: 37). It was only later, after Edison's invention of the phonograph in 1877, that Bell developed the wax cylinder technique for phonographic *re*production, something that had been a problematic aspect of Edison's tinfoil system. This will be discussed in the next chapter.

Another acoustic writing device was the *Phonodeik* ("sound-show"), designed by Dayton Miller in 1908. It was a photographic device that could write sound waves onto paper in a more nuanced manner than was possible before. This is the seismographic sound waveform discussed by Varèse (see the Preface), and indeed the harmonic synthesizer used by Miller was an early approach to the resynthesis of sound from the analysis of its harmonic components (Miller 1916). Other inventors of visualising sound in the nineteenth century were Eli Whitney Blake, who made an optical system to visualise sound vibrations in 1878; August Toepler, who created a stroboscope and other visualisation techniques in the 1860s; and Krigar-Menzel and Raps, who used photography in 1891 to improve the work of Helmholtz on violin strings (Beyer 1999: 157).

In 1906, Edward Wheeler Scripture came up with a system he called *speech curves*, as part of experimental phonetics research. Like Bell, Scripture was interested in phonography as a visual accompaniment and aid in studying speech and sound in general. For Scripture, it was important that the apparatus would be able to reproduce the recorded sound, as a way of judging its reliability (Scripture 1906: 14). But he foresaw other uses too: "from curves not like those of known instruments or vowels we may expect sounds representing musical instruments that do not exist" (Scripture

1906: 54). Again, the idea of using phonography for recording music was not current at the time, although the inventors often demonstrated their machines with song (Scott with "Au Clair de la Lune," and Edison with "Mary Had a Little Lamb" – song choices that clearly demonstrate how these inventors were not serious in promoting their inventions as musical media technology). Indeed, Jonathan Sterne says in an interview with the *New York Times* that there "is a yawning epistemic gap between us and Léon Scott, because he thought that the way one gets to the truth of sound is by looking at it" (Rosen 2008), and not considering that it could actually be played back to our ears at a later time.

Crossing the epistemic gap

Above, we read about the inventions of phonography, about sonic writing in the form of actual waveforms etched onto material substances. However, none of these early inventors took the conceptual leap of implementing the reproduction of the sounds they recorded until many years later. This is quite remarkable considering that we typically find, in futuristic literature or sci-fi, the seeds of technologies to come, a certain prediction or vision that later materialises, albeit in different forms. A famous example is William Gibson's conception of cyberspace in his 1984 cyberpunk novel *Neuromancer*, a vision that anticipates the popularisation of our current networked computer systems with connected devices and virtual reality headsets. In the area of sound, Douglas Kahn has written about "improbable musical instruments" (Kahn 2014) in relation to Guillaume Apollinaire's short story, *The Moon King*, from 1916, which describes a cave with an organ on which each key would play real-time sound from a particular location in the world. We think of networked performances, *musique concrète*, MIDI controllers, etc. Of course, like Gibson, Apollinaire is not writing in a vacuum: the *Telharmonium* had been invented a few decades earlier, streaming music through telephone lines into people's homes, and carbon microphones had existed since David Edward Hughes's invention in 1857.[4] Similarly, the electric loudspeaker had been patented by Bell in 1876 for his telephone, and the musical keyboard was a centuries-old invention. In light of the inertia of inventors to actually see the potential of reproduced sound, it is quite a wonder to behold Cyrano De Bergerac's science fiction novel from 1657, originally entitled "The Other World: Comical History of the States and Empires of the Moon," in which he describes a box that could play pocket-sized audiobooks, like today's mobile devices:

> As I opened the Box, I found within somewhat of Metal, almost like to our Clocks, full of I know not what little Springs and imperceptible Engines: It was a Book, indeed; but a Strange and Wonderful Book, that had neither Leaves nor Letters: In fine, it was a Book made wholly for the Ears, and not the Eyes. So that when any Body has a mind to read in it, he winds up that Machine with a great many Strings; then he turns the Hand to the Chapter which he desires to hear, and straight, as

from the Mouth of a Man, or a Musical Instrument, proceed all the distinct and different Sounds, which the Lunar Grandees make use of for expressing their Thoughts, instead of Language. (de Bergerac 1889: 195)

In April 1877, a French poet, Charles Cros, influenced by Scott's invention of the phonautograph, described an instrument he had built called the *Paleophone* (or "sound from the past") to the Académie des sciences in Paris. Twenty years after Scott had done the same for his recording device, Cros's invention was different in that it enabled a playback of the recorded sound. A stylus pin would etch sound waves onto a disk, but now the process was reversible: the pin could also read from the disk and emit sound.[5] Cros was of limited financial means and his paleophone was never built. In 1877, the same year that Cros introduced his invention to the French academy, Thomas Edison began experimenting with phonographic recording as part of his research into the telegraph. Edison did not imagine that a booming music industry would result from his invention: he was simply trying to come up with a method to record telegraph messages onto a disk of paper. Then, during one of his experiments, he had the idea of creating a tinfoil cylinder that would be written onto by a stylus moving from the power of actual sound waves. Edison saw the power of his machine mainly in preserving speeches, for example by politicians: "The main utility of the phonograph [is] for the purpose of letter writing and other forms of dictation" (Katz 2012: 13). He also envisaged that these documents could be preserved in a "well-stocked oratorical cellar" (Katz 2012: 13). In his 1878 article for the *North American Review*, "The Phonograph and Its Future," he presents the following list of possible uses: "Letter-writing, and other forms of dictation books, education, reader, music, family record; and such electrotype applications as books, musical-boxes, toys, clocks, advertising and signaling apparatus, speeches, etc., etc." (Edison 2012: 33). The idea of musical consumption, of mass-produced albums for repeated listening, was alien to the inventor. Edison writes that with some improvements, the plate onto which the sound is written can be made more durable, such that it can be replayed many times and indeed "render the record capable of from 50 to 100 repetitions, enough for all practical purposes" (Edison 2012: 31). One of the main identified uses of the phonograph for Edison was that, when played back at a slower speed, it would be much easier for the transcriber to write down the stream of words.

Edison's phonograph recording onto tinfoil was impractical. The sound was of poor quality, and the lifetime of the record was very short. Edison did not improve his system, reputedly as he had agreed to dedicate the next five years to work on the New York City light and power system (Newville 2009). It was up to Alexander Graham Bell, who had invented the telephone in 1876, and two collaborators to improve Edison's technique by replacing the tinfoil with a wax cylinder. They began this work in 1879, after Bell's father-in-law, who happened to own Edison's patent, encouraged Bell to work on the project. Bell subsequently got the patent for recording in wax in 1886 (Newville 2009). The Volta Laboratory Associates company developed the *graphophone,* and one of the main features was the zig-zag, horizontal movement of the engraving needle, as opposed to the hill-and-dale system developed by Edison.

However, having seen Bell's system, Edison improved his invention, now called the "perfected phonograph," and both systems were put onto the market in 1888 (Schoenherr 1999). What follows is a fascinating story of inventions, innovation, patenting wars, and legal settlements. This is the story of the origins of the music industry proper, where music becomes mass-produced, commodified, and shipped around the world, resulting in record companies, recording studios, printed musical media, radio, TV, and more.[6]

Conclusion

It might surprise us that the inventor of sound reproduction technologies did not foresee its foundational role in the way we write, disseminate, and consume music today. However, as McLuhan points out, "the 'content' of any medium is always another medium" (McLuhan 1964: 4). The analysis of how a new medium is shaped by an older is picked up in a more extensive study by Jay David Bolter and Richard Grusin (1999), where they analyse the mechanisms in which an older medium is *remediated* in a newer one (e.g., early television drama remediating theatre or the CD-ROM remediating the book). One of the key elements in understanding the inventors' problems of conceiving of the reproduction of recorded music might be that the phonograph is not really a remediation of the score: it is writing something completely different, the audio *signal* itself, as opposed to the *symbols* for its generation. There was nothing to remediate: applying the concept of ergomimesis, we can study the transmission of tacit knowledge, the way that skills and techniques are translated from one domain to a new one. This involves the ergomimetic process, and importantly, how learners misinterpret, mis-learn, anti-program (Latour 1991), de-script (Akrich 1992), or even reject what is being transmitted, thus resulting in new practices and inventions. The difficulty we have with the false notation-phonography transduction is that there are no prior ergomimetic equivalents: the compositional practice is essentially different in the transition from symbols to signal, the social forms involved in musical notation (composer, performer, audience) do not map directly to the new roles that emerge with the advent of recording technology; and the role of the listener has been transformed too, moving from active engagement (playing the piece or attending a concert) to passive consumption. I should note that I do not consider passive consumption to be negative in this context, but it is interesting to behold how the active engagement of musical participation diminished during the twentieth-century age of recording. This will be explored further in the fourth part of this book, but in the next chapter, we will study how phonographic technologies became adopted in musical practice.

10

Recording

For composer and ethnomusicologist Béla Bartók, phonography was the optimal form of sonic writing: "The only true notations are the sound tracks on the record itself" (Bartók 1951: 7). Not fully satisfied with notation in the transcription of music or objective approaches as attempted by people such as Charles Seeger and Mantle Hood, Bartók praises the objectivity of the machine in writing music. For anyone accustomed to studio recording practice, this is clearly a problematic statement, as there is no "true" or "objective" capturing of sound: our auditory perception depends on environmental and psychoacoustic factors. More technically, the choice of microphones, their location, the mix, the mastering, and other sonic writing processes also frame the sound to a particular effect. What, then, is the ideal technical and environmental context to reproduce the work? Can any listening situation be of the quintessential fidelity?

The appeal of phonography for Bartók is that the composer has full control and final decision in the production of the music. However, contrarily, there are other perspectives on the quality of notated musical work in which the idea of a "perfect performance" is perhaps irrelevant and the musical piece should always adapt to new contexts and cultures, over different time periods and in different geographical locations. Even within the same geographical continent, we might ask: why would an Irish, Portuguese, or a Russian clarinettist interpret a musical piece the same way? Their musical culture differs, as do the spaces they play in, the climate, and their natural language. Oboist Christopher Redgate, in an interview with the author, discussed how different mother tongues affects the way wind instrumentalists play; the sounds they can make on the instruments are different and their intonation too.[1]

The seeds of an industry

Bartók's comment is clearly of philosophical importance, and if there is a particular effect that electronic and digital media have bestowed on music, it is to change its ontological nature, or perhaps better, its "onto-epistemo-logical" status (Barad 2003). The epistemic nature of music technology is increasingly moving from the traditional *know-how*, or procedural knowledge, to a new form of *know-what*, or declarative knowledge: from tacit skill to explicit technical knowledge (Kirsh 2009: 397).[2] The ontological change introduced by studio recording in the twentieth century shifts the concept of the work from an idea expressed on paper, to an idea expressed on tape or disk (equally wax or digital).

Susan Schmidt Horning describes in her book *Chasing Sound* (2013) how little time musicians were given in the studio in the early days of studio recordings (between 1890 and 1920), where they would have to shout into a horn, loud enough for the needle to actually write the sound onto a master disk. This was clearly not ideal for quiet music, but acceptable for jazz, bluegrass, and rock. There was also the four-minute time limitation of the 78-rpm 10-inch disk, which defined the duration of the musical pieces. The early recording studio was seen as a place for documenting music, for writing it onto a disk for mass distribution, but it was not considered to be part of the creative process itself; it was merely a place in which music was documented. There would be no more people in the studio than the artist and the engineer. Most of the engineers in the early music recording studios were not acousticians and had a quite limited understanding of the nature of sound (Schmidt Horning 2013: 23). Their contribution to the development of improved recording technologies was from actual experience and experimentation, evidencing how practical know-how can contribute to technological progress. For example, engineers would make use of a horn into which the singer would sing, or instrumentalist play, and different-sized horns would fit different instruments for optimal recording (e.g., the lower the pitch of the instrument had, the larger the horn would have to be). This type of knowledge would not be found in a theoretical sound engineer manual, but would rather be passed on from one engineer to the next. Early studio engineers did not possess detailed knowledge of acoustics. The research conducted by Helmholtz and others in the nineteenth century, which helped some of the early work in the development of recording technologies, was not widely studied by practising recording engineers. Indeed, many of the early record companies were furniture makers who made phonograph players (and the phonograph player was a staple piece of furniture in those days), and who entered the business of music production primarily because they wanted to sell more furniture (Schmidt Horning 2013: 29).

The late 1920s brought a significant improvement in sound recording, with the advent of the electrified music studio. New technologies, often sourced from wartime military inventions, would be put into operation in smaller studios, where the "recording horn, diaphragm, and cutting stylus of the acoustic system had been replaced by the condenser microphone, vacuum tube amplifier, and electromagnetically powered cutting stylus" (Schmidt Horning 2013: 36). This new technology would bring better bass and high frequencies into the recording, as well as the "room" itself, as the microphone would be sensitive enough to pick up reflecting sound waves. This made the recordings richer and warmer in character. Furthermore, the studio became a space in which musical composition took place. In addition to the engineer there was the producer, a new professional role describing someone who would assist the composer or the band both technically and artistically. This new industry affected all styles of music, from the blues to classical music. Stravinsky, writing in 1930, saw the potential of the phonograph as a compositional medium: "it would be of the greatest interest to produce music specifically for phonographic reproduction, a music which would only attain its true image – its original sound – through the mechanical reproduction. This is probably the ultimate goal for the gramophonic composer of the future" (Stravinsky in Adorno 1990: 58).

Phonography as production method

In his 1922 essay "Production – Reproduction", László Moholy-Nagy, a proto-multimedia artist and a Bauhaus professor, theorised about the phonograph's potential as an instrument for production rather than the mere reproduction of music. He envisioned a system in which the "grooves are incised by human agency into the wax plate, without any external means," and this would constitute "a fundamental innovation in sound production (of new, hitherto unknown sound and tonal relations) both composition and musical performance" (Moholy-Nagy 2017: 482). A year later he wrote his text "New Form in Music: Potentialities of the Phonograph" (1923) in which he called for a "groove script alphabet," and this would supersede "all instruments so far" (Moholy-Nagy 1983: 291). He also argued, albeit from a different perspective than his contemporary musical machine composers, that the composer will now be able to write the music directly as sound, bypassing the performer, who "was in most cases able to smuggle his own spiritual experience into the composition written in note form" (Moholy-Nagy 1983: 291). It is evident how the fascination with the machine, so common in the 1920s, exalted the mechanical performance over the human interpreter (Patteson 2016). Moholy-Nagy optimistically argues that this new technology of the direct writing of sound waves will re-establish amateur musicianship, since the technology allows for people to engage directly with the sound, without having to gain skills on musical instruments.

It is unclear how Moholy-Nagy imagined that human gesture might etch sound waves onto the groves of the wax disk. Human movements are very slow compared with the movement of sound and vibrating objects, so there would have to be some mechanical transducer between human movement and etched sound waves. Flapping our hand back and forth really fast, we might be able to generate a 6 Hz sound wave, but as we know our hearing does not perceive a sound lower than about 20 Hz. Some kind of a synthesizer would need to be inserted into the equation here, but that work was never done. Moholy-Nagy abandoned his ideas of the groove script alphabet when a new method of making sound appeared in film in the early 1920s. The new "talkies" or "talking films" had an optical soundtrack next to the picture track (which is where the current usage of the term "track" for a musical piece comes from). This optical soundtrack used the method of recording sound waves in the form of light intensity written onto the film's soundtrack. This new technology was ground-breaking, with far-reaching repercussions including live musicians losing their role as accompanying performers, and films made in smaller language areas lost their appeal (this, for example, greatly affected the prolific Danish and Czech movie industries). However, the talkies also created the new profession of film composers, sound designers, and audio engineers (Murch 1995). In terms of composition and ergodynamics, the technology itself was revolutionary: instead of the defined tracks on a cylinder or disk, tape was a material that could be cut and pasted, reversed, hand drawn, and was generally more manipulatable.

Moholy-Nagy moved on from the groove script, picking up on this quality of the film and fantasised about pure synthesis, by means of which we can "write acoustic sequences on the soundtrack without having to record any real sound. Once this is achieved the sound-film composer will be able to create music from a counterpoint of unheard or even nonexistent sound values, merely by means of opto-acoustic notation" (Moholy-Nagy in Levin 2003: 48). And in 1932, having introduced his ideas widely across German lecture halls, he wrote: "today, thanks to the excellent work of Rudolph Pfenninger, these ideas have been successfully applied to the medium of sound film. In Pfenninger's sound-script, the theoretical prerequisites and the practical processes achieved perfection" (Moholy-Nagy in Levin 2003: 49). Pfenninger was one of the founding pioneers of abstract film, together with Oscar Fischinger, and both of them were instrumental in developing the techniques and aesthetics of abstract audiovisual cinema. Thomas Y. Levin (2003) describes how Pfenninger's *Tönende Handschrift* (Sounding Handwriting) came about as a result of his not having the financial means to hire a studio and musicians, instead deciding to write the soundtrack by hand. This, however, led him into the research domain of "acoustic writing," formed by Bell and Scrivener decades earlier, where the waveforms of the sounding phonemes of human language were studied and encoded. Through practice he was able to write different sounds as waveforms, by hand, and this is the birth of written sound synthesis, which we find implemented in multiple forms in computer music systems. Critic R. Prevot was both impressed and concerned by what he witnessed during a premiere in 1932. He claims that while what has emerged is a "new technological art," his "music-loving ear did go on a strike."

> Was this still music? ... rarely have we felt so clearly the inner difference between live art and technological construct. One heard piano and xylophone-like sounds, others which seemed to come out of a steam whistle – all of them crafted together with great precision, much as if someone were to build a tree out of a thousand pieces of wood, which can look deceptively real and yet will never bloom! (Levin 2003: 55)

Critical studies of the phonograph

Philosopher and composer Theodor Adorno had concerns about the phonograph as a medium, claiming in 1934 that it is "an artistic product of decline" and indeed the "antithesis of the humane and artistic, since the latter cannot be repeated and turned on at will but remain tied to their place and time" (Adorno 1990: 58). Place and time! Acoustics, weather, social context, mood, and other encompassing features of musical performance cannot be boxed, according to Adorno. This book argues that we are entering an era of new media technologies where the circumstantiality of musical performance and experience is being explored again to a renewed intensity, where music can be individualised and instantiated in ways unique to time and place. In addition to the emphasis on live performance resulting from a screen and networked media fatigue, we also embrace new technologies in which we create live systems, generative algorithms, sometimes applying environmental factors, interactivity, and statistical data streams, for example biometric data, as input into the music. From Adorno's perspective, writing long before digital media, recording was simply a "petrification" of music, turning what is alive and fluid into stone. Adorno, entrenched as he was in the tradition of scripted music, did see phonography as a form of writing, of preservation:

> There is no doubt that, as music is removed by the phonograph record from the realm of live production and from the imperative of artistic activity and becomes petrified, it absorbs into itself, in this process of petrification, the very life that would otherwise vanish. The dead art rescues the ephemeral and perishing art as the only one alive. Therein may lie the phonograph record's most profound justification, which cannot be impugned by an aesthetic objection to its reification. For this justification reestablishes by the very means of reification an age-old, submerged and yet warranted relationship: that between music and writing. (Adorno 1990: 59)

Adorno's critique derived equally from his compositional practice as from his critical theory. He points to the complex relationship between music and writing, as if lamenting the shift from the writing of symbols to signal, as the new ways of writing that phonography offers are not that of symbolic signs in the form of notes, but through the actual energy of sound as vibrations of air, transduced analogously to shapes, which are readable by a needle, from the disk. This is therefore not symbolic writing but the writing of energies, of signals. Sound synthesis, something we consider to be a normal practice today, was eerie and inhuman for Adorno: "the possibility of inscribing music without it ever having sounded has simultaneously reified it in an ever more inhuman manner and also brought it mysteriously closer to the character of writing and language" (Adorno 1990: 60). The problem for him was the withdrawal of the personal, the situated, and the individuated through the use of objective machinery. While some might agree with this perspective today, it is unlikely that we would use terms such as "inhumane." Phonographic technologies are human in the twenty-first century. The issue Adorno is struggling with is that the phonographic machines are writing machines, not of musical symbols, like the typewriter of alphabetic symbols, but of actual sound waves, of signals, and this prevents the reinterpretation and contextualisation of the music. It challenges the traditional literacy of music of writing and interpretation. Adorno did not live to see the proliferation of digital audio synthesis, where sound is not written with transduced energies, but rather with algorithms written in programming languages describing these energies. Considering the symbolic and literate foundation of computer music and live coding, it is tempting to think of him warming to such musical writing, one of symbolic instructions for machine interpretation. Later we will touch upon the question of the "inhumane" in future technologies of computer music.

Writing sound

The early twentieth century was thus a revolutionary period of musical technologies: there were mechanical musical instruments, synthesizers, and notational languages that allowed for more organic control, and phonographic technologies that enabled the recording and dissemination of sound. Patteson (2016) gives a good account of what was perceived as a rift in the early twentieth century between organic and mechanical music. The former extended the body, and allowed for control and expression, whereas the latter was machinic, automatic, and did not lend itself to human control. This dichotomy is problematic, however, as we also find human expression in mechanical music – for example, Rachmaninoff's recordings onto pianola music rolls, later to be perfectly executed by the "machine." It is interesting to consider the use of the term "organic" here, as its etymon is "instrument," referencing equally the body itself and technological extensions of the body.

Moholy-Nagy made the film *Sound ABC* in 1933, in which he presented the sound and shape of the alphabetic letters as well as other symbols on a 35mm film. He was applying the soundtrack drawing techniques developed by Pfenninger and Fischinger, later to be continued by Norman McLaren, who named his hand-drawn sounds projector "moviola," a new musical instrument for which one can draw direct sounds without any symbolic intermediary. Patteson discusses these as "media instruments," whose character is *notational transparency*, that is, there is no mapping or translation between the sign and the sound, thanks to its analogous nature, or what I have described as signal writing (as opposed to symbol writing). McLaren became one of the key proponents of synthetic sound, but experiments of drawn sound applied to animated music were conducted all over the world, for example in Russia by Nikolai Voinov, Arseny Avraamov, and Evgeny Sholpo (Smirnov 2013). In the 1940s, John and James Whitney created their abstract *Five Film Exercises* between 1943 and 1945, using a pendulum device to generate an entirely synthetic optical soundtrack. In the late 1950s, Daphne Oram began working on a similar technology called the *Oramics Machine* (Oram 1972; Grierson and Boon 2013), where synthesis notation was drawn directly onto 35mm film, read by light sensors and transformed into sound. The machine is an impressive piece of studio equipment, renovated in 2009 by Mick Grierson and team, and displayed in 2012 at an exhibition in the Science Museum in London.[3] An extensive survey on audiovisual work and synaesthesia, including and contextualising the work of Oram, can be found in Margaret Schedel's "Colour is the Keyboard" article (Schedel 2018).

Above we have discussed recorded and synthesised sound using optical reading heads, and how composers began to apply these technologies directly in their compositional process. The conception of the productive role of phonographic media in musical performance begins to appear in the 1930s, for example in Cage's *Imaginary Landscape 1*, written in 1939. However, the narrative that avant-garde composers like Cage, Schaeffer, Henry, Ferrari, or Parmegiani were the first to manipulate phonographic recordings is rebuked in Feaster's article "The Forgotten Origins of Phonomanipulation" (Feaster 2011). Feaster shows how the techniques of slow or fast playback, reversed playback, segmenting, mixing, and sampling were all applied in early uses of phonography, between 1890 and 1920. This was used for all kinds of extra-musical purposes, such as writing down speech, studying language, rearranging information, or simply for entertainment. Admittedly, the use of the etched disk as a directly productive, rather than reproductive, instrument, as articulated by Moholy-Nagy, had to wait six decades, when Grandmaster Flash and others instrumentalised the turntable. They did so by conceiving of the sounds of vinyl as something they could source in their own music, pulling in samples from diverse origins, scratching the vinyl and mixing sources, thus creating a whole new mode of musical performance. These ideas of sampling and using "found sounds" can in some ways be traced back to Schaeffer's musical objects (and Schaeffer is sometimes called the father of sampling), but there is a pronounced difference in that the turntablists and DJs were using existing music, mixing it together, and demonstrating virtuosic performance in operating the DJ equipment. In many ways, this corroborates the analysis of SCOT (Social Construction of Technology) scholars, who have conducted sophisticated analysis of how people redefine technology for new use, and also the findings of media archaeology scholars who re-interpret and reframe old technologies in new contexts in their study of residual media (Acland 2007).

Sound as compositional material

New media technologies inevitably spark new aesthetic ideas, and yet they may not lend themselves well to the execution of those ideas. Sometimes we need to wait for another technology to arrive for things to be practically achievable. A good example is Moholy-Nagy's ideas of gestural sonic writing, or "opto-acoustic notation", which is now perfectly achievable with sensors translated to electronic signals, or productive phonography (as opposed to reproductive) which became possible with tape recording technologies. Although the magnetic wire recording technology was invented by Valdimar Poulsen in 1898, it was not used much, and certainly not for artistic compositional purposes. It was with the invention of the magnetic tape in the late 1940s that we find the practitioners of *musique concrète* applying tape as a compositional tool, constructing new aesthetic theories of sound and music that were directly offered by the new medium. In the early days of tape, due to the relatively high cost of the necessary equipment, it could only be found in labs and recording studios, for example in radio stations. Pierre Schaeffer, an engineer at the Radiodiffusion Française, began his well-known work with *musique concrète* in that studio. Having access to tape machines and studio time, he was in an ideal situation to experiment with tape as compositional technology, one where pure sound objects became the musical material, instead of the more abstract musical notes. Again, we witness how the new sonic writing technologies of the twentieth century emphasise and celebrate the signal over the symbol: the musical event is not symbolic, it is the pure signal itself, the sound and its abstract qualities.

Schaeffer was inspired by Husserl's phenomenological method of bracketing, as developed by his philosopher contemporaries Merleau-Ponty and Sartre, and applying their phenomenological methods he proposed that we ignore the sound's origin in the composition and listening of music, and rather focus on its pure qualities as a sonic event. Founding the GRM (Groupe de Recherches Musicales) in Paris, with affiliated composers and studios, a new method of composing began to emerge, one which would be distinctively different from the IRCAM (Institut de Recherche et Coordination Acoustique/Musique) method that would continue note-based composition, or from the scripted music, *écriture*, as IRCAM's director Pierre Boulez would express it in a paragraph that reads like a direct criticism of GRM:

> I understand that the dialectic of composition better contents itself with neutral objects, not easily identifiable ones, like pure tones or simple tone aggregates, having no inner profile of dynamics, duration or timbre. As soon as one shapes elaborated figures, assembles them into "formed" complexes, and uses them as first-order objects for composition, one is not to forget [...] that they have lost all neutrality and acquired a personality, an individuality which makes them quite unfit for a generalized dialectics of sonic relations. (Boulez in Hoffman 2009: 63)

We detect a tension between the theoretical notion of music as a literary abstract form and the more situated and practical idea of music as sound: embodied, dirty, and contextual. If IRCAM celebrated the symbol, GRM worshipped the signal. From the genealogical tracing we have done in previous chapters, it becomes evident that we are yet again experiencing an explicit tension between *episteme* and *techne*, the Greek separation between theory and practice that we find in early Greek music, and which, as we have seen, has been a consistent thread throughout Western musical history. As the key site of high modernist music, IRCAM studies abstract knowledge and computer music manipulations of notes and this is where the well-known Max programming environment originates.[4] GRM, on the other hand, was more focused on the actual sound, on phenomenology, psychoacoustics, and processes to manipulate sound.[5]

Real or synthetic sound?

Since we are exploring twentieth-century aesthetic factions, it merits mentioning that much noise has been made of the tension between the French *musique concrète* and the German *Elektronische Musik*; initially there were contrasting ideas regarding the source of the sound, where the former applied any recorded sound in compositions, but the latter claimed to be able to synthesise any sound imagined. However, according to Schaeffer, this friction, which lasted for twelve years (Schaeffer 2017: 2), was resolved when composers began to understand that they were engaging with music at the same compositional level, one beyond notation, at the signal level, where the focus is on the actual sound itself:

> In addition, all aesthetics apart, these two types of music – if we may provisionally call them this – displayed worrying anomalies: the former was not written down; the latter was written by numbers. By going too far or not far enough, they did more than challenge traditional notation: they did without it altogether. The former was to abandon it when faced with a sound material whose variety and complexity eluded all attempts to transcribe it. The latter made it anachronistic

through a rigor so absolute that the approximations of traditional scores paled before such precision. (Schaeffer 2017: 3)

We cannot transcribe concrete sound, and Schaeffer is here pointing to the new compositional languages emerging in the middle of the twentieth century. But what does he mean, writing in the 1950s, when he says that electronic music was written using numbers? At the time, analogue synthesizers and tape were mostly in use, and the design of sounds would require calculations of frequencies, for example in additive synthesis (where partials are added to the fundamental frequency). The control interfaces of the new machines had numbers on them and their schematic notations would include numbers (of frequencies, filters, reverb times, and so on), but he is not talking about computer code. In systems like this, the compositional environment, the instrument, and the performer merge into one technological object assemblage, of which the musical score loses its central position. Schaeffer questions the established place of notated music: if there was any characteristic shift that the new electronic technologies of the mid-twentieth century introduced, it was one that transformed the composer-score-interpreter transmission to one where composers wrote directly with the technologies that executed the music. This was a shift from the symbolic to the signal: with electronic gear and tape, we were finally able to compose at the signal level, partly realising Moholy-Nagy's dream, although gestural control of synthesis had to wait for some time. As already mentioned, just as photography arguably liberated painting from the constraints of representationalism, phonography liberated musical composition from symbolic representation and opened up spaces for the exploration of the concrete (real sounds) and the abstract (synthesised sounds).

Phonographic writing as the condition of jazz

The twentieth century represented an upheaval in all musical practices. In the first decades, musical automata, such as player pianos and music boxes, entered a space previously occupied by live musicians, for example in bars, restaurants, and cinemas, but also in people's homes. With the boom in the music industry after World War I, the physical format onto which the musical work was inscribed moves from the musical score to the physical medium of the recording. The work-concept is transformed by this move, because instead of the work being a blueprint that affords interpretation by different performers at different times, locations, and cultures the published work now becomes the *actual sonic writing* that constitutes the piece. A new pattern of musical composition emerges in which musicians might have played music in various versions for a live audience, but once it has been recorded and mastered in the studio, it gains an ontological status as the work's original: the genotype of which all live performance phenotypes are constrained to express (Kania 2006). The studio recording becomes the work; it is the mnemotechnical object (whether stored on vinyl, CD, or hard disk) that listeners reference when they hear a live performance, an alternative version, or a cover version of the piece. And curiously, the studio recording becomes normative to a higher degree than what we have been accustomed to with prescriptive musical notation. This applies in particular musics of oral tradition, such as folk, jazz, and later rock and pop. Here, a live performance, where a band performs works from a studio album, will always be compared to the original and listeners will typically express their disapproval if the music veers too far away from the original. There are musical cultures such as some forms of jazz, free improv, or folk that explicitly reject this ontological model of music. Many individual bands that clearly operate within the framework of recorded music also reject it, for example the Grateful Dead, whose improvisational approach and innumerable bootleg releases (encouraged by the band) have resulted in countless versions of their music.

Recording technology has also affected the realm of classical music in that performers begin to reference other people's interpretations of notated musical pieces, a famous example being how conductors would gravitate to the recordings of Herbert von Karajan's interpretations of the Beethoven symphonies, recorded four times, and with each iteration limiting the scope of other people's interpretations.[6] In George Lewis's criticism of the Eurological tradition, he defines the reliance on the written score as the epitome of the work ontology. With the advent of recording technologies, this ontology is sharpened: for the culture of musical notation, the recording becomes an instance of the written piece, one possible rendering. And in aural cultures, or those that do not adhere to the literate canon of symbolic musical writing, but still use diverse technologies of sonic writing, the recorded piece can take on the status of the true original. This is evident in rock and pop, and certainly in the long tradition of jazz.[7]

For Bernard Stiegler, a theorist of writing and mnemotechnics, there is a difference between the traditional folk cultures whose music extends down the centuries and the twentieth-century phenomenon of jazz, which "belongs to the written tradition," of phonographic writing:

> jazz performances bear the signature of the musician and become works (*œuvres*) in the formal sense that is given to that term in the modern West. This is to say that the style is profoundly individuated, marked from a perfectly singular idiomatic difference in conflict with the tradition from which it comes, and recognized not only as an interpretation, but as a composition. (Stiegler 2014: 73)

Stiegler's argument in his analysis of Charlie Parker is that, by means of recording technology, his improvisation immediately becomes a composition. Indeed, the "work becomes the memory of the work" (Stiegler 2014: 75). For Stiegler, phonography, by virtue of jazz musicians composing in real time onto tape, becomes the *horizon* of the music. Of course, there is jazz music that is never recorded, but it is the technological infrastructure of its recording, dissemination, and listening that constitutes the possibility of its status as musical work. It is the reproducibility of the music that makes it producible:

> if jazz proceeds essentially, in its own unique way and in some way retroactively, from this analog recording and its real-effect, then the real time of the recording is

vital not only for the repetition of what has been recorded: it also makes possible the very thing that had been recorded – the essence of jazz, the space where jazz itself is captured. (Stiegler 2014: 75)

Ontology of the recorded work

In his *Musical Works and Performances*, Stephen Davies (2001) engages with this new form of musical ontology that emerges with recording technologies. In Davis's account, rock music ("rock" is used in a very wide sense here and could include other genres such as pop or heavy metal) is of an ontological type that could be defined as work-for-studio-performance, whereas classical works are of the work-for-live-performance kind. In "Making Tracks: The Ontology of Rock Music," Andrew Kania (2006) criticises Davis's view on rock music being for the studio, and he points to the role of club gigs, playing in the garage, and so on. The rock audience wants to see a performance of the music, and not a CD playback or other types of miming, when the band is on stage. However, the audience typically has a strong expectation of hearing a version that is a reproduction of the recording. There is another problem with the work-for-studio-performance in that very often the work is composed through the process of working in the studio. A simple example would be how Teo Macero, the producer of Miles's *Bitches Brew*, practically composed the music using an assemblage of recordings from Miles's improvisational studio takes. Kania does acknowledge the ontological difference in the musical score versus the recorded work, and argues that rock performers are trying to instantiate the work as composed in the studio: "live rock practice is dependent on recorded rock, but not the other way around" (Kania 2006: 404). Like Kania, Stiegler forcefully states that recording technology constitutes the nature of jazz music, just as the notated score is the foundation of classical music.

Kania further defines the ontology of rock as one based on tracks, upon the layered recordings of instruments onto the multitrack tape or DAW software. He applies Gracyk's (1996) terminology of "thin" and "thick" sound structures, where the model in classical music would be described as "thin," since the information encoded into the musical score is almost of a mathematical and abstract nature, whereas the "thick" sound structure of recorded music contains all the rich information of the instruments, the room, the human expression. For Kania, rock tracks are "studio constructions: thick works that manifest thin songs, without being performances of them" (Kania 2006: 404). Kania thus articulates a distinction between the rock song and the rock track, where the former is a "thin" structure, and the latter "thick."

The new forms of operation established with phonography prompt us to ask the following question: if the tape studio, and now computer music software, offers composers the chance to perfectly express their music, what then is the role of musical notation? Why would not all composers work like Glenn Gould in the recording studio, spending hours on retakes, cutting and splicing in order to manufacture the perfect rendition of the piece? Why would they not compose out of an abundance of recordings like Teo Macero, or treat the studio as an instrument in its own way, like Lee

Scratch Perry or Brian Eno? Like the early twentieth-century dream of machine music, to remove the intermediary of the interpreter, composers might thus bypass the problem of poor interpretations and get a version that is exactly what they hear in their "mind's ear." There are many possible answers to that question, one being that the musical score is about building a *relationship*, about creating a social context in which magical things can happen, which opens up new dimensions of the work. For many composers, the work is a framework consisting of diverse human actors (composers, performers, instrument makers, event organisers, media, audience), as well as the effect of a new context (the acoustics, the time of day, season, city, country, new culture of listeners). The work moves through time, and it is rewarding for composers to see their work adapt to new contexts. Another reason for composing with symbolic notation is that playing music is a very enjoyable activity, as we see in the number of amateur musicians (*amateur* being, of course, the French for "lover of") who play their instruments at home, for example, in order to relax after a long day at work. This role of performance in our musical culture is then clearly manifested when electroacoustic composers of the Western "art music" tradition, who primarily composed in the studio, became interested in the performability of their music, for example by including acoustic instruments as part of their electronic work, or by notating its execution, for example in terms of its formal structure, sound, or spatialisation.

Conclusion

Bartók's statement about the perfect phonographic score is therefore more problematic than it seems. He is right in that we can work our music in the studio, compose it with our advanced studio technologies, but by doing so we reduce music to a static linear product, stripping it of multiple other functions it has. Recorded music sacrifices the function of the reinterpretation, adaptation, revivification, and re-contextualisation of music as it can happen over time and in a specific place.[8] This chapter has explored the technological conditions that support a new form of musical work, and its accompanying work-concept, as discussed in Chapters 6 and 7. It is evident how new technology spawns unforeseen practices in which musicians apply it for creative purposes; practices in which the technology itself becomes the constitutional element of the work, and without which it would not emerge. The advent of recording, where music is written as a signal as opposed to symbolic musical notation, brought forth changes so drastic that we now find it difficult to think about music outside this framework. Across the world, recorded music is played via radio, cassettes, or CDs, from village market places to multimedia concert halls. New computational media support this format, and we carry extensive libraries of music on our mobile phones. The next part of this book will look into how new media changes this ontology of music, through its computational, interactive, and generative powers. First, however, we need to look deeper into the possibilities offered when machines begin to listen to and analyse sound, because analysis also implies the possibility of synthesis.

11

Analysing

Seeing sound is an old dream. Aristotle discussed the colour of musical ratios, as did Newton in his work *Opticks*. Synaesthetic approaches – those that combine or fuse the senses – have linked colours with frequencies, and musical instruments have been made throughout history that integrate the visual into their play function. Colour organs, pyrophones, light organs, and many other instruments have been built, particularly in the past two centuries (Peacock 1988). Dance, musical theatre, opera, and puppetry are also visual forms that involve music. Here, sound is an essential part of the work and we are reminded that historically music has always been part of an extra-musical context, for example in social ritual, sacrifice, work, dance, war, and so on. In the early twentieth century, Kandinsky, Kupka, and Klee worked on synaesthetic visual art, and it is clear that Scriabin, Debussy, Messiaen, and Schoenberg were composers informed by synaesthesia. Messiaen, for example, complains that "[o]ne of the great dramas of my life consists of me telling people that I see colors whenever I hear music, and they see nothing, nothing at all. That's terrible. And they don't even believe me" (Messiaen in Brougher and Mattis 2005: 211). Visual music is a field that became strong with film and sound, for example, with the work of Disney, Pfenninger, Fischinger, McLaren, and others. With the advent of computers, audiovisual experiments took on a new form with Harry Smith, John and James Whitney, and Steina and Woody Vasulka.[1] However, the relationship between sound and visuals is far from simple, ranging from artistic interpretations of the sound to direct representation of sonic parameters. In this chapter we will look at the latter, as it serves as an ideal basis for artistic explorations of sound.

Interpreting signals

Until the invention of phonography, the history of sonic writing has largely been one of symbolic inscriptions: musical actions, whether performed by human or machine interpreters (such as player pianos), are expressed through symbols and subsequently transduced, via interpretation, into the relevant sound signal. This notation can range from a generative theory, as we find in medieval times and many Asian or African musics, to note-by-note prescription, like the one that begins to emerge in Europe with the Renaissance. This transduction is metaphorical, carrying meaning from one systematic domain over to another, involving a process of interpretation, in which the receiving context is never the same. Therefore, when the technologies of signal inscriptions appeared (of carving into grooves, arranging magnetic values on wire or tape, storing numerical values of sampled amplitude in an array), a transformation took place in the way we wrote sound and music, moving from symbol to signal.

The imperfection of notation gives it the quality of being relevant to people in different cultural contexts at different time periods. For the student of music, notation also serves as a source for inexhaustible interpretation, just like the work of Dante or Beckett. With sound recording the focus of interpretation shifts, and there is suddenly an abundance of music from all over the world of different musical subcultures that do not necessarily engage with a symbolic writing of music. This poses problems for the analysis of music by musicologists and critics, that is, if they want to continue the same theoretical engagements with the music (see, for example, Covach 1997). In previous chapters we explored how it became clear to ethnomusicologists of the first half of the twentieth century that a comprehensive transcription of the world's musics into Western notation was impossible. What was lacking was a technology that would transform the signal into symbols, equally benefiting the musicology and ethnomusicology of oral musical traditions.

> In employing this mainly prescriptive notation as a descriptive sound writing of any music other than the Occidental fine and popular arts of music, we do two things, both thoroughly unscientific. First, we single out in the other music, what appears to us as forms that resemble structures familiar to us in the notation of the Occidental art and we write these down. In so doing we ignore the musical parameters for which we have no symbols, as they might not be important in our culture, but essential in the other culture. Second, we expect the resulting notation to be read by people who do not carry the tradition of the other music. The result can be only a conglomeration of structures part European, part non-European, connected by a movement 100 percent European. To such a riot of subjectivity it is presumptuous indeed to ascribe the designation "scientific." (Seeger 1958: 186)

The ethnomusicologist's dream of objectively notating non-European musics is old, but it strengthens considerably in the late nineteenth century as a result of the global trade and communication that resulted from colonial powers bringing instruments and music from faraway lands to Europe. Machines for "objective" writing would be a godsend and ethnomusicologists closely followed the development of Scott's, Bell's, and Edison's work. However, ethnomusicologists were interested in descriptive symbolic notation, and in the 1920s Milton Metfessel developed a technique he called "phonophotography"; its purpose was to notate non-Western musics whose character did not lend itself well to transcription using Western musical notation. Metfessel was aware of the unavoidable hermeneutic situation of the ethnomusicologist trying to transcribe the music of other musical cultures, a situation which led him to develop an objective technique where sounds can be "photographed on motion picture film" (Metfessel 1928: 28). Metfessel's "pattern notation" can be found in his 1928 book *Phonophotography in Folk Music*, but this notation technique measured the photographic waveforms of the sound, using zero crossings to calculate pitch for melodic transcription. Metfessel was studying folk music, in particular African American music, and he found that his technique could transcribe musical elements that traditional notation was not capable of:

> valuable detail and necessary accuracy have been lacking in studies of folk music.... [But] by using the phonophotographic technique, the word "unnotatable" no longer needs to be used with reference to music or speech. It has been possible to notate all the twists, quavers, trills, breaks in the voice, quick slurs, erratic tempi, and other similar features, so often a part of folk singing. (Metfessel 1928: 20)

Objective notation with the Melograph

Charles Seeger, whose definition of descriptive and prescriptive notation we came across in Chapter 7, worked on the Melograph in the early 1950s (Seeger 1951, 1958), in the context of other research that had the development of similar devices as its goal, with one device built in Norway (Dahlback 1958) and another in Israel (Cohen and Katz 1960). Like Metfessel, Seeger was interested in the objectivity of the machine as a notational device, as opposed to the subjective hearing of the ethnomusicologist (Pescatello 1992: 212). These were analogue devices, so there was no possibility of applying the computationally intensive Fast Fourier Transform to analyse the spectral content of the sound. Instead they would apply techniques of low-pass or band-pass filters, where the fundamental harmonic was filtered out and subsequently drawn. Another technique was to measure the number of zero crossings over a period (how often the energy of the bipolar signal would step between negative and positive values), a number that represents the signal's frequency or pitch; the higher the number of repetitions per second of the repeated pattern, the higher the pitch. Seeger wrote:

> We are working our way, for the first time in history, toward a bona fide universal technique of music-sound-writing. All that is now necessary, besides perfection of the equipment, is to standardize: (1) width of chart; (2) square co-ordinates; (3) two pitch norms (A = 440 and the octave); (4) one time norm (the second, with decimal divisions). (Pescatello 1992: 213)

We might humour Seeger's "all that is now necessary ..." optimism, as we are still far from finding a suitable system for describing the multiple parameters of sound such that it is readable and representative of the signal's complexity. The past decades have taught us that the paradigm of the transcription is equally as important as the transcription itself. What are we listening to and for, and where from? Why are we listening to these features? How do we use them?

Seeger worked hard on the invention of the Melograph and garnered interest in the project from Bell Labs, but he eventually developed and financed the project himself. He collaborated with Mantle Hood, whose book *The Ethnomusicologist,* a key early

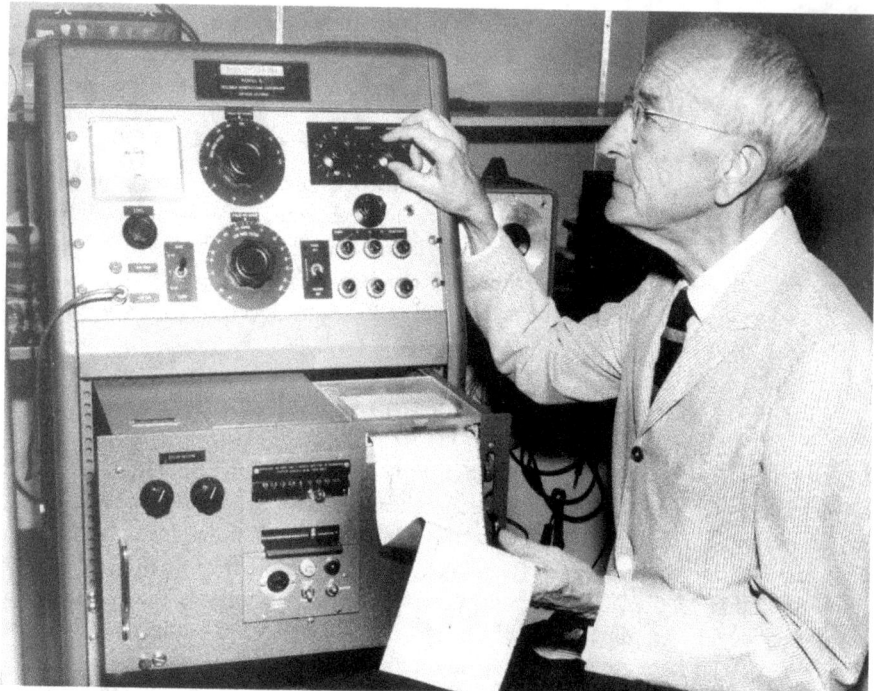

Figure 11.1 Charles Seeger working with the Melograph Model B in the late 1960s. © UCLA Ethnomusicology Archive.

work in the field of ethnomusicology, contains many illustrations of sounds in the notation format written by the Melograph. Hood was knowledgeable about various historical and cultural notational practices, yet he sought a more objective analytical tool for musical study that would describe the sound beyond the theoretical enclaves of a specific musical culture. Hood had a strong belief in this technology as something that might serve as an objective and analytical notational language for sound, or what Seeger called descriptive notation. With the Melogram,

> the ethnomusicologist will be able to supplement the symbolic language of the indigenous notation with graphic displays of pertinent musical practice known only in the oral tradition.... [It will enable the ethnographer] to compare whatever indigenous or extra-cultural theories have developed in relation to the tradition with the precise musical happenings that actually occur in performance practice. Such an application, need I add, is also pertinent in the study of Western music. (Hood 1971: 97)

The enthusiasm for machine phonography evident from the mid-nineteenth to the mid-twentieth centuries is understandable. Finally, theorists of music had gained objective means of writing down sound for description, analysis, study, and general understanding. This type of sonic writing went beyond indigenous notational systems in that it was agnostic about the music it notated, so it could equally represent a Bach cello sonata, American blues, or Australian didgeridoo. Furthermore, it solved the problem of the subjectivity of human transcribers (a musicologist specialising in classical music would hear different things in early jazz music than, say, a specialist in Irish folk music). The Melograph Model C was seen as the ideal machine for transcribing music, and contemporary ethnomusicologists became convinced that this notation would continue as a key analytical technology for ethnomusicological studies. Indeed, it was a fantastic tool: the spectral energy of the sound (the timbre, or the colour) could be studied, there was a nuanced description of amplitude, incomparable with traditional notation, and the fine-grained denotation of pitch was also of a standard that had not been seen before. Questions emerged, however, regarding the role of these devices: how are they to be used beyond simple demonstrations, and what precisely does the representation of an analysed sound add to the recording itself?

Analysis of music through visual means

Writing descriptive scores from a violin performance, for example an Irish folk tune, is easy to do as we know the character of the instrument, and we know that a good interpreter, embedded in the tradition, will be able to read and even play the score relatively well. But with electronic sounds, *musique concrète*, and other non-instrumental and non-notated music, creating descriptive scores becomes a difficult problem. We might perform a spectral analysis of the sound, and whilst looking at its partials, stacked on top of each other, can be useful this visualisation is not very descriptive, not even of pitch. A performer cannot even play back the music so richly represented in the spectrograph. For this reason, there has been considerable work conducted in electroacoustic music, whose primary compositional tools are computers, to reverse the representation: what they so elegantly provide in prescriptive notation they lack in descriptive or analytical notation. The rationale here is to provide listening scores that would aid the composer, the performer (of other instruments), and the musicologist or general listener in understanding the music, through representing it graphically in two- or three-dimensional space. Since electroacoustic music's primary focus is on the sonic events themselves as evolving dynamically through time, with the focus on timbral development as opposed to the numerical series of pitched events in symbolically written music, this representation requires a more advanced language than notes on a five-line staff.

Sound analysis is also important for electronic music, particularly when synthesising sound, in that through analysis we can get an idea of the source sound we want to replicate. The ability to study the sound's partials, amplitude envelope, and natural filtering is very helpful for deciding what type of synthesis design might be used. For example, if we wanted to synthesise the sound of a vibraphone (whose metal bars are

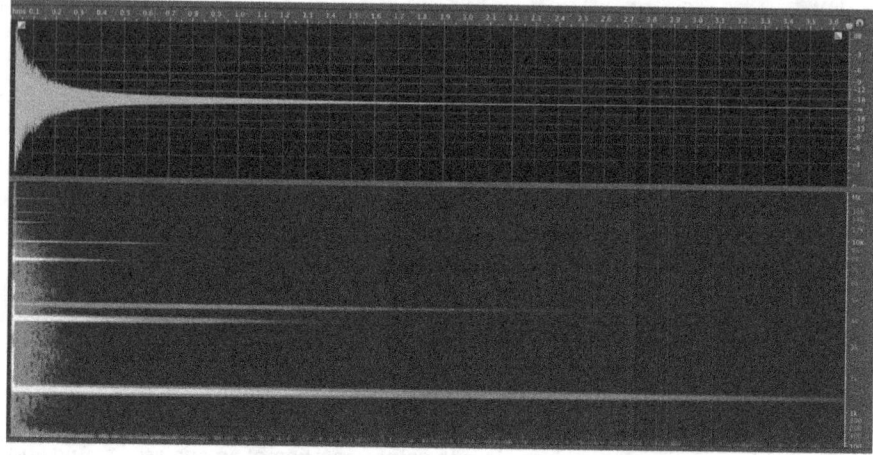

Figure 11.2 A spectrogram of a vibraphone note. At the top amplitude levels are drawn in time, but the lower image shows the partials of the sound after an FFT analysis. The partials are inharmonic (not multiples of whole numbers) and of different amplitude and duration. Photo © Thor Magnusson.

inharmonic, typically consisting of something between five and fifteen clearly identifiable partials), we might analyse the sound and note down the frequency, amplitude, and duration of each partial. These can then be reproduced with sine waves, with five to fifteen oscillators, each at a different frequency, amplitude, and duration; we can thus generate a very realistic simulation of the original vibraphone sound. This would be much harder to do without the visual representation, as it requires a very sensitive ear to be able to sense all the overtones of a vibraphone: their individual frequency, amplitude envelope, and duration.

Many of the techniques that underpin our understanding of sound can be attributed to the work of the mathematician Jean-Baptiste Joseph Fourier (1768–1830) and physicist Hermann von Helmholtz (1821–94). In 1822, Fourier presented an algorithm that demonstrated that any sound can be made out of sinusoidal waves. With the Fourier Transform, any signal, for example sound, can be deconstructed such that we can calculate its energy, depending on sample rate and other parameters. The Fourier Transform shows us the timbre of the sound; its partials, overtones, or spectral energy. The Fast Fourier Transform (FFT) and versions of it, such as the short-time Fourier transform (STFT), allow us to find the timbre of a sound within a specific time period of the signal (or window size), so a series of these will represent the time aspect of the sound signal. This is what we typically see displayed in a spectrogram in sound software. Before the digital computer, it would have been practically impossible to conduct such spectral analysis of a signal, and other means were used, as discussed above. However, it is beyond the scope of this book to delve into the specifics of the FFT algorithm, but readers are referred to Curtis Road's

Computer Music Tutorial (Roads 1996) and Nick Collins's *Introduction to Computer Music* (Collins 2010).

The notational quality of the sonic spectrum

Modern sound editors and digital audio workstations include diverse methods of representing sound, both for working with the sound and as an aid to the ears, where visualisation of waveforms, frequency spectrum, amplitude levels and more can help with understanding the complex layers of a recording. These software packages typically default to the depiction of sound in the standard waveform representation, similar to that found in seismographs or Scott's original phonautograph. Most of them also have a spectrogram view that draws the energy of the sound in a three-dimensional view, where the horizontal axis is time, the vertical axis is frequency, and the colour is amplitude. Sound editing software packages such as Audacity or Audition have good examples of these. Digital audio workstation software then typically has plugins showing the spectral energy in real time, which can be helpful when applying filters to the sound of individual tracks or the whole mix. The Sonic Visualiser software is also a powerful sound analysis tool that implements onset detection, spectral spread, dynamics graph, frequency cepstrum, and so on, visualising diverse audio descriptors (or identifiable features within the sound) that we might be interested in studying. This type of software gives the researcher a new perspective into sound and a drastically improved understanding of its qualities.

The types of scores that have been used to augment, and at times explain, electroacoustic and other musics have often been called "listening scores" or "aural scores" (*Hörpartitur* in German), and a very elegant and well-known example of this type of work is Rainer Wehinger's descriptive graphic score of Ligeti's electroacoustic piece *Artikulation*, from 1970. With digital computers, more sophisticated mathematical operations on audio signals become possible. Spectral FFT techniques enable both the analysis and resynthesis of sound. An example of this approach is GRM's *Acousmographe*, from 1989, developed by Olivier Koechlin in collaboration with Francois Bayle on an early Macintosh computer. In their 2004 paper describing the Acousmograph, Yann Geslin and Adrien Lefevre present the software as useful for annotating, analysing, and representing non-traditional music: "In the case of non-written music (ethnic, electroacoustic or improvised), the description and analysis of the musical content is not easy. A graphical medium is essential and has to be realized: this representation is not a score, but a sound transcription" (Geslin and Lefevre 2004). We note how their general ideology and description of the system is similar to that used decades earlier by Metfessel, Seeger, and Hood.

The FFT technique was much studied and improved in the 1970s using the digital computer, and it influenced practitioners both at IRCAM and GRM. In Paris, during this period, "spectralist" music emerged as a prominent compositional methodology, exemplified by the work of Gérard Grisey, Tristan Murail, and Kaija Saariaho. In this method of composing, the primary musical material is the sound itself, conceived of as energy in the spectral domain, as opposed to the discrete and abstract note events on a staff page. In other words, spectralism, which should be seen as an approach to composition and not an aesthetic, a genre, or a philosophy, is the first music that is purposely *composed in the domain of the sonic signal* and not the symbol. Some of these pieces were composed as tape pieces, and Jonathan Harvey's *Mortuos Plango, Vivos Voco* is a good example, while other pieces were notated for ensembles or orchestras. For the early composers applying this method, that meant they had to be creative in their use of notation, as the symbolic language does not fit perfectly with the spectralist approach. Grisey's *Partiels*, from 1975, exhibits many of these techniques. In this work, whose inspiration is a spectrogram analysis of the trombone (Finberg 2000: 116), we hear instruments morph into each another, making it difficult to perceive where one musical event ends and where another begins. The piece is a good example of an aesthetic that is inspired by the scientific understanding of sound.

Much electroacoustic music is based on these mathematical techniques of analysing and generating sound. Fourier's thesis was that all sound can be defined as a collection of sinusoidal waves, which implies that any sound can be synthesised that way too, and that is the idea of inverse FFT synthesis. Considering that electroacoustic music belongs to the lineage of written musical culture – the one of conservatories, musicology, concert halls, institutions, funding bodies, and media access – composers working in this domain, following on from Schaeffer and Henry, were accustomed to the musical score, written communication, and the composer-interpreter paradigm. This "critical music" was self-reflective, modernist, and progressive to the degree that practically every new piece represented an innovation in methods and aesthetics. The new language was mirrored in contemporary instrumental music, which, inspired by the explosion of sonic parameters to compose with, resolved to practices where many composers would create a new symbolic notational system for practically every new piece. This has to do with the concept of the work, the author, the product, and the intellectual transmission of the piece beyond the sound only. In Chapter 13 we will study how the instruments of the written tradition ended up in the hands of the composers of aural culture.

Instruments for seeing sound

Analysing music involves studying the compositional parameters at play. When the compositional material is sound itself and not symbolic instruction, a question arises as to which parameters are of relevance. There are innumerable features we could listen for when applying machine listening to sonic streams of music. Some might be parameters directly relevant to the composer, but the analyst might also be interested in features that were not of an explicit concern to the composer during the creation of the piece. Pierre Couprie's EAnalysis software[2] is a good example of a tool that aids in the analysis and reading of non-notated music, such as electroacoustic music. The software supports the use of diverse analysis plugins that return data in the form of audio descriptors (e.g., fundamental frequency, spectral flatness, loudness, spectral sharpness, inharmonicity, and zero crossing rate). These are epistemic signifiers for us to "listen in" to certain elements in the music, something that we can pay attention to and get an augmented perception of, via the power of the visual. Those signifiers are called "semantic tags" in the technical language of machine listening. Such spatial and visual representation of sound will help to get a temporal view of how the parameters progress over time, thus often giving nuanced analysis that traditional musicological tools might not be able to offer.

Software such as EAnalysis provides an analytical technical correlate to the new language of spectralist music, for example of spectral morphing, dynamic envelopes, partials, microtonality, formants, spectral envelopes, modulations, accelerations/decelerations, process, interpolation, rhythmic distortion (Finberg 2000). The spectromorphology of Dennis Smalley (Smalley 1997) is useful here,[3] and Manuella Blackburn's work of notating in the spectromorphological domain proposes symbolic elements to represent spectral signal features (Blackburn 2011). Signal inscription software like EAnalysis thus serves an important purpose in familiarising people with the parameters composers are working with in new musical work. This type of analysis software not only provides the visual means to study these parameters, but also supplies a language, a vocabulary that might suggest new avenues of study to the musicologist (Couprie 2016).

Another analysis software project, *TIAALS: Tools for Interactive Aural Analysis,* has been developed at the University of Huddersfield, by Michael Clarke and collaborators. The software is created for general-purpose analysis, but has been applied to specific analytical contexts in which the authors have performed thorough musicological analysis of electroacoustic music of work such as John Chowning's *Stria* from 1977, Barry Truax's *Riverrun* from 1986, Hildegard Westerkamp's *Beneath the Forest Floor* from 1992, Trevor Wishart's *Imago* from 2002, and Natasha Barrett's *Hidden Values* from 2013; a good selection from the electroacoustic canon. Using the TIAALS software, Clarke, Dufeu, and Manning introduce a new method of musicological analysis that offers an interactive engagement with the work, in terms of synthesis techniques, exploring of the musical shape of the work, understanding the decisions taken by the composer, and viewing or listening to the composers' own accounts of their work. They argue that this is a substantial addition to what traditional musicology performed through text only can achieve (Clark et al. 2016). The power of TIAALS is that, unlike EAnalysis, the sonogram image is a linked depiction of the FFT data, which means that the sound can be manipulated via the image. Events in time and frequency can be isolated for focused listening, or transformation. Selections of sonic events can be placed on analytical charts which is both visual and aural. Clarke and his team are currently working on a new project called IRiMaS, which aims to use interactive musicological tools for understanding music through manipulation, study, and tests. The premise here is that symbolic analysis in musicology is not sufficient any longer, thus presenting a "significant conceptual challenge to the traditional textual bias of much musical research and leading to new enhanced modes of musicological knowledge."[4]

Conclusion

Visualising music is not merely a method used in the analysis of music, but one that can equally be an aesthetic part of the work, to such degree that the music would not stand independently without the visual element. This is different from music videos or film music. Visual music, or audiovisual music, is a multimedia approach to composition. For the greater part of the twentieth century, the tools needed to compose music and make a film were a completely different set of equipment, requiring different skills, studio access, and belonging to bespoke institutional infrastructures, equally in terms of education, arts councils, theatres, media coverage, and so on. Today, however, composers operate with technologies – for example the software on their laptop – that are just as much an editing studio for a movie as they are a recording studio. The result is that the artificial division between the art forms is fading, and with new media playback devices, such as mobile phones, there exists an excellent framework for experimentation, development, and dissemination of audiovisual content. This collapse between the distinctive art forms will be studied in the next part of the book.

12

Machine Listening

Musical instruments are technologies of inscription: they incorporate musical ideas defined by the instrument maker, but when we play them, we "write" sound waves into the air. The musician listens to and experiments with the ergodynamics of the instrument, but at times a new notational language is required to get more out of it (Craenen 2014). Musicians *listen to* instruments and they *listen for* particular features, sometimes not satisfied with the available instrumental palette, which prompts them to make new instruments, software, or sound algorithms. It is as if we are hunting for a sound, probing the ether with our sonic antennae, our instruments, and eventually we find what we are looking for, or discover something new. As explored in the previous chapter, machines can also listen to music, and *listen for* particular qualities in the music. Using computer music algorithms, we can apply machine listening, or music information retrieval technologies, to analyse sound or musical patterns, we can detect similarities or differences to other work, and this is done at diverse levels from pitch, spectra, rhythm, structure, etc.

What to listen for?

Working with machine listening forces us to be explicit regarding what we are listening for or paying attention to. Until now, machines have analysed music by defined algorithms, such as FFT or filter bank pitch detection. The results of this analysis enable us to work with sound in new ways, for example representing it graphically for deeper knowledge and application in composition and critique, or as visual artistic accompaniment to sound. Inversive FFT enables us to re-synthesise sounds that might have been manipulated in the visual domain, for example by amplifying or attenuating partials, changing their frequencies or temporal qualities. These signal analysis technologies began to emerge in the 1960s as common scientific tools, but we had to wait until the 1980s to see them in computer music software. Even then, it was not until the late 1990s that personal computers became fast enough for real-time audio recording and manipulation. The first decade of this century saw an explosion in sound analysis technologies, and with machine listening, deep learning, computational creativity, and digital humanities reading and listening techniques, our machines now have ears that serve as extensions of our own; ears that can hear better and more than we do, informing us in ways that can be extremely helpful. However, to hear sound and to understand sound are not the same, and we have yet to reach a stage where machines exhibit general musical perception on par with humans.

But what happens when machines begin to listen? An objective definition of sound would typically describe it as the movement of molecules within material form (air, water, solids)[1] and it is tempting to say that the computer can hear what we hear. There are clearly problems with such statements though: the computer's ear, via the microphone or other sensors, is a subjective opening up to the world, dependent on the mic's build and its location. Provided we are satisfied with how the sound enters the computer, and its resolution there, we can then apply algorithms to actually *listen* to the sound. Machine listening, sometimes called "computer audition," is used in diverse areas of sound, for example in ecoacoustics (Sueur and Farina 2015) where soundscapes are analysed to assess and investigate ecological status (Eldridge et al. 2018), acoustic surveillance, lie detectors, monitor machines in factories, as well as in various areas of music. The novelty here is that the computer is "understanding" the sound in the sense that it is attentive to features within it: we ask it to listen in on certain features in the audio, applying semantic tagging, and to provide us with the results in ways that suit the particular context.

In previous chapters we have seen how machines hear and write sound by applying mechanical means of transcribing physical phenomena (sound waves) to another format (e.g., etched grooves, magnetic forms on tape, or binary numbers). This is mindless transcription without understanding, as if we were to listen and write down a language we don't speak with the letters of an alphabet. Pierre Schaeffer talks about four listening modes, which he relates to abstract/concrete and subjective/objective modes of listening to sounds. The first is to listen (*écouter*), a process in which we point our attention to something; we focus on the sound, its location and source. The second mode involves our aural perception (*ouïr*) of the sounds around us, in the periphery. The third is to hear (*entendre*), which means we become selective to the source of the sound and try to deconstruct what the sound consists of. And the fourth mode of listening entails understanding (*comprendre*) the sound, and that is where the sound becomes known, we can place it and relate it to extra-sonorous objects (Schaeffer 2017: 83). Shaeffer creates a fourfold matrix and further divides this into objective modes (*écouter* and *comprendre*), which are concerned with the object of perception, and subjective modes (*ouïr* and *entendre*), which are listener-centric. He also divides these into concrete modes (*écouter* and *ouïr*), which focus on the sound without trying analyse it or retrieve meaning from it, and abstract modes (*entendre* and *comprendre*), which involve us trying to understand and contextualise the sound. Schaeffer was, of course, interested in listening to (*écouter*) and perceiving (*ouïr*) sounds in his music without a semantical or source understanding of them. Sounds should be treated as raw objects, found objects, and this in itself is an aesthetic ideology. The goal was reduced listening (*écoute réduite*), a phenomenologically inspired method where the sound-in-itself becomes a self-referential object that we bracket out from its contextual background. Michel Chion builds on this approach to listening and identifies three listening modes: "causal listening," where you seek to understand the source of the sound; "semantic listening," which applies a code or a language to interpret the sound and its meaning; and finally, "reduced listening," a phenomenological mode of perceiving sound without seeking to find its source or meaning (Chion 1994).

Machine listening

How, then, might a machine listen to sound? It could do all of the above according to how we design its listening mechanism. Historically, machines have listened to human musicians via two methods: symbolic data and signal data. With MIDI or other control communication protocols, we can get direct information about the musical event, such as note, velocity, pitch bend, and so on. The protocol makes it easy to separate notes, and here polyphony is not an issue, as voices and instruments can be placed on different channels. This is equivalent to the symbolic musical score and we can add all kinds of symbolic level metadata, for example regarding timbre.

But the machine can also listen to the audio signal itself. But problems arise: what are relatively simple tasks for humans, such as identifying the instruments in an orchestra, turn out to be very difficult for a machine. With new developments in music information retrieval (MIR), semantic audio, machine listening, and deep learning we are beginning to see new modes of machine listening. These new listening technologies can analyse audio signals and extract meaningful information (what is specified as meaningful in the particular context), offering new techniques for musicians, composers, and musicologists to understand, classify, create, recreate, and manipulate sound (Collins 2016).

The use of trained neural networks in machine learning is a technique that is transfiguring the field of music technology. Although computationally expensive (it might take hours, even days or weeks, to train networks on fast computers, depending on the size of the networks and the data), these techniques can be applied in the analysis of music, its representation, and eventually recreation. We can apply them in our instruments, which can then learn about our individual behaviour and adapt to it. They can also be used in music production software, assisting in mixing and mastering; for example, applying styles analysed from the user's favourite track, album, artist, or producer. Machine listening might also work for listeners who get new music recommendations depending on their listening habits, and whose interactive input to the music can be guided. A good example of a machine listening project might be the performance and analysis of an Indian raga. A raga is never played the same way, as improvisation plays a crucial role in Indian music, and yet connoisseurs can typically identify which raga is being performed. This has to do with the scale used (the raga), and the general mood (*rasa*) of the performance. With machine learning, the system can be set to listen to different performances of the raga, and without any one-to-one pattern comparisons, it can identify the particular raga. Machine learning systems often perform tasks like these better than human specialists (see, for example, Waghmare and Sonkamble 2017).

There are various techniques available for machine listening. With the Fast Fourier Transform, for example, we get a map of the sound's spectrum at a particular period in time (over a short window, say from 16 to 1,024 samples, which is 5.8 to 23.22 milliseconds at 44,100 Hz sample rate). This map is stored as a two-dimensional array of numbers for each window frame, with an analysis of the amplitude and phase of each frequency band in each frame.[2] With a series of these over time, we can detect changes in timbre that are hard to write down by hand, for example how the individual partials of a piano note fade out at different times in the note's lifetime, or how a trumpet mute filters out the higher harmonics, effectively working as a low-pass filter. These techniques allow us to write sound in a new way; they concurrently provide a method and a language with which we scientifically understand the nature of sound. Theoretically, any sound can be re-synthesised if the phase and frequency information exists for each of its partials, applying a technique called additive synthesis. With such a numerical representation of sound, stored as numbers in arrays, it is therefore possible to apply all kinds of mathematical applications to it, for example pitch shift, time stretch, filter, amplify, or attenuate frequencies, reverse, inverse, and so on. Signal processing research has developed methods that enable us to detect "audio descriptors" in the signal, which include formant frequency, RMS energy (root mean square), intensity, dissonance, inharmonicity, centroid, onsets, spectral peaks and valleys, and spectral flux. Applying onset analysis, rhythm descriptors can also be populated with BPM (beats per minute) information, tempo changes, or novelty curves. Sound effect descriptors can also be applied to identify the effects used (e.g., reverb, echo, filtering, mastering), thus gaining information about the space in which the sound was recorded.

In common music software we find various spectrum-based features in use. MARSYAS (see http://marsyas.info) is an example of an early software framework that implements the analysis and annotation of sonic features such as the spectral centroid, spectral rolloff, spectral flux, bandwidth, RMS energy and zero crossings, as well as MFCC feature extraction.[3] An example of a practical project applying these techniques includes the *Million Song Dataset* (Bertin-Mahieux et al. 2011), in which a million songs by nearly 45,000 artists were analysed into fifty-five distinct fields or audio descriptors. The database is provided for other researchers to experiment with. But in general, the audio feature extraction parameters in any music can be stored in a stream of feature vectors, which can then be looked up, compared, classified, and rearranged, according to the application. With spectrum-based features like these, we can begin to classify the genre of the music, detect instruments, get an idea of room size, compare performances, and develop algorithms for tempo tracking, automatic transcription, score following, structural analysis, song identification, and so on. These audio descriptor algorithms themselves are relatively easy to design and implement. It is harder, however, to figure out what to do with them: how do we apply this analysis to get what we are interested in? This is where now machine learning is entering the picture among other AI algorithms.

One of the critical points with domains such as music information retrieval is that it considers music primarily as a sound-based information structure. We therefore get optimistic claims that machines can become musically creative since the data is all there (the machine can even perform an analysis of how music changes, thus possibly able to predict future developments). This is problematic, as we only hear what we listen for, and the machines will be conditioned by what we instruct them to do, and thus potentially marginalising crucial elements of musical expression which might go ignored if they are not registered for processing in the system. Examples include extra-audio signals, such as human gestures, performer relationships, communication with audience, and many more, all features that we would certainly define as being musical.[4] Some of these signals have been studied with MIR methods, for example machine analysis of performance gestures (Tzanetakis et al. 2007). This is important, as music is more than sound: it is a set of practices that come together in a social activity that Small calls musicking (Small 1998). Musicking includes fashion, dance, politics, drugs, media, merchandise, economics, and many more areas of human culture that constitute the musical environment. Live performance can also be seen as a cybernetic system, or what Simon Waters and others have talked about as performance ecosystems (Waters 2007). Considering the process of analysing, abstracting, representing, and processing the descriptors in audio and gestures, it is interesting to attempt the same with extra-auditory signals, although clearly the machine will not (any time soon) reach the level of a human in their creative thinking. However, the author of the MARSYAS music information retrieval framework is optimistic about the potential of information retrieval in musicology: in an article called "Computational Ethnomusicology", Tzanetakis et al. raise the prospects of using computers in ethnomusicological research, furthering the programme that Seeger started half a century earlier. For them, *computational ethnomusicology* is the "design, development and usage of computer tools that have the potential to assist in ethnomusicological research" (Tzanetakis et al. 2007: 3).

Cutting sound up into databases

Around the new millennium, increasing computing power allowed for music information retrieval techniques that analyse audio signals into chosen audio descriptors (via a process often called "audio indexing") that were stored in a database for analysis and resynthesis. The ISMIR community (International Society of Music Information Retrieval) was formally created in 2000, roughly coinciding with the release of Diemo Schwarz's Caterpillar system (Schwarz 2000), Aymeric Zils and François Pachet's Musaicing system in 2001 (Zils and Pachet 2001), and Michael Casey's Soundspotter system in 2002 (Casey 2009). The common technique in these systems is often called "concatenative synthesis," as the sound is analysed into small units (from 10 to 1,000 milliseconds), with affiliated audio descriptors. These are stored in a database for real-time calculation and access to the sounds, based on their descriptor values which can be written, composed, or derived from gestural controllers. In a sub-technique of concatenative synthesis, often called "audio musaicing" (referencing the visual equivalent of the mosaic), the audio descriptors are generated from the analysis of existing audio, such that sound can be synthesised, for example by analysing an incoming audio signal, or applying some other generative algorithm. In Casey's soundspotting technique, the incoming sound from one instrument can be analysed and replaced by the sound of another instrument, a procedure in which matching sound descriptors provide the nearest unit of similarity.

Schwarz later developed a system in Max/MSP called *CataRT*, a system that implements a "[c]orpus-based concatenative synthesis [that] plays grains from a large corpus of segmented and descriptor-analysed sounds according to proximity to a target position in the descriptor space" (Schwarz et al. 2008). Schwarz's system enables the analysed audio descriptors to be visualised graphically on a two-dimensional plane where grains are distributed over a parameter space. The visual representation of the grains is coloured, so the third descriptor can be expressed on the colour scale. This allows users to perceive the quality of the source sound and navigate and explore the sounds in the database. Various projects have been made using the concatenative synthesis framework provided by CataRT, ranging from systems where the sound of one instrument is replaced with another, to new gestural interfaces that navigate the sonic parameter space with the visual supplement of the software.

One of the techniques made available with the use of feature extraction and fast database access is soundspotting. This is a method where a large source database can contain a large collection of sound materials, and a target sound is played and matched with sounds in the source database, to be replaced with the target sound by applying a synthesis technique called concatenative synthesis (Serra 2007). In this type of synthesis, grains from the incoming signal are analysed (feature extraction), mapped onto the closest match in the database, and replaced by that grain (Casey 2009). An example would be a guitar player whose guitar sound is matched to the analysed signal of a saxophone such that the guitar sounds like a sax, all in real time. For Casey, the author of SoundSpotter, the goal in machine listening is to make "computer systems that are capable of humanlike hearing and decision making" (Casey 2009: 449). Artists have been using these technologies for nearly two decades, but we are witnessing more general applications of this technology, as implemented in commercial software, for example Adobe's VoCo ("Photoshop for audio") software (Jin et al. 2017). Adobe has not released the software yet, as it presents some serious issues vis-à-vis privacy, security, politics, and media, at a time in which an increasingly relaxed attitude to facts is currently causing drastic global problems. However, vocal sound spotting is bound to happen at some time, and aside from social and political risks, the technology will eventually become beneficial in the film and game industries, where character voiceover, help for people with vocal degenerative diseases, computer vision sign language translation, sonification of lip reading, or game character voice synthesis are among the most promising applications.

With the advent of deep learning neural networks, computers can be said to listen differently than in previous computational techniques. Instead of coded features added to vectors that are stored in databases, we now have multidimensional artificial neural networks that listen to the sound. With audio databases and score databases, we have a "big data problem" – the analytic problem becomes scientific, equally involving musicology, computer science, and the digital humanities. The question when feeding a musical signal to the neural network becomes: which parameters the network should be trained with? Should it be the raw sample stream? Should we hand-craft features and ask the system to listen for those? Should we let the machine discover its own features, and come up with its own representations? (Wyse 2017). Varèse called us primitives in the early twentieth century in terms of understanding machine sound generation, and we can confidently state today that we are primitives in our understanding of machine listening, and by implication, machine creativity. The latter will be explored in Chapter 15, but this chapter has set the ground for that discussion.

Conclusion

Online music technologies listen, understand, classify, and organise musical signals (whether audio, gestural movements, or symbolic data), and they do this through an analysis of the sound itself. These technologies also analyse user behaviour when listening, recommending, searching for new music, tagging genres, sounds, instruments, and other musical features. We are therefore slowly witnessing the emergence of a technology that might reconstruct or become instrumental assistant in the creation of new music. Through the centralisation and clustering of services in the few information conglomerates that stream and sell music, a market for data analysis and quantification of creativity is beginning to emerge that will engender new production methods, largely based on big data, and artificial intelligence methods in applying them in new musical products. David Beer describes the social power of the algorithm, pointing out how it can shape our consumption of information and eventually our perception of the world. He writes: "The notion of the 'algorithm' is now taking on its own force, as a kind of evocative shorthand for the power and potential of calculative systems that can think more quickly, more comprehensively and more accurately than humans" (Beer 2017: 11). Investigations into the role of algorithms are also taking place on smaller scales through open access and open source projects, and there is a dedicated fascination and curiosity of working with machines that are at the same time our instruments, our notational languages, and our acute expert listeners. New musical systems and codes are thus being created once again, like the technique of plainchant, the fugue, serialism, or the blues, and here they are written into the technology we apply for music-making, in ways that remind us of the epistemic nature of our musical instruments: the cave flute, the lyre, and the synthesizer patch. The next part of this book builds on the previous chapters, and explores what happens when digital technologies become the primary operative logic in new musical creation.

Part IV

Digital Writing

13

Transductions

The previous parts of this book have explored three distinct modes of sonic writing. In Part I, *Material Inscriptions*, we looked at how we inscribe our musical ideas into the bodies of musical instruments, which in turn become epistemic tools for sonic writing, as if writing sound waves with sonic pencils. In Part II, *Symbolic Inscriptions*, we analysed the ways in which we notate music and express sounds or performers' gestures through abstract symbols that can signify all kinds of things, from pitch, timbre, and dynamics, to bodily movement, interaction with the instrument, or the definition of the instrument's state. In the third part, *Signal Inscriptions*, we examined the methods that become available with sound recording, such as engraving with a stylus on writable surfaces, ordering the magnetic flux of tape, or converting analogue signals to binary digits stored on disks. In digital musical technologies these modes are applied in various ways: we create our instruments from diverse materials; we program them, notate for them, design solutions to symbolically represent their state, and we synthesise, process, record, analyse, and classify their behaviour, performance, and audio signals. We make instruments with ergodynamics that derive from millennia of musical practice, applying instrumental, gestural, and skeuomorphic metaphors in our new designs. Digital technologies encompass a wide range of genealogies of developed techniques from the past millennia, and this is a fact we often forget.

The core quality of digital musical instruments is their computational nature, and this involves arbitrary physical materials, mappings, and sound synthesis. The instrument becomes an epistemic model into which we write our music theory, and we test that theory through active engagement and performance, equally in the personal lab of the studio as well as on the social lab of the stage. Musical instruments are models of music theory (ergotypes in practice) and models can be mechanistic, theoretical, analogical, metaphorical, paradigmatic, and representational (Bailer-Jones 2009). They are "thing knowledge" (Baird 2004), comprising a unique functioning-in-the-world, sometimes designed into them, but at times, as is the case with acoustic instruments, inherent in their bodies, where their function is largely given by nature.

Wherein lies knowledge?

Digital musical instruments are hybrid objects with a pronounced epistemic nature and are of such complex origins that we can conceive of them, via Hans-Jörg

Rheinberger (2016), as "experimental systems," in which individual objects are complex "spatiotemporal entities" whose origins we can only analyse as part of a larger social context. In his foreword to *Experimental Systems*, Michael Schwab argues that the "move from a single experiment to an experimental system is necessary since it is the system that provides the context *against which* an experiment carries meaning" (Schwab 2013: 6). A creative musical context has been emerging where each composer/producer/musician develops their own experimental system that produces the musical work as an outcome of a performative practice. This work derives from a particular software constellation, electronic equipment patching, self-built hyperinstruments, or live coding environments, all of which embody bespoke theories of music. As a unique construction, such a theory of music, in its material manifestation, offers particular ergodynamics to the performer that are investigated through embodied and aesthetic approaches. The instrument equally offers an "enactive landscape" to explore, through its affordances and constraints, and a musical outcome resulting from the musician's exploration of the instrument's performance (*ergos*) potential (*dynamis*). It is, therefore, fitting to think of an instrument's ergodynamic quality in the manner ludologists apply the word "gameplay" in the analysis of computer games.

After a few decades of remediation in which digital instruments simulated physical instruments, and software simulated the recording studio, we are now in a situation where the nature of the computational begins to emerge, not as a simulation of the physical world, but as design materiality on its own. This change in algorithmic thinking manifests in the move from symbolic logic and expert systems to machine listening, deep learning, and big data analysis and manipulation. We benefit from earlier work, where the language of new instruments contains ergophors from both acoustic and electronic music technologies and is augmented by design tropes from other domains, such as computer games, HCI, web design, and now machine learning. Digital natives are less concerned with the digital as a subject of focus, and so hybrid technologies expand, in particular with the ease of DIY and the hype of the Internet of Things (IoT). Into these new technologies we bring with us knowledge of the past, its music, and its methods, but we also forget tacit practices, abstract away unarticulated features, and add new features that are given for free by the nature of the computational.

Furthermore, the new musical instruments made possible with computational technologies reassert the ancient dichotomies between the theoretical and the practical. A key difference is where the types of knowledge reside, and we can, through some great generalisation, look at how people's roles have changed with the advent of the digital. Here we observe that *designers* of digital software or electronic circuitry need to know what they are doing, but *users* can then "mess about" and discover the musical intelligence embedded in the instrument. The converse is true when acoustic instruments are concerned: here *performers* need to know what they are doing (thousands of hours spent on musical education, both theory and practice), but *instrument makers* typically "mess about," exploring the qualities of their materials when developing new instruments, not necessarily knowing much about the theories of acoustics or music.[1] Using a sensor involves voltage and baud-rate specifications that need to be looked up in a manual and adhered to, whereas a piece of wood does not come with specifications. Wood has properties, of course, but they are not defined, specified, standardised, listed, or formally communicated. Digital designers build their instruments from simple, formally specified elements, while acoustic luthiers are trying to tame complex materials.

Instruments as media and objects of resistance

Marshall McLuhan (1964) famously stated that "the medium is the message," opening our eyes to how the qualities of a particular medium constitute the way we communicate through it. In terms of organology, this might translate to "the instrument is the music." These are hyperbolic statements, conflating medium, code, signal, context, and message into one thing, but the key thrust for McLuhan was to demonstrate how media, in shaping the form of what is presented, define audience behaviour and expectations. Decades later, Bolter and Grusin (1999) analysed how the evolution of new media involves a primary step of remediating the previous medium. Early newspaper websites serve as a good example: they remediated printed newspapers. The analogy with McLuhan prompts us to consider how all musical instruments colour our musical expression: while it is true that we express ourselves through instruments, they have never been a neutral channel relaying pure information. Quite the opposite: they frame our thinking in equal measure, whether we are composers, performers, instrument makers, or indeed members of that wider assemblage of people and institutions who contribute to the cultural establishing of the instrument in its place. The instrument does not channel the music, as a tweet channelling a text message; it is the *origins* of that music, its archetype (*arkhetupon*), the very source of the sound through which the performer expresses herself. That expression would be nothing without the instrument. Moreover, a phenomenological engagement with the instrument, in the spirit of Ihde (1990), would demonstrate that musical instruments are not simple transparent extensions of our bodies, like spectacles or hammers. They are technologies whose phenomenological mode constantly shifts from being one of embodied extension to that of a partner in dialogue: from invisibility to dialogic confrontation. The instrument equally enables and resists, and it is precisely this polarity that makes it fun to play.

Musical instruments are objects of mystery.[2] Even after years of practising, musicians discover new qualities in their instruments and new techniques of playing them.[3] Knowledge of the instrument's ergodynamics emerges from the continual dialogue and exploration of its constraints, while experiencing improvements in play and a deeper understanding of the instruments place – its history, evolution, and future potential. This depth and mystery is what makes a good instrument. But the instrument is not the same for all players: it emerges or comes to life differently in the hands of different performers, so ergodynamics operate beyond any subjective-objective dichotomy.[4] Varela et al. (1991) describe embodiment as an *enactive* stance towards the world in which we create our (not "the") world through active engagement. Enactivism is resourceful in explanations of embodiment, something we should bear in mind when analysing how musicians incorporate knowledge of their instruments through repeated practise. For the instrumentalist, a particular concept of music emerges via the ergodynamics of their instrument, defining ideas and compositional possibilities. The general rules of HCI, of designing seamless interfaces that disappear through user-tested functions, simply do not apply here. The violin, for example, is not particularly user-friendly. It takes years of practice to play it decently, and through the process, hands, neck, and shoulders, as well as the left ear (which is too close to the sound source) can become greatly damaged. However, the violin's infinite dimensions of potential investigation make it a focus of desire; it is a beautiful object with a remarkably rich history, and for the performer, establishing a relationship with the instrument is a profound physical and psychological experience. Heidegger's phenomenological terminology of "ready-at-hand" (*Zuhanden*), which relates to the instrument as a medium, and "present-at-hand" (*Vorhanden*), which conveys the instrument's objective qualities, are useful in the analysis of instrumental ergodynamics. In the former mode, the instrument disappears through its use, but in a *breakdown situation* the object suddenly emerges to us as present-at-hand, and we need to attend to its properties (Heidegger 1962). Heidegger used the hammer as an example, appearing as present-at-hand when the head becomes loose from the handle. The phenomenological relationship we have with the hammer shifts when it ceases to be an extension of our body, or part of our body schema (Kirsh 2013), and suddenly becomes an object of study. However, the difference between a tool and a musical instrument is that many present-at-hand experiences are *not* necessarily a breakdown situation,[5] but rather comprise what makes the instrument interesting to play, and what draws us to explore it for hours on end on the path to mastery.

A good musical instrument, therefore, exhibits the duality of shifting between being a means to an end and an end in itself.[6] Marc Leman argues that musical instruments should be transparent technologies that "give a feeling of non-mediation, a feeling that the mediation technology 'disappears' when it is used" (Leman 2008: 18), and we agree that this is a great state of affairs: one of flow (Csikszentmihalyi 1996), concentration, and expression of motor skills. But we also acknowledge, as Derek Bailey writes, that "[t]he instrument is not just a tool but an ally. It is not only a means to an end, it is a source of material, and technique for the improvisor is often an exploitation of the natural resources of the instrument" (Bailey 1992: 99). A musical instrument fluctuates between these modes, but its performance ecology (Waters 2007) forms a cybernetic system where performers listen to (and feel with their bodies) the instrument's performance, constantly reacting to its behaviour, with innumerable other parameters feeding into the cybernetic loop, parameters such as architectural acoustics, audience behaviour, fellow performers, and behaviour of the sound system. The concept of instrumental boundaries become complex in electronic systems, where the use of microphones and outboard effects, pedals, modular synth elements, mixers, and other accessories, make up an assemblage whose borders are vague, and whose individual elements can often be swapped out with others (Bowers and Haas 2014).

Embodied musial gestures

In his article "Beyond Art and Technology: The Anthropology of Skill," Tim Ingold (2001) explores the notion of skill as an embodied activity with a strong environmental nature, typically through the use of technology. Ingold rejects the Platonic dualisms of mind and body in instrument use: the essence of skill is "both practical knowledge and knowledgeable practice" (Ingold 2001: 20). But skill is not merely a technique of the body or a thing-in-itself; it is a "total field of relations constituted by the presence of the organism-person, indissolubly body and mind, in a richly structured environment" (Ingold 2001: 21). For Ingold, the study of skill demands a wider ecological approach. The dexterity of human movements does not lie solely in the precise control of motoric movements. All human movements are different, as any study in biomechanics or ergonomics demonstrates: the precise movements in hammering, drumming, or strumming are never identical in living systems. Kinesiologists, who study human movement and apply their science in sports, dance, martial arts, and rehabilitation, call this fluidity of movements "motor redundancy." However, the musical intention behind the act can be the same in all the cases and the measured *result* of the movement too (the beat was repeatedly on time).

Imagine a drummer on a steady beat. The body is moving, individual parts of the body are never in the exact same position, and the muscles of the arm are differently applied in each strike. Yet the drummer might be in perfect time. While the timing of the drumstick hitting the skin might be almost machinic, the performing body is a shape-shifter throughout. In this context, Ingold introduces neuroscientist Bernstein's (1996) views that dexterity lies not in bodily movements, but in how these movements respond to environmental conditions that are never the same from one moment to the next. Dexterity involves ad hoc "sensory corrections," a continual adjustment, or "tuning," into the ongoing task. For Ingold, "skilled practice cannot be reduced to a formula" that is passed from generation to generation. Skill is rather "a generative schema or program that is established in the novice's mind" (Ingold 2001: 21) by observing accomplished practitioners. Furthermore, in the NIME literature a distinction is made between sound-producing gestures and ancillary gestures (Wanderley et al. 2007; Jensenius et al. 2009), where the latter accompanies the former, but can be prospective (preparing for coming movements), communicative (to the audience or other performers), or generally expressive (of ideas and emotions).

Consider this example: in an important part of a piece, a pianist lifts her arms, head tilts backwards, and then suddenly a bodily force is exerted downwards and forwards, with accompanying facial expressions. The embodied preparation of playing a two-handed chord might have taken a few seconds, and this concerns not just the physical time of the clock, but also the subjective time of psychological tension and energy. Here, the wait and the build-up of energy might be more musically impactful than what could be measured in milliseconds or how the actual chord sounds (or looks on a spectrogram). A large part of the musical communication of the performer happens in the semiotics of the corporal space. This is observed by Godfried-Willem Raes, of the Logos Foundation, who has built an impressive orchestra of robot instruments over the past decades. For Raes, it is important that the robots give a visual indication that they are about to play, preparing the audience for the sound to come (Maes et al. 2011).[7] From the perspective of motor-image and instrumental theatre, what happens before the sound is emitted is of equal importance as the sound itself, as we, the listeners, are partaking with our mirror neurons – we feel the notes about to be played, even prior to hearing them (Matyja 2015).

Ergomimetics

The shape is now emerging of what I am exploring as *ergomimetic* memory, or the imitation of gestures or work practices, in *identical* (say a student of the guitar learning from a teacher), *divergent* (a saxophonist applying their skill when playing the clarinet), or *transduced* (e.g., where a pianist plays on a touchscreen) technologies. Ergomimesis happens when one generation initiates the next by introducing them to new environmental and technical contexts in which the technique or skill is applied. However, it is not that simple, as this initiation might not be through direct apprenticeship, but through "distance learning," as we find with British musicians imitating American blues in the early 1960s or Palestinian rappers making hip-hop music today. There is a level of noise and misunderstanding that infiltrates the signal, conditioned by the environmental context, and this becomes an essential part of the creative process. In divergent ergomimesis, the performer adapts to new technology with the knowledge or intentionality they bring with them. The process of learning the new instrument (say, the clarinet, which has a different fingering than the sax) is one of adapting to a new technological context by engaging in a dialogue with the instrument and discovering what kind of ergodynamics the new tool offers.[8] Transduced ergomimesis can be explained, in the spirit of Heidegger (who, like Nietzsche, enjoyed philosophising with hammers), by archaeologists demonstrating that the *act of hammering* existed long before the hammer, in the form we know it today. But since the action existed (e.g., using a rock to crack a nut), the potential existed for developing the technological object itself through iterative and evolutionary processes.[9] The behaviour is established in a space of functional action and semiotic gestures and designers subsequently implement a support for it in the new technological domain. Although the above discusses how human actions constitute the potential for new technologies, the converse might be more common, that is, where a new technology, built for a particular purpose – often imitating some other technology – interrupts the function it was built to accomplish, yielding new unforeseeable behaviours, or is adopted to a completely new and unforeseen use.

The question of trained musical motor memory is of importance in the design of new instruments. This science ranges from HCI to psychology and sociology to musicology. For example, in the Augmented Instruments Lab at Queen Mary, University of London, Andrew McPherson and colleagues are exploring this question of ergomimesis through diverse projects, or how skills identified in one technological domain are implemented in another domain. The idea is to leverage the skills of trained musicians in their playing of new instruments. In many ways, this is a postdigital approach, where the design is a complete hybrid of acoustic, electronic, and digital elements, and where the designers equally enrol the affordances and material qualities of each of these domains. In a recent research project, Jack, Stockman, and McPherson proposed the concept of a *bottleneck* in defining how rich biomechanical gestural space is reduced in the design of an instrument: "The instrument takes that input and projects it down via the sensors of the instrument which capture only a small subset of that gesture, constituting a bottleneck in the interaction. From this bottleneck the instrument then projects out again to an expanded set of sonic and kinematic features and behaviours" (Jack et al. 2017). The bottleneck concept is useful to describe the interface aspect in ergodynamics, that is, what is projected as the instrument's affordances for interaction, as its expressive bandwidth, and how those offer an exploration of the instrument's constraints. McPherson's Magnetic Resonator Piano (MRP) is a good example of such an instrument, where the performer can play the instrument as a normal piano, with the addition of soft onsets, slow envelopes, pitch, and timbral change (McPherson 2012). This is a distinctly new instrument, becoming individuated (in Simondonian terms) and stabilised (in actor-network theory terms) with its unique ergodynamics, which offer new compositional and performative approaches. However, an instrument is never a fixed thing; just as any acoustic instrument is perpetually open for interpretation, an "emerging" instrument is bound

Figure 13.1 The Magnetic Resonator Piano. The electromagnetic frame has been installed into a regular grand piano. © Andrew McPherson.

to be too. McPherson explains: "What I've found is that pianists and composers come to disagree about the optimum identity of the instrument. Some see it as an extended piano, rooted in traditional piano technique with an additional layer of control, while others see it as a wholly new instrument and even seek to avoid engaging with the traditional hammer action, and still others see it somewhere in the middle" (personal communication, 2018). It is this exploration of a new instrument's ergodynamics that characterises much contemporary music: it is a search, or research, for new sounds, new sonics to "torture" out of the instrument, just as Bacon and other seventeenth-century scientists used their instruments to "torture" truth out of nature. McPherson has studied the new sonic language of the instrument by inviting composers and performers to explore the instrument through playing, encouraging them to come up with their own notational systems or improvisatory practices, as there are features in the instrument that he, as the inventor, will not have discovered, and might never do.

The halldorophone is another example of an instrument in the process of innovation; it too leverages acquired performer expertise, in this case the embodied skills of the cellist. The halldorophone is more than a cello imitation: it is deliberately made for feedback that contains interface elements to control it, so in many ways, a classically trained cellist will not necessarily have an understanding of the dynamics of feedback which constitute the instrument's functionality. The shaping of feedback can be likened to a potter shaping clay, but it has to do with amplitude, instrument position, bodily movement, resonating strings, and many other non-linear parameters that are hard to have full control over. It is therefore interesting to compare the distinct approaches to the performance of my Sussex colleagues Alice Eldridge and Chris Kiefer, who, inspired by the halldorophone, and collaborating with Halldór Úlfarsson in the building of their

Figure 13.2 Halldorophone. An instrument incorporating acoustic, electronic and digital elements. © Halldór Úlfarsson.

own feedback cellos, have a very different way of playing the "same" instruments. It is quickly evident that these instruments are different beasts, even if physically similar, as the internal, electronic structure that processes the sound is different: Eldridge's cello is electroacoustic, while Kiefer's is a digital electroacoustic hybrid. It is further intriguing to observe the differences in how Eldridge, who is a trained cellist, plays with her instrument, compared to Kiefer, whose background is in computer music and, in particular, feedback. They bring different experiences to the instruments, with extremely varied results. This family of feedback instruments represents a new approach to the delegation of an agency, in which the instrument itself becomes an actor in the performance ecology, and the performer shapes, moulds, and influences the instrument, as well as playing it in a normal manner. Many of the new instruments we see appearing these days have this character: the aim is not to gain full control and mastery over the object, but rather become collaborative and conversational with the instrument in a musical dialogue. Decades of interactive technologies are changing our ideas of what instruments can do. This is explored by Andrew Johnston and colleagues in what they term as a "design for conversational interaction," where "musicians are engaged in a kind of musical conversation with the virtual instrument as if it were another musician" (Johnston et al. 2009: 209).[10]

Ergomimetic interfaces

The section on epistemic tools, in Chapter 3, articulated how objects (from a rock to a rocket) can suggest use, symbolising their function through appearance and design. The argument was that many digital technologies are of a higher conceptual nature than acoustic technologies, as evidenced by menus, sliders, buttons, and knobs, most of which are labelled with words signifying their function and numbers signifying their state. However, and this is an observation made by many developers of digital instruments, even if the interface itself is symbolic, with use of screens, numerical knobs and sliders, and menus of all sorts, we, the users of the instruments, still gain motor skills through repeated entrainment. Even the live coder will have a motor memory of writing a for-loop in their fingers. These skills are typically ergomimetic; they are reused motor patterns that might come from a different field, and can easily be applied in making music. The movement of the mouse, the turning of knobs, the pressing of plastic buttons, or soft rubber pads are all actions that are applied in diverse use cases beyond music. Computer game interfaces are increasingly used as musical interfaces, and even the gameplay is translated into a musical context, both in terms of performance (as in animated notation) or teaching (as many instrument learning apps nowadays include machine listening and can give learners points for playing the right note at the right time, and generally measure their progress). Instrument makers therefore increasingly design for embodied skills from other domains, applying human–machine interactions that are now becoming universal. If new music technologies apply interdisciplinary skills, from operating computer interfaces to game controllers or apps, this is bound to have implications for musical education.

The question of musical education is relevant. Just a century ago we had more time, greater attention spans, fewer entertainment distractions lying around, and a musical instrument in a household was of focal significance. Interface skills were also less diverse, as there were fewer objects to "play" or control. Thus, motor skills gained by practising a particular instrument were typically exclusive to the domain of the particular instrument family. A cellist would be more familiar with the double bass and the violin than a percussionist (consider the divergent ergomimesis discussed above). Musical instrument "interfaces" were unique and did not derive from other technological domains. In fact, the opposite is true: musical instrument design is often influenced by the design of newer technologies, as the case of the typewriter demonstrates, where we began "[p]laying the literary piano" (Stingelin 1994: 71) in the new profession of typists and secretaries, when typewriters became popular in the 1920s (Kittler 1999) (and this would be an example of a transductive ergomimesis). What makes this change possible is the efflorescence of interfaces in our daily life through electronic and digital equipment. Our lived experience now largely happens through an interface: making coffee, buying a train ticket on a phone, taking an elevator, working in front of a screen, cooking a microwave lunch, or flipping through TV channels with a remote. Interfaces are a control layer that maps movement in one domain to different movements in another, and modern existence implies engaging with the world via interfaces most of the time and not through direct application of energy (Galloway 2012; Gill 2015). Interface critique is therefore due (Andersen and Pold 2018) and in music we may ask if the experience of future generations will be that mapped interfaces (as in digital instruments) are more natural than direct play surfaces (acoustic instruments).

Conclusion

New digital instruments pose the problem that not only are they fluid, potentially with unique mappings for different pieces, but, with the implementation of machine learning, they may also evolve with the performer through their relationship. Machine learning changes the mass-produced interface into a unique object, as no two instruments will be the same. Here digital objects, which have hitherto been generic and exchangeable, become unique instances, like violins, guitars, or analogue synth modules. Instruments with machine learning capabilities will learn from their players, who "train" them to adapt to their playing, so no instrument will be the same. Players can, of course, share their training data and black box models, and thus play each other's instruments. It is unlikely that an instrument trained by another player will be fully satisfying to someone who did not train it, but perhaps the sense of alienation with a differently trained instrument might become a source of interesting and rewarding creative engagement?

Once again, music technologies can therefore serve as the experimental platform or laboratory for our embodied being-in-the-world, for how we act, create, share, collaborate, and communicate through our new networked computational technologies.

The digital redefines musical practices and understanding. New instruments can be single user, collaborative, networked, audience-participative, or multitemporal. They can be simple and direct, or involve complex artificial intelligence and machine learning, where the instrument shapes itself around the behaviour of its user. The new instruments can be systems that analyse our behaviour, share with others, consult from databases of musical works and performances, suggest or predict, all in a bespoke way tailored around users and their musical preferences. The intriguing question will be to see how those preferences are formed when we become a cog in a larger machine-learning musical intelligence system.

We therefore get a techno-aesthetic or ergodynamic materiality that is established through a relationship of playing that is unique to each performer. This prompts the question: with instruments like these, what is the destiny of the composer? What is the instrument that is being composed for and how would one notate for it? Equally, how will studio producers perform live on stage the pieces they composed in the studio? With new music technologies the idea of the musical piece is transformed, and the next chapter will engage with how new digital media drastically reposition the people, objects, and institutions of music, reframing old practices and shaping new ones through the emergence of musical systems impregnated with new ideas.

14

New Notations

[A] new and notable trend in instrument design has attracted my attention. Instruments themselves are now being fashioned in the shape of notes and other musical symbols. This is an interesting attempt to cut out the intermediary score and grapple with notes directly. (Nilson 2016: 5)

Before notation, repeating musical pieces was the only way to preserve them: the music would be written into our memories, but memory is a medium that quickly degrades, so the music would have to be repeated at regular intervals for it to survive. Therefore, in oral cultures, an emphasis is placed on the accuracy of the performance, on the fidelity of reproduction, although all musical cultures contain distinct clusters of ideas regarding the concepts of reproduction, fidelity, and authenticity. No performance of a piece can ever be the same, but there tends to be a "family resemblance" of rules regarding how a musical piece is to be played. And these rules differ across musical cultures. With notation, there is a sense in which music is liberated from the weight of tradition. Music can now be preserved by writing it down, and consequently other avenues of exploration emerge. This explains, to some degree, the explosion of innovation in musical composition that followed the advent of notation and, in particular, the printing press. With the press, music became easily transmitted geographically through score books, which quickly became a commodity. Thus, the press instilled new notions of authorship into musical culture, for attributing the right author could now make a pronounced difference for the composer's livelihood.

Writing on fluid formats

In previous chapters, we explored how the advent of notation and recording, and symbolic and signal inscriptions, changed the way we compose, perform, disseminate, and enjoy music; we also explored the associated changes in the notion of the musical work. We are currently entering yet another epistemic shift in music technologies, where our notational languages further diversify with the advent of programmable machines, resulting in methods such as machine listening and computational creativity, and musical media formats such as streaming music, games, apps, and virtual reality worlds. Here the audio signal can be manipulated through a hierarchy of languages and implemented interfaces (abstractions written in programming languages) that analyse, understand, sculpt, shape, synthesise, or recreate music. Subsequently, the way we engage with music changes too, from the previously defined roles of composers, performers, listeners, etc., to new functions where these mantles are less defined, more prone to merging and fluctuating, even within the same musical system (and here I use the word "system" deliberately to signify a piece or a work – *systema* is the word used in ancient Greek for a musical scale). Computational media are different from previous media in the sense that we will not arrive at established formats for algorithmic, generative, interactive music, but rather enjoy a state of constant innovation and exploration. This would have caused problems for earlier models of musical commercialisation (imagine the record shop if there were no standards for LPs or CDs, with every piece of music coming in different packaging formats), but with digital systems, which are increasingly software- and browser-based, the question of distribution standards becomes less important with the gradual disappearance of physical storage and transport. However, this melange of formats can clearly cause confusion regarding language and expectations in regular listeners, critics, and musicologists.

In light of new notational practices, we might ask how sonic writing is constituted in the current networked computational multimedia. We are witnessing the formation of a new type of oral culture (or rather *aural* culture), with improvisation and live coding on new instruments, one that is established through electronic media of audiovisual documentation. This is a culture where the text, picture, video, tweet, Snapchat pic, Instagram (what an appropriate name!) is speedily disseminated, but without being carefully composed, thoughtfully consumed, or archived. It is a culture of fleeting media exchange, more reminiscent of how our body language communicates in social settings: "the world of electronic media has put on centre stage oral cultures or immediate communication; writing, the elaboration of unheard universes, the results of reflexive rational speculation thus seems to be put in question" (François 2015: 67). It is not writing as such that is put into question, but older practices of writing that depend on media of scarcity (tablets, papyrus, skin, paper, wax, wire, tape, vinyl, film, etc.). Writing is not disappearing, it is simply drastically different, in its evanescent omnipresent nature and with the addition of extra-textual channels, as exemplified by Snapchat or YouTube. With the excessive writing in text, pictures, and video we lose the interest in archiving, and enter a world of new orality. For us, the question becomes: how do sonic writing practices form in new computational media, where the process is more important than the product, and how does this redefine established notions of instrument design, notation, and phonography?

A piece or an instrument?

Digital musical instruments are epistemic tools par excellence; they are an assemblage of technical elements, software and hardware, that express a vision of how a music could be thought, composed, and performed. Everything is explicitly designed in a digital instrument: there are no "freebies" emerging naturally from the materials, as we find with the flat bone bridge of the sitar or the detuning effect resulting from the difference in temperature of elements inside an analogue synthesizer (Pinch 2007).[1] The instrument often becomes a piece in itself, as the creator of the instrument has some musical purpose in mind, often quite specific. The boundary between a piece and an instrument is deliberately vague, and it can fluctuate.[2] One of the leading digital luthiers, Perry Cook, once listed the principles of digital instrument design. Among his sound advice, we read: "Make a piece, not an instrument or controller" (Cook 2001). Cook qualifies his statement by pointing out "that huge control bandwidth is not necessarily a good thing, and that attempting to build a 'super instrument' with no specific musical composition to directly drive the project ... yields interesting research questions, but with no real product or future direction" (Cook 2001). It should be stressed that Cook is not talking about a notated score or some conceptual schemata: he is talking about a physical object embodying the piece itself. Such an object represents a complete fusion of the conceptual and the material with a bespoke purpose, bespoke ergodynamics, and a bespoke playability. This tenet was articulated by Gordon Mumma in the 1960s, when he said that "I consider that my designing and building of circuits is really 'composing'" (Mumma in Nyman 1974: 91; see also Born 1995: 60).

A musical instrument is a vehicle of theory: it is a model of *ergos*, one that implements the instrument maker's ideas, the performer's movements, and the audience member's understanding of how the sound was generated. In its encapsulation of knowledge, it demonstrates theory and upon its encounter the performer explores its potential or *dynamics*. The way we build the instrument, how it sounds, how many holes we put on the flute or strings on the lute, how we tune the instruments; all this is constituted by a relationship between theory and practice as manifested through history. La Monte Young's *Well Tuned Piano* is an example of a piece that requires the instrument to be in a certain tuning: through its unique configuration an improvisational performance is made possible, and the piece itself somehow emerges from playing on the piano. There is no notation for the *Well Tuned Piano* except for the tuning, and this has been kept secret. Typically, it is Young himself who performs the piece, although he has taught his collaborator, composer Michael Harrison, the tuning and how to play it. But it is a concern for contemporary musicians what happens to their music when the instrument is given to other users. For different reasons, the field of new music is therefore experiencing a change in practice, one where the composer's performance is centred and the notation, or the recording, becomes peripheral to the piece. In an interview I conducted with composer Alex Mendizabal, he epitomises this mindset, stating that "[m]usic is away from the record itself." Music happens elsewhere in a much more complex techno-aesthetic and social context, and a recorded documentation cannot be the music. Mendizabal adds that he has problems with the authorship concept implicit in musical media: "It feels to me grotesque to think there might be copied records in different sites, a will of omnipresence" (personal communication, 2018).

"Make a piece," says Cook, but what does he mean? What kind of a piece? The tendency in the performance and composition of new digital music is to advance beyond the work-concept. Looking over the fields of NIME, SMC (Sound and Music Computing), ICMC (International Computer Music Conference), CHI (Computer–Human Interfaces), DIY instrument making, and open source music software, as practised in studios, hacklabs, and research labs across the globe, what we find are conceptions of music that are very different from the Romantic understanding of the musical piece as a unique work with an ideal interpretation. We are witnessing the emergence of a drastically new notion of a musical system as an *invention* in the understanding of Baroque music, a discovery of a musical idea that can be developed through disposition and style. This invention can be an instrument, an app, or a new rule set, for instance. Affiliated with this change, we observe how the musical score is decentralised, and the instrument or the system take up its former position. To support this statement, we observe how performances with new musical instruments or live coding languages often do not have work titles: it is the instrument or the system itself that is listed after the performer's or ensemble's name in the concert programme. The performer is also typically the designer of the instrument or the creator of the software. This is also yet another example of how the distinction between the roles of composer, instrument maker, and performer blur in this context, and more updated terms might be "producer," "designer," or simply "musician."

Music as invention

The hypothesis I am putting forward here is that within musical culture, practices are beginning to form that resemble those that existed before the twentieth-century commodification of music, which was enabled by recording technologies, but to some extent also those before the nineteenth-century establishment of the work-concept and the standardisation of notational practice. These new practices transcend the linearity of the musical score and recording, not by rejecting them, but rather by including and opening them up, introducing interactivity, agency, interpretation, exploration, collaboration, and environmental factors. The new works are typically unique expressions, not simply in terms of content, but also in terms of form. This emergent cultural formation can be found in the examples of musical automata, innovation in instrument design, reinstitution of pre-work ideas of musical performance, formulations of new music theoretical systems, incorporation of improvisation and indeterminacy in scores, of the effacement of performers' names or titles of pieces, and so on. The relationship between the work and the author has a long history and is analysed in Foucault's essay "What is an Author?" where he explains how once scientific theories were only considered truthful if attributed to their authors, while anonymous stories, folk tales, epics, and tragedies were judged as true by virtue of their age (Foucault 1984). With modernity the justification by power of the author is reversed: scientific theories do not acquire their validity because of who the scientist is, and literature, art, painting, and music tend to be strongly attributed to the creator. Today, once again, these tectonics are shifting around.

In Chapter 6 we read about how, in the sixteenth century, musical composition came to be considered an art form whose ontogenetic structure shares that of other arts, such as rhetoric, where invention, disposition, elocution, memorisation, and delivery are key to the successful creation and execution of a work. Bach famously used this structure of rhetoric when he wrote the pieces of work known as "Inventions." The invention is not a musical genre, but a concept that signified the thematic idea that underlies a musical composition. The invention precedes the instantiated written composition and Bach wrote variations of his inventions, as these can be developed in different ways. The invention is "more than a static, well-crafted object, but instead like a mechanism that triggers further elaborative thought from which a whole piece of music is shaped" (Dreyfus 1996: 2). In the first century BCE, rhetorician Cicero writes in *De Oratore* that the orator "must first hit upon what to say; then manage and marshal his discoveries [*inventia*]" (Dreyfus 1996: 3) into elegant execution through arrangement, style, memory, and delivery. The issues explored in contemporary works of notation and new musical media are bringing us back to pre-work ideas of music, more akin to these earlier ideas of invention, and questioning the author function as articulated by Foucault. However, by questioning the work and the author function, this does not mean the effacing of artistic signature: musicians still bring their system, vision, aesthetics, and personality into the music, it is just that the notion of what constitutes the identity of a musical work is changing. For much contemporary musical work, the invention is not the notated piece, but the theoretical structure that supports it, whether instrumental, notational, or computational.

In an article from 2000, Jonathan Impett investigates the concept of invention in his practice of interactive music: "In my work, I needed a name for a particular software construct: a unit of musical behaviour which encapsulates materials and behaviours from multiple sources which is formed by the interaction of several dynamical systems.... The *invention* in this context is the locus of materials, behaviour and relationships" (Impett 2000: 29). Impett applies the concept to his swarm-based musical system, but it is also relevant to embodied musical systems, such as Laetitia Sonami's sensor glove, or software-based systems, like live coding environments. Invention is therefore not a musical work, but something that supports it by engendering action and interaction. Impett has a good reason to apply the term to his practice, but for the current context I refrain from defining it as precisely and narrowly as he does. Rather, it should also include how invention is defined in the twentieth century – namely, as something that has been put together, not discovered, but actively created through combinatorics or out of nothing (as in *poiesis*). Examples include generative compositions, new musical instruments, audio games, live coding systems, animated notation pieces, unit generators in audio programming languages, verbal scores, virtual worlds, and so on. These are theoretical systems that "contain" the music through their affordances and constraints – objects, in other words, that generate actions through their ergodynamic potential. The contemporary act of composition and performance in these new media is therefore a search, a *ricercare*, where we explore the ergodynamics of the instrument, its playability, sound, and musical structures. Thus, the questions often asked, in twentieth-century language, whether the work is the instrument or the musical piece, or whether the live coding language is a system or a composition, become irrelevant. By applying the concept of *invention*, we transcend the work-concept and the conceptual problems that have lingered as linguistic remains of a musical culture we have now surpassed.

Semiotics of notation for new instruments

Notations for new musical instruments and systems are therefore a complicated issue. This could be explained through the different nature of what we have defined as nineteenth-, twentieth-, and twenty-first-century instruments, studied in previous chapters. But in considering it as a problem of notation, we could do well to reflect upon the distinct semiotic notational conditions in the three instrument paradigms. This problem relates to the question of mapping, which is the key difference in the way musical instruments work. An acoustic instrument does not have an arbitrary interface that maps to some functions. It *is* the interface. And yet we do not tend to use such terminology, since these instruments are not "mapped." With electronic and digital instruments, the question of mapping becomes essential, and subsequently the question of ergonomic design: where to place the knobs, sliders, screens, how they should feel and function, and how they are grouped together (perhaps from the perspective of Gestalt theory, in order to create a good workflow and playability). When these considerations are combined with the historical continuity, remediation, and musical potential of the instruments, we get their ergodynamic character. To understand this problem of mapping we can apply a semiotic perspective, and use the Peircian trichotomy of the sign, which divides signs into the types of *icon*, *index*, and *symbol* (Peirce 1868). Briefly explained, the *iconic* sign is one which resembles, imitates, or reflects the qualities of the signified object. A statue, iconograph, or onomatopoetic word is iconic. They resemble (visually, sonically, etc.) the signified. The *indexical* sign does not necessarily resemble what it stands for. However, it is directly connected to it: for example, footprints in the snow, smoke, a ring tone. These are learned signs, but they are contiguous with their origin. Finally, *symbolic* signs are arbitrarily assigned structures where the signifier and the signified might bear no relation at all. This is based purely on convention, which is determined by a population of users. A green traffic light, skull and crossbones, a red cross: these are all symbolic signs that are learned and highly cultural. Peirce notes that these signs often overlap, stating, for example, that a symbolic sign might contain an iconic element.

Our musical instruments can be subjected to this semiotic model in order to make explicit and understand a certain unease musicians report when describing the qualitative differences between acoustic, electronic, and digital instruments. The model can also inform our concept of notation for diverse instruments. From the perspective of semiotics, *acoustic instruments* are of iconic nature: the string on the guitar functions at the same time as the sign, the interface, and the sound source. There is a direct and necessary relationship between interface and sound, one that is based on acoustics or physical laws. *Electronic instruments* can be defined as indexical. There is a contiguous link between the sign and the signified (e.g., between the filter knob and the filter behaviour). A voltage-controlled low-pass filter works a certain way, and its behaviour is clear. We might, however, re-wire the knob such that it increases the cut-off frequency when turned to the left and decreases it when turned right. This is a convention, an index, but it is not arbitrary, as the behaviour is still based on the principles of electronics. *Digital instruments* are clearly symbolic. The mapping between the interface element, whether screen-based or physical, is arbitrary: there are no natural laws that limit our design options. In the design of digital instruments metaphors are often used at the interface and interaction levels. We need to base our understanding of the technology on something familiar.

When we study the structural nature of our instruments from this semiotic perspective, we realise why it is so much easier to notate for acoustic instruments: here, a symbol on a page can be directly mapped to a resulting sound (e.g., in five-line staff notation) or a gestural action of the performer (as in tablatures). The composer knows the ergodynamics of the instrument and is able to instruct the interpreter to perform an action with a specific result. With electronic instruments, however, it is harder to create prescriptive symbols for their behaviour. How might we notate an amplitude envelope, a pitch change, a timbral morphing of a sound, panning between speakers, a filter movement, and so on? This can be done, of course, but it is likely that each notational system would be specific to the instrument, if not an individual musical piece: an invention therefore. Furthermore, electronic instruments are unstable, so the symbolic notation cannot refer directly to a sonic result. On modular synths, people used "patch sheets" to store the state of the modules (which cable was connected where, and the state of the knobs), a practice that is now typically done with a digital camera, but even this could not preserve the patch fully. It is well known in the modular synth communities that in a complex system, you can never arrive at the same state again. A solution would be to apply more imprecise notation for these imprecise instruments. This trouble with the state singularity of complex systems triples with digital instruments. They change like the wind; a change of a number value in the code could result in a drastically different instrument, one where the sound engines can change as well as the mapping engines. What is then being notated for? The composer thus moves beyond writing notes, harmonies, or metre, and focuses on general design, where the notation becomes the structure of the instrument itself, for example, expressed as an invention, a compositional system, in a Max, Kyma, or SuperCollider patch, for example.

Composing for digital instruments

The argument is that the composer–performer paradigm is being transgressed in new work, but that does not necessarily signify a total rupture or discontinuation of older practices. Established music behavioural patterns are persistent, and they are evident in the various laptop orchestras, where the setup is quite traditional: the composer, the written score, the conductor, the orchestra on stage, and the audience – the close resemblance of eighteenth-century practices. It is evident that new technologies will not destroy older musical cultures, and unlike the Futurists a century ago, no one is calling for this destruction. But new technologies do bring with them a new episteme, which becomes the ideological signature of the age. And threads will cross the diverse epistemes – for example, a composer writing a piece on paper for a new musical instrument, or a pianist working with brain-controlled robotic hands. From an abstract viewpoint, two general compositional approaches have emerged for composers in this new musical paradigm: the "idiomatic approach" (Vasquez et al. 2017) and the "supra-instrumental approach" (cf. Cadoz 2009). In the former, the composer writes work specifically for the new instrument, thus supporting its *raison d'être*, its continuity, and development. This establishes a tightly knit and rewarding relationship between the composer, the instrument maker and the performer. The problem surfacing here is that new instruments are never in a solid state – they resemble fluid processes – and so a composition written for an instrument in 2020 might not be realisable five years later unless specific actions are taken to preserve the state of the hardware, software, operating system, protocols, and physical gear and cables. Will, for example, the Bluetooth standard even be supported in computers in ten years' time? Will the software work on the next update of the operating system? The supra-instrumental approach, on the other hand, represents an approach where the composer writes more general notation that could be for any instrument, high-level musical gestures, and instructions which can be further explained through natural language. Writing supra-instrumentally can involve quite nuanced definitions of sound (e.g., synthesis type, spatialisation, change in filtering, sonic mass, granularisation, sound sources), but neither the software nor the interface is specified. Such a compositional approach is reminiscent of Renaissance music where composers did not write music for specified instruments (Schulenberg 2001), at a time when innovation in instrument design was booming, and where performers were more than mere interpreters of music.

How do we compose for, and with, instruments and systems that are in constant flux? The traditional composer-role is described above with the idiomatic and supra-instrumental compositional approaches. But perhaps the question itself is flawed, as it is symptomatic of a thinking that defines the composition as a series of linear events in time, symbolically notated in paper format. With digital technologies, the potential is opened up for various new conceptions of the work, such as real-time scoring, interactive pieces, game narratives, and virtual reality worlds. Clearly, many of these forms could have been possible in pre-digital times, for example composing in real time on staff paper using an overhead projector to project the real-time score for the performing musicians. We could have involved the audience in our performances, and many Fluxus artists did, creating immersive musical theatre experiences, all without digital technology. However, technology embodies the cultures that emerge with it, and the fact is that it is often *because of* the "digital" that these ideas are picked up and explored. The idea of the instrument as a piece emerges and articulates very clearly in digital technologies, and the paradigm has now affected ideas, language, and development in analogue and acoustic instrumental practices as well. This is not unheard of in pre-digital practices, and the resurgence of interest in the work of Harry Parch, who built bespoke microtonal instruments for his new music (Parch 1979), is a good indication of that. However, the digital paradigm means that Parch's approach, which might have seemed eccentric half a century ago, has become rather ordinary today.

Figure 14.1 Performance on a tangible score by Enrique Tomás. © Enrique Tomás.

New notations

The instrument incorporates notational elements, but conversely the notational is becoming increasingly instrumental and systematic. Animated notation is an example of a practice that investigates the potential of computer-generated scores, where performers follow instructions in real time. Considering the multimedia qualities of the computer, such scores need not be on staff lines, but could take the form of abstract shapes (e.g., 2D or 3D animation) and could be displayed on screen, tablets, or virtual or augmented reality headsets. In this form of musical practice, the ideas of playing music and playing a game converge, since the score (not the game score) is projected for the audience to watch as well; it emphasises the aesthetic features of the score, inviting the audience to follow, and partake in the music. The audience can easily judge the performers' fidelity to the score. Critics have often voiced the opinion that this type of playing becomes highly reactive for the instrumentalists, as they do not know what happens next, and they have not had time to familiarise themselves with the music. The performance becomes one of waiting for a visual trigger to play a note that puts the performer in a role more akin to a gamer shooting objects than an interpreter of a piece. This is set against the more common practice of the interpreter studying the score in its spatial layout, practising its parts, and finally rendering the whole piece – a piece in which the score has typically been inscribed into the performer's bodily memory (ergomnemonic memory) and the score itself is merely a memory aid.

However, such criticism might fall short, as who is to say that a performer necessarily needs to know the piece they play or whether, indeed, it is not possible to gain an in-depth understanding of an animated piece by performing it many times. Animated notation critically engages with the ontological notion of the musical piece, leaving the piece open and flexible. The work of Ryan Ross Smith, Cat Hope, and Lindsay Vickery is exemplary here, as well as the miscellaneous experiments of the SLÁTUR composer collective. The collaborations of Hope and Vickery have been influential (Hope 2017), as has David Kim-Boyle's (2018) notational practice. In Smith's work (an example of which is on the cover of this book), the performers are given the freedom and flexibility to interpret the score as they see fit, and experiment with different interpretations (by establishing their own rules) of how to perform the piece. This relates to the approach of Cardew (1967), whose graphic scores required the ensemble to discuss and decide how to interpret the piece, thus delegating the creative responsibility or authorship (concepts that were being critically questioned) to the ensemble itself. The manner in which abstract animation inspires musical form has a long history, of course, and in many ways the animated notation practices refer to live piano music or the abstract music of composers like Pfenninger, Fischinger, Smith, McLaren, and the Whitney brothers (Abbado 2018).

Animated scores have a game-like event-response dynamic, where players are often unprepared and follow instructions that appear in real time. We are familiar with such practices for example in karaoke or computer games such as *Guitar Hero*, but this kind of musical instruction can once again be traced back to Guido of Arezzo,[3] whose technique, known as the "Guidonian Hand," was a sight-reading system that enabled the composer or person knowing the music performed to indicate by pointing at parts of their hand which notes were to be sung. This type of pointing, indicating, and instructing is an essential element of music, the role of the orchestral conductor being a good example. A new style of musical direction emerged in the 1970s, conflating improvisation with composition, under the name of "soundpainting" (Thompson 2006), where the conductor composes the music in real time with the orchestra in front of her. This language of musical communication has been used in various settings, with Nadine Couture applying skeleton gesture recognition with the Kinect sensor to control flying drones that create a sonic swarm (Couture et al. 2018). This is done by gestures functioning as "letters" in an alphabet that enable the creation of a string, or a "word" which turns into a series of commands. Julio D'Escrivan live-codes soundpainting, by applying computational thinking as part of his instructional sets: "I have remarked on its similarity to live coding as you have the chance to instruct direct actions that are either single events, or streams of events. These can be instructed to interact with each other or depend on each other" (personal communication, 2017). Dynamic scores have existed in various forms throughout the ages; one could trace such methods from Guido's "Hand" to John Zorn's instructional piece, *Cobra*, but it is the real-time nature of the computer that makes this practice easily achievable. With computational multimedia, real-time generation of musical scores of all kinds becomes increasingly common (Hajdu 2007; Eigenfeldt 2014). Pedro Rebelo and Franziska Schroeder have worked on networked dynamically-generated scores, with performers in different physical locations (Rebelo 2010). Dynamic scores can be precomposed or generated in real time, and then streamed to the performers (and sometimes audience) through laptops, tablets, or phones; they are often projected for the audience to see, for example in Rebelo's work, where the scores are visually stunning. These scores can also be generated with input from machine listening systems where the composition adapts to what the performer is playing.

Live coding

The idea of considering computer languages as notational systems is fundamental in the practice of live coding. Here, the internal mechanism of software is exposed and rewired through live, direct, and exploratory use of custom-made programming languages (Collins et al. 2003; Magnusson 2014). Practitioners perform on stage by writing code that generates audiovisual work; it is a form of real-time notating or scoring music, visuals, dance, or robotics. When performed live, the screen is typically projected, enabling the audience to follow the development of the code as it progresses during the performance. Since the computer is interpreting the code in real time, every evaluated edit is immediately reflected in the musical or visual end result. This practice presents an ideology that approaches the production and performance of computer music from the core of what the computer instrument is, namely an algorithmic machine. Working across musical styles and venue types, live coders talk to machines in their native language of code, projecting their screens to the audience, who can follow the performance of the piece as it is written and heard. This is a very honest form of computer music, where the problem of mapping is exposed and the causality of the musical events is clear, even if members of the audience are not familiar with the programming language used. For many live coders, the graphical interface represents a concealment, an abstraction, obscurantism, as it closes off possibilities. Conversely, interfaces also "suggest" action, where good design helps the user to navigate possible musical ideas. Clearly, the language of code offers an incredible width, depth, and flexibility, compared with other compositional systems; it presents an opportunity to think musical practices from the ground up.

Notation, instrument, audience, performer, composer, piece, agency, authorship, participation – all these concepts are up for grabs in live coding practice. As an example, PowerBooks UnPlugged is a live coding ensemble whose system, the Republic, is in square contrast to the framework we see established in many of the laptop orchestras that subscribe to nineteenth-century hierarchies of the composer, the conductor, the orchestra and the silent audience. During PowerBooks UnPlugged performances, the players sit amongst the audience, composing in real time by writing code and sharing it with each other such that it is not clear who is doing the composing or who is playing what. Established musical hierarchies are abandoned (Rohrhuber et al. 2007). Networked collaboration in the form of code sharing of code is common amongst live coders, and the removal of power hierarchies in

Figure 14.2 Alexandra Cardenas live coding at the GEIGER Festival in Gothenburg in 2014. © Ruud Gielens.

networked music is a focus we find in the work of Shelly Knotts, who has composed various pieces in which established musical roles are rethought – for example with democratic voting systems, mixing algorithms that make sure performers have equal play time, and other playful algorithmic means of challenging established conventions.

Related to live coding, algorave (Collins and McLean 2014) is an emerging subculture in electronic music (see www.algorave.com). Here performers typically apply methods of live coding although not strictly, as graphical user interfaces can also be algorithmic. The music in algoraves is beat-based dance music, on repetitive rhythms in a night club context, where projected code lights up smoke-filled dance floors. This genre is about movement: that of algorithms and human bodies. Like live coding, algoraves have become a global phenomenon, with club nights popping up in over eighty cities around the world. This club performance context means that the coder needs to be able to write quickly and keep the attention on the audience. For this reason, new live coding systems have been created in the form of micro-languages, such as TidalCycles (McLean 2014), Sonic Pi (Aaron and Blackwell 2013), ixi lang (Magnusson 2011), and Gibber (Roberts and Kuchera-Morin 2012); languages which are often built on top of domain specific languages for music, typically SuperCollider, ChucK, and Extempore; or more general interpreted languages, such as Python, Haskell, JavaScript, Ruby, and Lua.

Characteristics of postlinear work

As part of the research programme for this book, I organised a symposium on new notations together with Frédéric Bevilacqua and the Sound Music Movement Interaction team at IRCAM.[4] The speakers at the symposium presented their unique approach to the concept of notations in the twenty-first century, and their presentations can be found on the project's website, as well as IRCAM's. A common theme involves how new notations often question the concept of linear time in music and open up for alternative understandings of time and how to structure musical events. This also relates to the hesitance of defining a new notational standard. Claudia Molitor, a presenter at the symposium, is an apposite example of a composer for whom the idea of the musical score has been emancipated from the tradition of *écriture*. Although Molitor works with computers as part of the workflow, her work is not "computer music," and it is more the conditions of the computational than it is a specific format that defines the characteristic compositional mindset of her work. Thus we find pieces composed for particular locations, like London's Waterloo Bridge, or a journey, like in the *Sonorama* piece, which exists as an app that tracks the location of the listener and plays music composed for the particular land or cityscape through which the train travels. Another piece, *Entangled*, explores the concept of touch and how the score can physically manifest in space. Here three percussionists pull strings looping around the stage and on encountering a knot on the string, they play. In many ways this is similar to the animated notation approach described above, where the score is visible and performers "react" to the score more than interpreting it. For Einar Torfi Einarsson, a musical score can be a spatial installation, and one of his pieces consists of the internals of a piano stretched out in three-dimensional space. The physicality of the score is also explored by Enrique Tomás, another of the symposium's presenters, who composes "tangible scores," or physical objects that combine the interface and the musical score through notated graphical shapes carved into wood or embossed on paper and other materials. The performer plays the score through interacting with the tangible score which contains contact microphones, such that hand movements on the score are translated into sound (Tomás 2016). The tactile score gets the form of a body suit in Bhagwati's musical notation: the notation is communicated through vibro-tactile information by means of a bespoke suit (Bhagwati et al. 2016). Composer Sarah Angliss has an extensive history of work: her compositions range from writing staff notation for performers to designing robots that collaborate with her on stage. Her pieces are highly diverse, comprising of dance choreography, installations, robotic carillon, and opera. In times of multimedia and audiovisual documentation of performances, there is no need for composers to limit themselves to musical expression on staff notation.[5]

In the compositions, writings, and discourse of contemporary composers, we gain an awareness of the diversity of their practice. There is a sense that, unlike their predecessors from a pre-recording culture, the music of contemporary composers will not persevere and be played for centuries to come around the world in the same way that we have experienced with the performance of Classical and Romantic music in the

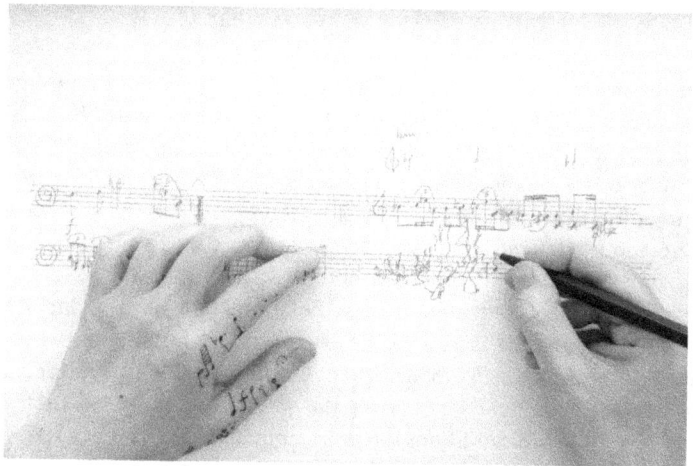

Figure 14.3 A still image from a piano/video piece by Claudia Molitor for pianist Zubin Kanga, called *You touched the twinkle on the helix of my ear* from 2018. © Claudia Molitor.

nineteenth- and twentieth-century concert culture. Current musical work is more evanescent, site-specific, based on the technology and cultural context of the time, and it is more likely that the documentation of the work, in audio or video form, will preserve it for future generations. In improvisational or site-specific pieces, this can never give a satisfying picture of the work, but it might be better than nothing. In this respect we are witnessing a movement back to practices of the Renaissance period, that is, before composers began to write idiomatic works for bespoke musical instruments.

Conclusion

What, then, is the point of composing music for live performance? The written musical piece – whether that is a score, instrument, audiovisual piece, interactive game, or a virtual reality world – is a particular object in the world. It opens up new ways of thinking, and is a space for exploration that has much more potential for engagement and interactivity than an instance of the work recorded in the studio. For composers of written music, the rationale for writing music differs, but for many it has to do with providing the condition for engagement, communication, a world to explore, an offering of gameplay, perhaps not so different to what computer game designers do. As such, the notated score has an ergodynamic: it can be studied, played, investigated, navigated, and reinterpreted at different times in different places. When composers realise that the instruments of today might not be with us in a decade, the compositional work becomes more about shaping sonic and human relational processes through time, about creating situated social contexts and narratives that provide apertures or clearings for the interpretation and creation of new worlds. Composers are interested in written music as objects that generate social formation (e.g., Harris 2013), in the form of people – composers, performers, event organisers, media, critics, and audience members – coming together around and exploring a piece, a situation, a *habitus*. Contemporary composers are moving beyond the linear mentality imposed upon music by the qualities of notated and recorded music, and re-conceiving music as a form of collaborative social practice that emphasises presence in time and space, mindfulness and focus. When recorded music is freely available and in abundance on the Internet, many practitioners do not see the point of adding another piece of recording to that speedily growing collection, and so the musician's focus is directed back onto the social dimension again – that is, on musicking.

15

Machine Writing

Music technologies have always been at the vanguard of changes in human culture, used to explore new modes of expression, communication, economic models, and social development. For decades, practitioners and philosophers of computational creativity have discussed whether computers can really be creative and compose something original. This has been a revelatory pursuit, as it has uncovered our assumptions of creativity, humanity, and the nature of artworks. Here, we might note that the idea and injunction of originality are phenomena that begin to emerge in the eighteenth century; they experience a heyday in nineteenth-century Romantic music, as well as in twentieth-century avant-garde and experimental music, but are subsequently questioned in postmodernist theory and practice. A prevailing view of the state of the art in computational creativity is that computers cannot create original music in the sense we attribute to humans. Below, we will explore this question in more detail, but there are many issues with such a statement, as a "computer" is only what a human has programmed it to do, so we must be careful not to state that humans cannot create original music if they notate with code, but can if they notate with symbols on paper. Such statements can easily be questioned, but the answers are far from simple, and we might also take note of the fact that in our current post-postmodernist culture, the concept of originality is somewhat marginalised, and is not the holy grail it was in twentieth-century modernism.

Transforming rules

Cage's silence and Merzbow's noise has made anything in-between these extremes musically possible. Today, we can therefore ask why, if Duchamp was able to place a urinal on a plinth and frame it as art in 1917, we might not simply record any sound and call that music? Cage has already done this to some extent, and we are happy to have him masterfully turn on a few radios and call the resulting cacophony music, but we seem to have problems with the idea of machine-created music, even machine-assisted music. The first problem we encounter is that the output of machine music is often not particularly great, but neither is the output of many human musicians. The question then shifts to that of creativity, where we might ask how creative composers or music producers really are when they apply the rules of the fugue, the sonata, the serialist system, the blues, or the house track. How is the case of a human following the rules of these compositional systems different from a machine following the same rules?[1]

Many argue that real creativity in music, art, or indeed any human activity, is to transform the rules, or at least to explore them from such a new perspective that the content produced is novel, surprising, and valuable. This is the argument of my Sussex colleague, Maggie Boden, a pioneering researcher of computational creativity, and in her ground-breaking books, *The Creative Mind* (Boden 1990) and *Creativity and Art* (Boden 2013), she defines three different types of creativity: combinatorial, explorative, and transformational. Boden's account is quite formalistic: the rules of an art form are "search spaces" that we navigate by *combining* elements within the space, *exploring* the space, searching for new constellations, or *transforming* the rules of the space. The difficult question of computational creativity is really whether computers can be said to be creative in the transformational sense as they can clearly combine and explore search spaces of rules, and this has been explored by many (e.g., Ritchie 2006; Wiggins 2006; Saunders 2012). Astute readers will recognise how combinatorial creativity derives from seventeenth-century combinatorics and transformational creativity can be seen to borrow from twentieth-century avant-garde ideas. Limitations of scope prevent me from entering the philosophy of creativity here,[2] but suffice it to say that there are many other aspects which relate to creativity, such as the social context related to the 4 P's (person, process, product, press) as presented by Saunders (2012), and also simple and relatively unstudied factors, such as misunderstandings, mistakes, errors, and general noise in the system.

Machine as a human extension

The above concerns might seem rather peculiar when one considers that music-making machines have been with us for thousands of years. Greek water organs, Chinese wind chimes, medieval aeolian harps, Arabic mechanical robots, and European clockwork musical automata are all examples of an attempt to delegate music-making to non-human processes (Koetsier 2001; Collins 2018). These were often projects that anthropomorphised the mechanism and were perhaps also designed to imbue the machine with a "soul," a situation that comes as no real surprise given the prominence of the concept, in eighteenth-century Europe at any rate, of music as a tool to express the subjective inner life of humans. In the nineteenth century, pianolas – mechanical machines that read piano rolls – could be placed in front of normal pianos, performing from punch-card pieces that might have been played by expert pianists and composers, such as Rachmaninoff, Joplin, Debussy, Mahler, Scriabin, and Gershwin. Later the playing mechanism was implemented inside the piano itself, and the player piano was born (although this, too, is sometimes called the pianola). Before the advent of recording, player pianos would be very popular in Europe and America, and there were specific jukeboxes that played the piano in bars, restaurants, and clubs. A good example from the first decade of the twentieth century is the Mills Novelty Company's musical mechanism called the *Violano-Virtuoso*, which included two violins and piano strings that gave mechanical control over all three instruments. A paper music roll contained the notation to control the player mechanism. These technologies were expensive, so clubs and bars would lease them from the manufacturer, whose additional revenue stream was producing rolls with new music. Predating sound recording, pianolas and music boxes were popular replacements for live musicians in clubs and bars that were often open twenty-four hours a day. The popularity of these machines largely ceased in the 1920s as result of high production costs and in conjunction with the increased popularity of recorded music.

The idea of machines contributing to the *creation* of music, as opposed to merely playing it, has not been as prominent as that of musical automata, perhaps mostly because the articulated distinction between composition and playing is only a few centuries old. Working as the programmer of Babbage's *Analytical Engine* (a mechanical computer) in 1842, Ada Lovelace wrote about the possibility that the engine "might compose and elaborate scientific pieces of music of any degree, complexity or extent" (Roads 1996: 822). Note that here the machine would not *originate* anything new and the attribution of originality should go to the programmer (Saunders 2012). Here, Lovelace immediately connects to the ideas of combinatorics in music introduced in Chapter 8; for example, Guido of Arezzo's algorithmic system of composition, Kircher's *arca musarithmica* compositional device, or dice used as compositional aids by composers such as Haydn and Mozart. Indeed, in the 1950s, various experiments inspired by information science began to pop up that experimented with the automated creation of music. The *Push Button Bertha* pop song in 1956 (Ames 1987) was one of these early examples, or the more contemporary music sounding *Illiac Suite*, written by Hiller and Isaacson in 1957, using algorithms, such as Markov chains, that are still used and well known in the field of computational creativity.

Generative music

Generative techniques have been used in computer music since its birth. This use could have been expected, considering that the origins of the modern computer come from the work of researchers exploring the combinatorial arts, formalisms, and abstract machinery that might become programmable; and music is a prominent feature in all these areas. Lovelace's statement, on the prospect of "scientific music" that is calculated by an analytical machine, is a natural element of this theoretical mentality, as music was an essential part of the liberal arts. Research in informatics and computer science has resulted in a canon of pieces using various techniques for computational creativity. A few examples will suffice. In 1972, James Anderson Moorer used heuristic techniques and generative grammar for composing music, a technique later used extensively in David Cope's well-known *EMI* (Experiments in Musical Intelligence) system (Cope 1996). Kemal Ebcioğlu's 1988 *CHORAL* software was a knowledge-based heuristic system of 350 rules for generating choral melodies, and uses a generate-and-test method with intelligent backtracking, resulting in pieces that sound very similar to Bach's compositions (Ebcioğlu 1988). Genetic algorithms have proved a successful technique of generating musical data from set rules of defined search spaces (Horner and Goldberg 1991) and so have other artificial life algorithms, for example, cellular automata (Miranda 1993). Artificial neural networks have been used for a long time to learn models of compositional or performance styles successfully (Todd 1989; Bharucha 1993), and performers have made use of computational systems to contribute to their performance.[3] A lively academic research strand deals explicitly with the theoretical problems of computational creativity, where pioneering authors, shaping the conceptual framework of the field, include Maggie Boden (1990), Graeme Ritchie (2001), Simon Colton (2008), and Geraint Wiggins (2003). International conferences, such as Computational Creativity or Musical Metacreation, demonstrate the liveliness of this research field and a recent book explicitly expounds on the compositional and philosophical aspects of the field (McCormack and d'Inverno 2012).

The history of computer-based generative music systems thus extends over the past sixty years, with Hiller and Isaacson generating the *Illiac Suite* in 1957, but many of the interesting academic research projects have not reached the public in the form of music or actual systems; they are introduced through reports, whether in academic journals or in the media. The problem with a piece like the *Illiac Suite* is that when we listen to it, there is nothing that points to its origins as an algorithmic piece. It is distributed as a score and performed by human players. For the realisation and experience of this piece as a computer-generated work of art to happen, the actual *system* of the music should ideally be shipped and executed by the listener. Below is a selection of a few notable projects of generative music that have been released into the public domain and that exemplify certain trends and technological frameworks that are evolving. They have many things in common, but it is evident that the lack of protocol has created very diffuse experiments that can be seemingly hard to group.

A form without a format

Brian Eno's 1996 release of generative music written for the Koan software is one of the earliest examples of distributable generative music. In the early days of the World Wide Web, Koan was a tool for designing and generating music on various platforms, including a browser plugin, so that the music could be stored online and executed in the user's browser. Eno's floppy disk was relatively expensive (£45) and did not achieve widespread popular reception, although the media were interested.[4] Around this time, Eno gave important talks and interviews on the topic of generative art, acknowledging the inspiration of the emergent field of artificial life on his thinking, and influencing many artists who attended his presentations or read about them in the media. Antoine Schmitt and Vincent Epplay released *The Infinite CD* in 1999, distributed in the form of software on a CD-ROM for the Mac and Windows operating systems. The authors state that their reasons for making the work were to search for new ways of composing and thinking music, where the computational process of the machine can be part of the piece's rendering. A similar project, mentioned in Chapter 8, is *The Morpheus* CD-ROM from 2001, which was initiated by John Eacott and included works by five composers who all wrote their music as SuperCollider code (Eacott 2001). The album contained a runtime version of SuperCollider that would interpret the compositions in real time. This release was a highly successful experiment but suffered from the fact that SuperCollider was, at the time, only available for the Mac, and it will now only run on Mac OS 9. In 2003, Runar Magnusson and I collaborated on a generative music project called *SameSameButDifferent*. Two versions of the project were released: *v.01 – humans and insects*, and *v.02 – Iceland*, by the ixi label in 2003 and *Rafskinna DVD Magazine* in 2008 respectively. The first piece focused on similarities and differences between humans and insects, on intelligent swarm behaviour in these species, while the second consisted of field recordings from Iceland, rendering new schizotopic generative soundscapes at every play.

As a result of the lack of standards, the formats of these pieces become essential factors, and in 2016, Spyros Polychronopoulos created a generative music product *Live Electronic Music (Sound Object)*, a small black box, the size of a matchbox, with only a USB power source and a headphone plug. Running a Pure Data patch, the music changes with every play. Other distribution formats housed generated releases, such as the electronica band Icarus's release of a CD of generative music in 2012, on which all the pieces were generated by software, but mixed down to a fixed-media audio file, when the album was purchased. Another approach was *Rand()% Radio* by Tom Betts and Joe Gilmore from 2003, which was an internet radio station that would stream generative music created by a large group of composers, running their software in real time.

Regarding the establishment of mobile apps as formats for generative music, the progress is slow but stable. Brian Eno's *Bloom*, Björk's *Biophilia*, Radiohead's *PolyFauna*, or Massive Attack's *Fantom*, and perhaps Lady Gaga's *ARTPOP* apps are examples of how innovation is often driven by, and dependent on, well-known early adopters of new technology (see Pinch and Trocco 2004). There is a diverse array of distribution formats, with a range of physical and online shops, as well as streaming services (but there is no generative music category, for example, on iTunes, Google Play, or Spotify).

In the current age of cloud computing, generative music work is shifting in favour of browser-based systems. The Web Audio API is a standard system for a cross-platform browser environment, enabling both high-level synth graph patching and very low-level sample-based manipulation. The music is written in the congenial and straightforward language of the web, JavaScript, but it gets compiled to bytecode that runs on a JavaScript virtual machine, so it runs very fast on the phone, tablet, or laptop. There is a myriad of experimentations in the Web Audio API, for example, Patrick Borgeat's interactive music applications (www.cappel-nord.de/webaudio/) or generative music in *This Exquisite Forrest* (http://labs.dinahmoe.com/theme/). Collaborative music environments, such as Dynamoe's *Plink* system, also question formerly established musical roles. Finally, the Bronze environment is a compositional framework that supports generative composition, with a generative player (www.bronze.ai/player/example), proposing a standard for artists to compose and release their music in a generative form.

The examples above demonstrate the problems of establishing a consensus for a format of generative music. These problems derive somewhat from its heterogeneous and non-standardised media formats. The lack of protocols or standards is bound to continue, as composers and developers use different programming environments depending on artistic goals, and write for one or more of the many end-user playback platforms (operating systems, device types, etc.). The field can be characterised by its spirit of experimentation, investigation, and innovation, and it is questionable whether there will ever be a standard for generative music at all, as the technology moves fast and musicians, coders, and other innovators are quick to explore the novel features of any new technology, protocol, or language as new material in their work.

Systematic musicology of systems

This lack of a standard or familiarity with generative music is not necessarily a problem. Generative music a fluid music in a fluid form, whose media format might never stabilise, like in older media, such as vinyl or CD, where the medium disappears and the consumer's focus is purely on the content. It might be the forte of generative music that the medium keeps surprising us, that the music is playfully critical of its format, potentially providing a fulfilling experience with sound that is very different from the linear music on the radio. The focus is shifting from the linear work to the dynamic system, including playback systems, performance systems, and instrumental systems. Therefore, from the perspective of musicology, we ought to establish a discipline that builds up methods, concepts, and language to analyse algorithmically generated musical works. Such musicology would study algorithms as musical material, inventions notated through the medium of code, and apply diverse techniques, ranging from early symbolic AI, rule-based systems, and expert systems to artificial life and contemporary deep learning neural networks. The formation of such musicology could be seen as part of the larger and more scientific *systematic musicology* (Bader 2018).

Up until the 2010s, there were very few successful musical systems based on artificial neural networks. However, this is rapidly changing with the use of fast GPU (graphics processing unit) calculations and cloud computing technologies that enable cheaper and faster customised deep learning designs. The ground-breaking part of the approach is to get all of these elements to work together. Deep learning has become one of the key current focus areas within computer music in the late 2010s as it offers an incredible power in domains ranging from synthesis to high-level compositional structures.

But what are our chances of understanding music that is composed from a source repertoire that we might not have easy access to, say a system that listens to Indonesian gamelan, Norwegian death metal, and Italian Baroque music? What can we know about the deep learning "thought process" that led to the piece? With deep learning and the applications of neural networks in the creation of music (Roberts et al. 2018), it becomes decidedly difficult to establish where the musical creativity comes from, as scrutinising the code is not sufficient: the system actually learns, and the musicologist, or any other individual wanting to study the music deeply, would therefore need to have access to the training data too. There is nothing to be observed in the networks

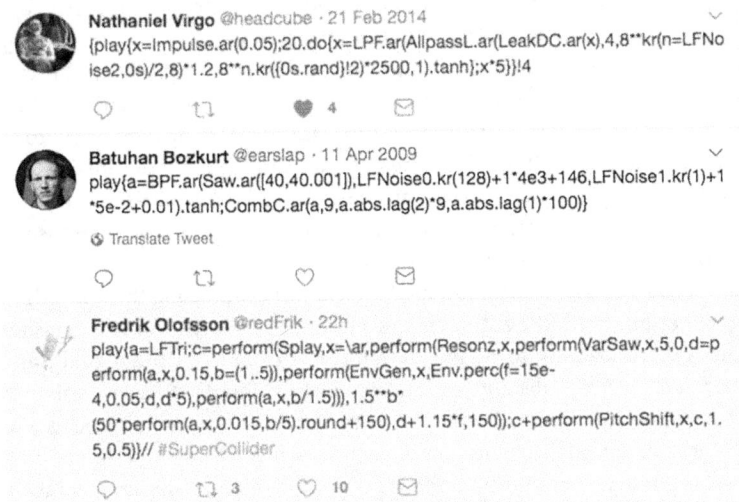

Figure 15.1 A screenshot of three SC Tweets (SuperCollider code). Here miniature pieces were written with the constraints of 140 characters, until it was raised to 280 characters.

other than their structure – the description of the neural connections – so exclusively focusing on the code would be like poking at the grey matter of a musician's brain to try to find music in it: the music is not there, as the knowledge is learned by the network, through demonstration, training, and reinforcement.[5] For this reason, it is ideal for musicologists if training data and other supporting sources, as well as learning context (such as how the network is designed, how it was trained, or the aesthetic reinforcement bias by the composer), are made available. Open source software, as well as open access to source data, becomes important: if musicologists and other interested individuals are not able to inspect the source code of the system, and now the source repertoire, a large part of what musicologists have traditionally used to analyse musical pieces (such as the score, sketches, music theory, biographic data, or social context) will not be accessible.

Composers as programmers

The composer and live coder Nick Collins has explored computational creativity in music since the late 1990s. His projects include automatic orchestration for human instrumentalists, breakbeat cutting (with the well-known BBCut algorithm), stochastic synthesis, and automatic electroacoustic music generation (submitting unheard pieces to computer music conferences and festivals), culminating in a system that was called, before threatened with a lawsuit, *Android Lloyd Webber*. This compositional system was designed for composing work for a West End show, as the "world's first computer-generated musical" (Collins 2016). Collins's musical output is diverse, ranging from automatic pop tune generators to electroacoustic music and pieces for an ensemble. His current research involves music information retrieval (MIR) techniques to analyse large corpora of data, whether they be MIDI files or audio data, in order to develop both automatic critics (Collins 2016) and composition systems (Collins 2017). Collins has coined the term "corposition" for music resulting from processes that use data to generate new pieces, an excellent term that combines "corpus" and "composition," but also "positions" the piece as part of a corpus.

Collins explores the uncanny in new music; namely, the feeling that something awry is going on not only with the authorship of machine-generated music, but also with its reception: who is listening to it? For example, in the spirit of the digital humanities where texts are read via "distant reading" methods (Moretti 2013), we might suggest machine listening as "distant listening." However, this does not work well as a term, as the machines can indeed listen in a very nuanced and precise way, in some circumstances even more accurately than our human ears can, although humans are still much better at tasks such as stream segregation or rhythm detection. The question of *machine critics* emerges, as the machine can listen to all of the released music in the world, which is something that no single human will ever be able to do.[6] The title of one of Collins's latest research papers gives an idea of the terrain he is exploring: "Towards Machine Musicians Who Have Listened to More Music Than Us: Audio Database led Algorithmic Criticism for Automatic Composition and Live Concert System" (Collins 2016).

Bob Sturm is another machine learning researcher and has collaborated with Collins on various projects. He has recently been involved with creating a machine learning system trained on a symbolic corpus of folk tunes from the UK and Ireland. This system can then generate new tunes in the same style. In this case the machine learning is not at the signal level of audio, but at the symbolic level of notes, parsing through large databases of text. This is an intriguing project in many ways, firstly as this kind of music follows some well-defined rules – so a machine learning system is likely to be successful in generating plausible tunes – and secondly because the project engages with a living folk tradition, and some of the most challenging questions come from the fact that Sturm is engaging with a community of practitioners who have a variety of opinions on the role of a computer in their practice. Sturm, whose background is in both computer science and music, has this to say:

> Anyhow it is important to realise that no matter the level that you're working at in a computer, whether it's the acoustic realm, or the symbolic realm, that it is still not music. Music is a cultural activity; it is a human activity steeped in all of these rich sets of experiences that humans have, in their daily sonic life or in thinking about music, reflecting on music, the feelings they had. Those data representations are really impoverished when it comes to music as an experience. So these models that we built aren't necessarily understanding music, they are really good at building plausible patterns that can be turned into music. But all this is still worlds away from the experience of music that we have. (Personal communication, 2018)

Many similar questions have been asked by François Pachet, formerly of the Sony Computer Science Laboratory in Paris, but currently the director of Spotify's Research Lab (I mention this to signal the emphasis key players in the industry are giving to machine listening technologies). Pachet has for many years been involved with projects of computational creativity, and his *Continuator* project is well known (Pachet 2003). Funded by the European Research Council, Pachet worked on a symbolic creative system project called *Flow Machines*, which resulted in a piece simulating the Beatles, called "Daddy's Car" with much acclaim. Music tech companies are catching up with contemporary AI research and the paradigm shift AI will bring to future music practice, something Google recognised in 2012 when they hired synth designer, composer, and futurist Rey Kurzweil as the head of AI to work on machine learning and language processing projects. Another example of a system that is becoming a prominent player in the field of computational creativity is AIVA (Artificial Intelligence Virtual Artist), which applies a deep learning algorithm to music composition. Up to now, the system operates on the symbolic level, as opposed to machine listening level, and therefore outputs symbolic data (notation) and not signal (synthesised sound). AIVA was the first non-human to be registered for intellectual property rights with the French Société des auteurs, compositeurs et éditeurs de musique (SACEM). In 2016 an album was released with AIVA compositions played by humans. The results are convincing: there is nothing extraordinary about his music, but the fact that it is performed by human interpreters clearly makes it more listenable than many of the old AI examples we are familiar with where the creative system has been outputting notes that are then lazily realised by some hollow MIDI sounds.[7]

User-generated music

The work described above is taking place in research labs and universities around the globe. But invention eventually turns into innovation, and after decades of experimentation, we are now seeing how these technologies are being applied in commercial products, often by venture-capitalist-funded start-up companies. Artificial intelligence in music creation is becoming prominent in its most powerful manifestations. Companies such as Jukedeck, Amper Music, and Humtap enable people to generate music for their YouTube clips, family videos, advertisements, or films without having to clear copyright and pay for the use of the music. Through web services, people can select moods and instruments and define in more detail the kind of music they are after, without having the professional skills to compose a new piece. There is a plethora of applications similar to these where the music theory is embedded in the software and the user simply evaluates the output and decides whether it is fit for the purpose. In sound for the moving image, software is being developed that permits the music to be composed according to parameters that are specified on the film timeline. The *Filmstro* software is a good example, but it has three sliders ("Momentum," "Depth," and "Power"), which are deemed sufficient by the developers for the composition of film soundtracks. Imagine, for example, a scene in a movie that goes from sadness to hope to joy, and we need a soundtrack to express those emotions. For this situation, it is unlikely that we would find an already existing piece of music which would follow the path of this particular emotional trajectory, but a bespoke track could easily be composed quickly by non-musicians with such compositional tools.

Future systems might use biometric data (pulse, perspiration, blood pressure, breath) from test viewers, detecting their emotional state, for example, and this data can then be fed into the compositional system. This will clearly have to be under the final decision of the composer and eventually the director, as emotional parameters are complex in a film. Here, the use of sensor and biometric data as part of the compositional process, where all kinds of parameters can be pulled into the compositional pool and controlled over time (say sliders that define the amount of "sadness," "hope," "joy," and we could add "anger" and "danger," just for good measure), is a promising prospect that will equally aid professional film composers and amateur YouTubers. This music might not be oozing with genre-breaking creativity, but neither is most film music, and we are already at a stage where listeners would not be able to hear what music was made by a human and what by a machine, as long as either humans or machines play both.[8] It is likely that most game sound and music will be generated on the fly using the above technologies.

Deep learning in music

Machine-created music makes many people feel uneasy: suddenly the deepest expressions and innermost emotions of our soul, to use a Romantic language, are generated by a bit chewing machine! But this is not only happening in creativity: machine learning and AI have become integral to our instruments and tools, the potential of which can be seen, for example, in Adobe's VoCo software (described in Chapter 12). This area of research is now greatly stimulated by the recent power and progress in the field of deep learning networks, including approaches that involve the generation of speech and music – for example, with DeepMinds's *WaveNet,* which was ground-breaking in its sample-by-sample level neural network analysis and regeneration of signal, or Google's *Magenta* project, which focusing on machine learning in music, using the Tensorflow open source machine learning framework. Other projects include the image generating *Deep Dream* and DeepMind's game intelligence *AlphaGo.* Different methods are studied and implemented for bespoke purposes, such as generative adversarial networks (GANs), feedforward convolutional neural networks (CNNs), recurrent neural networks (RNNs), long short-term memory networks (LSTM), reinforcement learning, and many more techniques that can be applied equally in areas of synthesis, instrument design, and computational creativity.

An example use of deep learning can be found in an output of the Magenta research project, the *NSynth* (Neural Synthesizer)[9] which uses deep learning networks to analyse sounds, find their sonic characteristics (in terms of timbre and dynamic changes), and create new sounds that are not simply a mix between the two source sounds, or older forms of cross-synthesis FFT convolution (Collins and Sturm 2011), but an actual combination of the source sounds as perceived by the system. The results are promising (Engel et al. 2018). Using the *WaveNet* project, this has been done at the sample-by-sample level, not through oscillators or sound buffers of longer duration, as in the granular or concatenative synthesis, as implemented in the VoCo software. New deep learning technologies quickly demonstrated a strong and wide area of applications, ranging from voice synthesis, through automatic creation of sounds, melodies, and rhythms (e.g., Roberts et al. 2018), to its implementation in musical instruments. However, the analysis of such sounds requires extreme computing power and time using GPUs, which, once again, makes the cutting edge of musical experimentation the domain of the industry and university research labs, similar to when early tape, computer music, or wavefield synthesis first appeared. The output of the Magenta analysis is the NSynth, which is a system encompassing an instrument database that has been subjected to the WaveNet deep learning analysis of source sounds. Applying automated parametrisation, or self-directed learning, the audio features are not defined by humans but discovered by the network.

The new and exciting techniques of deep learning have been the domain of music from the start. Or rather: researchers in deep learning find music an ideal test domain for their research as music is intrinsically constitutive of our human nature; it is based on numerical and mathematical relationships, and we have strong intuitive knowledge of what makes good music. Indeed, many of the earliest proof of concept demonstrations have been in musical applications and the emphasis labs such as DeepMind or Magenta put on research in music and deep learning exemplifies that. To demonstrate the power, relevance, and potential of this new technology, music is seen as an interesting challenge that connects to people's imagination. Regarding challenges, there are no general solutions to the machine analysis of complex musical audio signals. We have seen how tasks that are relatively easy for humans, such as stream segregation might prove difficult for a machine listening system. For example, considering the richness of data in an audio signal compared with the simplicity of musical notation, the question becomes which representations or features the network should be trained with. Should it be the raw sample stream, hand-crafted features, machine-discovered features, MFCCs, or other spectral representations (see Wyse 2017)? Deep learning both excites and promises great changes in music technology, but this research is only in its nascent stages.[10]

Conclusion

The future challenges of deep learning are not simply technical: they will become aesthetic, epistemological, ontological, cultural, and ultimately ethical. For example, the question of the musicology of code and the problems machine learning poses for that. What do we mean by music analysis when there is no one way of inspecting the mechanism by which the music is composed, whether this mechanism is used in synthesis, instruments, or composition? A typical approach is to reveal the conditions that underpin the compositional thought. In what historical, technological, and music theoretical context does the work originate? What was the underlying research? What were the compositional problems? How does the work solve those? What is brought forward as new revelations? Which possibilities did the composer have in executing the work? How was the composer limited by the technology available (e.g., no GPU farm for training the networks)? What is the algorithmic structure of the system? Which pieces, or other data streams, were used as training data? What happens to copyright if a deep learning system is fed some albums by two distinct groups and generates new music from that?

Is it here the future ends? Is it true, as some people say, that we cannot look into the future in the same way as Cyrano De Bergerac could speculate about the equivalent of the iPod from the invention of mechanical clockwork? Some argue that we have reached a stage, as articulated by critic of neoliberalism Mark Fisher, where the future cannot be imagined, that cultural time has stalled, and we have difficulties grasping our historical moment. For Fisher, we are experiencing a flattening of cultural time: "While twentieth-century experimental culture was seized by a recombinatorial delirium, which made it feel as if newness was infinitely available, the twenty-first century is oppressed by a crushing sense of finitude and exhaustion. It doesn't feel like the future. Or, alternatively, it doesn't feel as if the twenty-first century has started yet" (Fisher 2014: 8). As a lecturer in music, I have performed experiments with my students and asked them to compose the music we will be listening to in twenty years' time. Surprisingly, but perhaps not, the music they have typically submitted sounds very much like something from the 1990s.

The diversification of musical technologies demonstrates how our musical imagination tends to be techno-poetic – that is to say, the creative process integrates with the tools we use, our cultural memory, which is always at hand as a database that we can reference, mix, and recreate from. If, in the past, we could metaphorically claim that the violin is our conversational partner, our future music technologies will include systems with which we will have actual discussions and whose internal workings might be as difficult to understand as our band member's mind (Bretan et al. 2017). This potential for intelligent complexity is why music and music technologies are such an exciting area of research: music continually keeps up with, and often engenders, the rapid changes in the development of material technologies and computer science. Apropos of today's music, there is no doubt that deep learning will continue to surprise, challenge, and scare us into new ways of making music that we find unimaginable at the moment.

16

Music in Multimedia

When our primary music technologies have moved from being instruments or studio equipment to simulations in digital software, a substantial portion of the musician's work happens on the computer. As a multimedia device, we plug in sound cards, screens, interfaces and controllers, externals, and all kinds of devices that configure the computer for a particular purpose, whether it is for admin, research, multimedia home entertainment, or a recording studio for music production. As an artist, when working on a computer, the exclusive focus on *sound only* begins to feel peculiar as we are really working in a multimedia studio.[1] The digital natives now leading musical development are used to thinking about music in as diverse set of contexts as streaming, music videos, movie soundtracks, adverts, social media memes, game sound and music, and installations. Music written onto a CD, for example, is just one of many manifestations of music and not to be considered as original or special in any way. Before the computer, practitioners would have to use completely different studios for photography, film, and sound, but today all this is available in devices we carry around with us in our pockets at all times. Feature films have been made on mobile phones, where the filming, editing, sound, and musical composition were all done on the device. The costs can be minimal when using open source development tools. People's mobile phones can also become part of VR headsets, for example using Google's *Cardboard* system. We carry production and consumption entertainment devices in our pockets, just like the book in the past, but now it is multimedia. Creative content increasingly becomes "environments" to explore, and interactive audiovisual material becomes the norm, involving tactile interfaces, audiovisual material, non-linear structures, and interactivity. The Greek Muses are all working in multimedia these days!

Computational thinking in multimedia

But there is another change that has been brought forth by computer technologies: they prime our mind to think in terms of distributed networks, cybernetics, complexity science, interdependencies, and non-linearity. Experimental musicians, as voyagers into the unknown, began exploring these concepts in the 1950s and 1960s, but it is with networked algorithmic information machines, the computers, that this type of thinking becomes fully achievable as artistic work (as an example, consider the effort of splicing the tapes in Pousseur's *Scambi* piece from 1957, compared with downloading an app or using the Bronze player). It has now become natural for us to see things around us as forming part of complex systems and this paradigm can be found in sociology (Latour 2005), informatics (Castells 1996), biology (Estrada 2011), aesthetics (Bourriaud 2002), and music (Barbosa 2003). It is not as if this manner of thinking was impossible before, but through the immersive theoretical and practical experience in computational thinking, with a networked internet of things, we begin to embrace computer science practices in other fields too. Indeed, as the notion of the postdigital (Berry and Dieter 2015) conveys, it is not of great consequence for us whether the system is operating through 0s and 1s, electricity, mechanics, or human movement: the postdigital brings with it a certain form of computational thinking that infiltrates all our operations (Wing 2006). A good example is the music of composer Masahiro Miwa, whose work involves writing software as scores that are performed by human performers. The results are fascinating as the human interpreters operate by following computational instructions, in the manner of a machine executing software. Miwa also works with musical machines, physical installations that follow computational logic in intriguing ways. This exemplifies yet another mode in which new notational practices are emerging, here as mechanical structures that generate music. The cybernetic work of Agostino Di Scipio is also prominent as he has composed a large array of pieces for feedback with complex environmental and instrumental elements. The halldorophone demonstrates how feedback is becoming part of instrument design – a feature of the instrument, rather than an error. There is no coincidence that cybernetics and Jimi Hendrix were born in the same decade.

Interactivity, networks, algorithms, generative patterns, protocols, standards, APIs, controllers, motors, lights, sound, interfaces ... tactile, haptic, sonic, audiovisual, olfactory ... intelligent, emergent, non-linearity, artificial life, genetic algorithms, neural networks, deep learning ... code, symbols, mapping, translation, transduction, media, movement, channel, location, GPS ... social context, immersive, augmentation, hyper, virtual, participatory, distributed ... open, open source, open access sonification ... screens, buttons, pads, sliders, knobs, accelerometers, biosensors, EEG, big data, personalised, individuated ... and so on. These are just a few ingredients, characteristics, and descriptors of new media works. It is no wonder, then, that the musicologist, the critic, the journalist, the shopkeeper, or the general consumer are a little confused. The technical system, and the ideological and aesthetic frameworks that accompany it, have exploded! How can new music of this nature be analysed such that there is a meaningful continuity with our musical history and aesthetics? This is where the concepts of ergomimesis, ergophor, and ergodynamics work well, as they explain *how* ideas, techniques, and methods are transduced from one system of practice to another (ergomimesis), *what* the new object has carried over (ergophor) from past practices into its new context, and the *function* and *nature* (ergodynamics) of the new object as a system of ideas, thoughts, practices, history, ideology, aesthetics, and technical solutions that need to be studied, questioned, explored, tested, and practised in order to find its creative potential, its *dynamis*, for musical expression.

We can, along with Jonathan Sterne, question mediation in instruments: "scholars have imposed a distinction between instrument and medium in their theoretical logic while musicians and engineers have long since bridged it" (Sterne 2007: 4). Sterne is primarily looking at how media reproduction devices, like turntables or tape, become instruments at some stage, as there is "no reproduction without the artifice of an instrument, and all instruments in some way reproduce sound" (Sterne 2007: 14). He references Brian Eno's influential essay, "The Studio as Compositional Tool" (Eno 1983), in which Eno describes the potential of the studio for performative actions. Although the obverse – how instruments mediate – is less explored in Sterne's text, it does encourage us to look at instruments as technologies of representation (Sterne 2007: 15). We have analysed the dual mode of mediation vs. artefact, agency as a mediation for performers' intentions, but at the level of the object, digital instruments are clearly media devices of agency and resistance: they are defined by their interface and interaction design, their ergonomics, as simulations of physical systems or explorations of the algorithmic nature of the computer itself. Information is communicated through their bodies via mappings, APIs, and protocols. Furthermore, the instrument's depth needs to be designed too: its imperfections, non-linearities, and quirks. For example, it is a well-known trick in subtractive synthesis to add one or more slightly detuned (by 0.2 to 2 Hz) saw wave oscillators to the primary one, thus producing a thicker and warmer sound when added together. It is the imperfection of the detuning of the oscillator that satisfies the human ear and gives it a "character" and "personality."

The invention is the work

In previous forms of media we rarely had to question the medium itself when enjoying a new piece of artistic work: reading a new book was about the content, and only exceptionally about the design or act of reading a book (exceptions include experiments with hypertext books or visual poetry); performing a new piece of music was about the piece, and not the notation (with an exception in novel notational practices such as graphic scores); listening to a vinyl album or a CD was a "ready-to-hand" format, to use Heideggerian language, and the object, the medium, disappeared out of focus when the music emerged (except perhaps listening to Christian Marklay's *Record Without a Cover*, which emphasised the faults of the medium). Contrarily, the experimentation and creativity in new music often emphasises the exploration of the instrument, the format, the media context (networked computational media), and the creator–audience communication channel. A certain alienation from the object is celebrated as a method of investigation and we find this equally in acoustic composition (such as Helmut Lachenmann, Sarah Nicolls, or Giovanni Verrando), electronics (e.g., Ewa Justka, John Bowers, or the Dirty Electronics Ensemble), and digital practices (such as Marcus Popp, Tristan Perich, or the Owl Project). The instrument itself is often aestheticised too, fusing its status of being an aesthetic object and the function of making one (the music). The microtonal sculptural instruments of Harry Partch come to mind, as do drone master Tony Conrad's "invented acoustical tools" sculptures, or the instrument art exuberance of Ken Butler and Tuni Panea. New instruments are not necessarily objects held and manipulated by a performer on a concert stage. Indeed, new works often reject the idea of the virtuosic player being the central focus of attention in a context where the audience is comparatively passive. They can be systems, or compositions, that are extended in space through networked technologies or extended in time through systems that evolve or adapt during their lifetime or performance.

In new digital media formats, the central position of the linear recording is transcended, although it becomes difficult to market such work. The shops (online and offline), the media, the critics, and the audience do not have the network of conventions that build up around repeatable products, such as a novel, a painting, or a pop song, and this removes the fluidity of commodification, as the work tends to be "present-at-hand." It is tempting to think that this was the case with vinyl, radio, TV, music videos, CDs and MP3s, in their early days, but the difference is that all these formats required standards that quickly became concretised as they needed set machines for production and playing. The computer, on the other hand, is not a medium that requires a standard of form or content (the computer is clearly built on all kinds of standards and protocols, but those relate to its function, not content), so the works we make using the digital computer will continually engage with the latest technology, whether it is new hardware, interfaces, programming languages, audio technologies, or other computer science topics ranging from human–computer interaction to machine learning. It is quite impossible to conceive of digital music, stabilising into formats that function like the vinyl album or the CD. McLuhan talked about hot and cold media, but in this context, we might talk about stable and unstable media. Indeed, already in 1986 the V2 media lab in Rotterdam published their "Manifesto of Unstable Media," well ahead of their time (http://v2.nl/events/manifest), praising the dynamic nature of computational media, the constant change and evolution of its condition.

The paradigm of twenty-first-century music is becoming one that Jacques Attali (1985) characterised as "composition," where people participate in the music, where the concept of the amateur is reawakened, and where the composer disappears into the networks that produce new musical experiences. Like Tomlinson, whose ideas we explored in the introduction, Attali sees music as an essential feature in establishing cohesion in human society: it is about taming noise, about establishing order and collaboration where otherwise there would be chaos. For Attali, there are four historical stages of music: *sacrificing* (the stage before monetary exchange; here music creates the possibility of civilised society); *representing* (concert halls, notation, printed music, musical services, patrons); *repeating* (recording, stockpiling, marketing, distribution, technology, industry, business); and finally *composing* (where music is not primarily about exchange value anymore, but about experience and pleasure with an increased participation of amateurs).[2] A large group of musicians work in this spirit, rejecting recording, documentation, and publication of their work, and rather emphasise the present moment, the social situation, and the spiritual attention we can give the music, there and then. A good example is Alex Mendizabal, whose work involves turning buildings into instruments by installing strings into them, cycling around cities with amplified sounds (using a car battery for electricity), or using hundreds of balloons as musical instruments that sound in concert. The work represents an opening for the audience, an invitation into the work, not to be documented, but to be experienced in real time and real place. In many of these new works, the distinction between production and consumption of music is effaced and a new term emerges, the "prosumer" (Toffler 1980), which describes how the lover of music is now applying the same technologies as the music producers, in making their own music and remixing, in interacting with, collaborating, or generally engaging with music both in real space and online.

There is no coincidence that in the twenty-first century we find a revival in instrument building, electronics, general DIY, makerlabs, fablabs, hackerlabs, open source, open hardware, and open data. The concept of DIWO (Do It With Others) is becoming prominent in the discourse and signifying the new forms of social behaviour. The open source practices of computer programming have certainly been inspirational, but those, in turn, built on the musical practices of sampling, sharing, and collaborating. The instrument is the music, the piece (as we revamped McLuhan's phrase earlier), and this is an approach we see in the work of instrument makers like Godfried-Willem Raes, Trimpin, Sarah Kenchington, Ajay Kapur, Marianthi Papalexandri-Alexandri, Gil Weinberg, Walter Kitundu, and Eric Singer, all of whom have built robotic instruments that border on being installations, musical pieces, and instruments for musical expression. Some of them have collaborated with other artists in making bespoke instruments for their work, as does instrument maker and roboticist Andy Cavatorta, who made the *Overtone Harp* with Molly Herron and *Gravity Harps* with Björk. To thematically focus on the harp, the Electromagnetic harp by Úlfur Hansson and the Light harp by the Bent Leatherband are other examples of how new musical systems are ergomimetically based on one of the oldest instruments, the harp. Our current language does not fit neatly for these objects, but essentially these are *inventions,* as conceived in medieval rhetoric, in Bach's pieces, and in Jonathan Impett's reflections on his practice. These systems are composed to be explored through an ergodynamic investigation.[3]

Musical automata have been documented in some of the earliest historical written accounts, but the practice of creating them was boosted in early modernity with clockwork automata, and associated cog-wheels and springs. Today we find automata implemented as machine mechanisms (see, for example, *Felix Machines*, Graham Dunning's turntable instruments, Miwa's *Thinking Machines*, Sarah Kenchington's mechanical orchestras or Sam Underwood's diverse array of instruments). Many of these pieces have been recorded and documented using video, but this medium does not capture the pieces themselves, as they are spatiotemporal installations that invite us to freely explore them as they are performed; their essence is beyond a documentation that is bound and guided by the subjective eye of the moving camera or microphone.

The boom in popularity of modular synth-making exemplifies in many ways this coalescence of composition and design, arrangement and performance. But this is more than mere technological design: artistry, craft, skill, invention, and composition are also involved. The instruments of Derek Holzer can be mentioned in this context. He builds little electronic synthesizers, most often in cigar boxes, and they feel like little pieces of music that one performs. These synths can be purchased from Holzer, but he also gives workshops where people build their own synthesizers, enabling participants to compose their own synths. This approach of experimental investigation into synthesizers can also be seen in the work of Ewa Justka, whose audiovisual performances are extremely noisy, powerful, and chaotic and where the technology looks like the sounds that it is producing. We find similar approaches in Martin Howse's *Substrate* project, which is a synthesizer made out of earth as a key material (resulting in the *ERD, Earth Return Distortion*); John Bowers's home-brew electronics (including the *Victorian Synthesizer*) explore redundant technologies and performance ecologies; Owen Green's *Cardboard Cutout*, shabby electronics made out of found materials; and last but not least, in the work of the godfather of dirty electronics, John Richards, whose influence has been wide-reaching and deep. For Richards, there is no distinction between a workshop, electronics design, a composition, and a performance: these stages are rather a highly integrated techno-social process of musical creativity (Richards 2011).[4]

New platforms for invention

Contrary to common ideas in popular music circles, musical multimedia are not limited to music videos or visuals behind the artist on stage. Musical multimedia express their core meaning when other elements than sound become integral to the work, often to the degree that it is unclear whether the work should be defined as music. Musical video games are good examples. Since the late 1980s, there have been sound toys, sound games, and other interactive environments that provide experiences which take place at the common borders of the computer game, music production software, and score playing. An early example is *Rez*, a 2001 Japanese "shooter" game where the user would create the music by shooting loops (composed by Coldcut), in gameplay that was exciting, expressive, and performative. In Rez, the distinction of playing music and playing a game collapses as the soundtrack changes dynamically depending upon the player's performance of the game. The game consists of flying in an interstellar space, collecting targets that can be locked and then terminated by releasing the "lock-on" button, on a controller that gives haptic feedback to game events. The music is fast and rhythmic, seamlessly representing the events in the game, and producing sets of quantised highlight notes and sound effects as part of the play mechanism.

Computer games are ideal environments for generative and interactive audiovisual media composition. Immersed as first- or third-person avatars in 3D worlds, or operating with signs and animations in 2D worlds, the player can interact with the music, or "play" the music, in innumerable different ways and forms. Japanese multimedia artist Toshio Iwai is one of the pioneers in this field, releasing the *Otocky* game in 1987, *SimTunes* in 1996, and *Electroplankton* in 2005. And with games such as *Vib Ribbon*, in the early 2000s, creators began exploring the role of sound as a functional element in the gameplay, as opposed to mere ornamentation. Vib Ribbon enabled the user to load in their own music, and the terrain in each level of the game would be generated from the musical events. This was done through a spectral, amplitude and onset analysis of the music. *Audiosurf* takes this idea into the third dimension. Dedicated research into audiovisual gameplay has been conducted by Tom Betts, aka Nullpointer, whose games *AVSeq*, *In Ruins*, and *Permutation Racer* are research projects into generative audiovisual artforms (see www.nullpointer.co.uk). Other recent games, such

Figure 16.1 A screenshot from Tom Betts's *AVSeq*, an interactive musical game from 2010. © Tom Betts.

as *Proteus*, *Panoramical*, or *Thumper* make use of sound as an essential gameplay element and, like Vib Ribbon, the sound and music have been composed specifically with the knowledge that the soundtrack will be constructed through the play. The generative nature of these games has demonstrably set their composers a challenge to which they have risen.

The mobile phone app is also an excellent context for exploring interactive musical works. The designer can use the phone's sensors (sound, camera, light, proximity, gyroscope, compass, barometer, etc.), GPS locations, and the user's social networks as elements of the musical work, which in turn can be shared, uploaded, and stored. Very soon after the iPhone launched in 2007, researchers began looking into writing musical apps that supported the affordances of the device. All kinds of experiments took place in labs around the world, but Ge Wang's projects have been particularly successful in exploring the potential of the iPhone as a musical instrument. Wang co-founded the company Smule, a pioneering business that has created instruments such as the *Ocarina* (Wang 2014), rap-making software called *Autorap*, and other apps that blur the boundaries between audio games and musical production environments. Another game, *Papa Sangre*, took the audio game format at its word and got rid of all visual elements, except menus and footsteps, so the game is sound-only. Locative media projects making use of geo-tracking have been created for "sound walk" projects where the listener walks around the city (or other spaces) and mixes the music. Projects such as *NoTours* (www.notours.org) or *UrbanRemix* (http://urbanremix.gatech.edu) enable composers to compose "in space," where the listener receives different sounds depending on their navigation in a city or natural landscape. The majority of research into the interactive and collaborative platform of the musical app has been done in academic labs or hacklabs around the world, but there are pioneering artists, such as Brian Eno and Björk, who pick up on these developments as they find that the new formats offer new ways of creating content. Businesses have formed around musical apps, and as an example, the electronica group Mouse on Mars have created a label for app software instruments. They recently teamed up with composer and programmer Jan T.v. Falkenstein in developing a music making app called *fluXpad* which is a graphic sequencer that allows the user to draw the notes and beats.

The Web Audio API is a standard proposed by key actors in the media computing industry (Google, Mozilla, Apple, the BBC, etc.) with the aim of making an audio content platform that is supported by all browsers, on all operating systems, on all computers. This would solve many of the technical problems that game creators constantly face. This standard is new, but there is already an academic conference formed around it (WAC – Web Audio Conference) and diverse user groups exist.[5] At IRCAM in Paris, Norbert Schnell led a project called *CoSiMa* (http://cosima.ircam.fr), which set out to create a platform that can be used by composers and musicians to compose bespoke pieces of music for interactive co-locative performances (Schnell et al. 2015). In some of their early pieces – for example, *Chloé X Ircam*, *Terminal*, and *CollectiveLoops* – they invited people to pick up their mobile devices (phones, tablets, laptops) and log onto a web page, which becomes peoples' instrument and notational score at the same time. The audience is then able to perform using the loudspeakers on their device, which at times has resulted in a multichannel audio piece of 1,000 distributed speakers. In this work, Schnell and colleagues ask why the audience has never been included in orchestral work: why this invisible wall at the front of the stage? They then proceed to enable this with technologies everyone has in their pockets.[6]

Music in other realities

Second Life and *Minecraft* have been used as environments for real-time music, but with the recent availability of both high-quality virtual reality headsets, such as HTC Vive and Oculus Rift, as well as cheaper cardboard DIY sets, we now find immersive 3D worlds in which the user can build modular synths, navigate sonic spaces, and generally be immersed in the music-making process via avatars that might include other people on the Internet. This offers new channels for musical creativity, dissemination, and education, as distance learning might also take place in these environments. Here, issues of embodiment, latency, distance, spatial arrangement, or first- versus third-person perspectives, become interesting research topics, as we get ready for being creative, playing, studying, and generally communicating in miscellaneous virtual worlds. There are already synthesizers built for VR, such as *LyraVR* or *Soundstage VR*, and new experiments in musical content are emerging; good examples of more popular facing experimentation would be Björk's *Notget* virtual reality music video or Brian Eno's *Bloom: Open Space*, where he turned his *Bloom* app into a VR environment. Virtual reality opens up new spaces for creativity that are non-linear, participatory, multi-user, and so on, but also for live events. As an example, Ash Koosha, an Iranian-born London-based producer, was recently banned from entering the US due to Trump's travel ban. However, on 16 August 2017, he performed for hundreds of people through the WaveVR (www.thewavevr.com) platform. Club nights, concerts, and other social events will take place in this new space for creativity. The night club scene from the sci-fi movie *Ready Player One* has been recreated in VR, and this was presented in the 2018 SXSW, in participation with dozens of VR arcades, where users can come and "jack in" (to use Gibson's ageing *Neuromancer* metaphor). MelodyVR is another new venture, promising "VR performances to please any crowd," with a line-up of various musical acts, ranging from KISS to the London Symphony Orchestra. The promised experience is (as this is written in the first half of 2018) a little implausible: "Sneak into sold-out stadiums, backstage sessions or underground clubs. Jump onstage with the band, go cross-country with the tour bus, or be the only one in the audience at a secret gig" (https://melodyvr.com). We are yet to see how this materialises, but it is clear that immersive and augmented media are capturing the imagination of creative artists, and it will be yet another platform for future audiovisual creative content.

In the above WaveVR and Melody VR examples, we witness a hefty process of remediation (Bolter and Grusin 1999), which is typical when an older medium is presented in a newer one. The rock concert becomes virtual! However, after a period of remediation (typically done by industry) and formal experiments (explored by academia and artists), the medium emerges as something of its own, with ergodynamic qualities that are not simply simulating what we already know, but investigating the medium's unique qualities. We should expect to see experiments with head movements, binaural audio, physical movement, virtual movement, speed and height, and any such work that applies psychoacoustic knowledge in composing specifically for our perceptual nature. It is exciting for artists to investigate human perception further (as they have arguably always done), applying scientific, experimental results in studies of surround and binaural audio, both in virtual and augmented reality. Google's *Resonant Audio* and Facebook's *Spatial Workstation* are projects that support developers in creating content which implements psychoacoustic technologies. Microsoft has spatial audio in the *Hololens* augmented reality headset, and Apple has implemented spatial audio in its ARKit.

As with any new emerging technology, musicians are exploring how virtual reality can be applied in the making of music. A team led by Stefania Serafin has explored what they call VRMIs (Virtual Reality Musical Instruments) and provided guidelines for their design (Serafin et al. 2016). Furthermore, at the 2018 international NIME

(New Interfaces for Musical Expression) conference there was a workshop in virtual reality audio that included the topic of VIMEs (Virtual Interfaces for Musical Expression), spatial audio in VR, issues of embodiment, and what it means to compose for this new creative platform. These are questions also asked by researchers at Melodrive, a company which promotes its "deep adaptive music" (DAM) to support a better feeling of immersion in virtual and augmented reality worlds. Building on knowledge of psychoacoustics, artificial intelligence, and generative music, their focus is to develop systems for user-tailored generative music or soundscapes that represent the world inhabited by the user "in a meaningful and smooth way" (Velardo et al. 2018: 9). A new media format like virtual reality or augmented reality will always be an exciting platform for artists to conduct their experiments; just as Nam June Paik or Steina and Woody Vasulka began to explore the expressive power of video in the late 1960s, we are only at the beginning of composing music in these new realities.

Sound art as music

Music is ubiquitous and continually infiltrates other media whatever form they take. Some define sound art, or sound, as a category that includes music, but also other practices that involve installations, art, sculpture, moving image and more. Abiding by the argument of this book, I suggest the opposite: music is a fundamental human practice, involving sound, dance, movement, art, poetry, narrative, drama, space, architecture, and more. It is the stuff of the Muses. However, there are some who rigidly define sound art as spatial and music as temporal, or sound art as necessarily relating to other media, while music is pure. But such definitions seem fuzzy and historically vague to boot. The history of sound art demonstrates diverse origins, for example with Duchamp's composition of music, and Cage's experimentation with sculptures and visual art, through practices that immaterialised the boundary between the arts. Alvin Lucier or La Monte Young's work is, at times, more related to our definitions of sound art than music, should we care about this distinction, but neither composer questioned their work as music. For sound art as a bespoke field, one that is distinct from sonic arts (Wishart 1996), names like Gordon Monahan (www.gordonmonahan.com), Trimpin, and others come to mind as early practitioners in America, but in Europe people like Christina Kubisch, Christian Vogel, and Bernhard Leitner took to sound as a unique material with which to make art, relating to architecture, sculpture, and happenings, and, in an educational context, establishing sound as part of activities found in art academies and design schools, in addition to the conservatories or university music departments. This has been a significant development, equally benefiting the arts and music. However, even if the origins are the same through both musicians and visual artists turning to intermedia practices, what explains the separate discourses that have emerged and set music and sound art apart is the respective institutional homes in music departments and art schools respectively. In the context of this book, the distinction between sound art and music is not of crucial importance, as I operate with a historical lens that sees music as a very broad fundamental social activity.

Figure 16.2 Picture of Semiconductor's *HALO* installation, interpreting data from the CERN experiment ATLAS on the Large Hadron Collider. © Semiconductor

With today's sound art establishing itself firmly as an integral part of the fine arts, increasingly merging with the other spatial arts (installations, sculpture, architecture, video art, media art etc.), we notice that for young people it makes no sense to separate sound out from the other arts when their laptop is their artistic laboratory and workshop. Today the field of sound is abuzz with activity, with people such as Carsten Nicolai, Kaffe Matthews, Robert Henke, Ryoji Ikeda, Samson Young, Cevdet Erek, Susan Philipsz, Janet Cardiff, Christine Sun Kim, and the sound art veteran Yoko Ono, most of whom have worked in music. Their work incorporates the use of instruments, sculptures, objects, video, landscape, and more. This fusion or expansion of the musical domain prompted composer Johannes Kreidler to ask if music is dissolving completely into media art (that was the topic of his "Music or Media Art?" lecture at the Darmstadt Summer School in 2018). Composer Kaffe Matthews has, for example, composed pieces for her *Sonic Bed*, where the audience an immersive experience lying down in a bed frame with speakers embedded on the sides. Another work by Matthews and her FOAM collaborators is the *Sonic Bike* where she uses GPS tracking to compose a piece specifically defined by the path the listener cycles through the "composed" environment. This work morphed into the *Sonic Kayak*, enabling perhaps a little calmer environment for sound than afforded by city bicycles.

The discussion of sound art, or the related field of sound studies, cannot be dealt with in any depth here, and other works define the field very well (see authors such as Michael Bull, Douglas Kahn, Seth Kim-Cohen, Gascia Ouzounian, Brandon Labelle, Salome Vogelin, and Jonathan Sterne). However, I do think sound art is relevant in the context of this book as yet another example of music breaking out of the formal and material constraints that the twentieth century forced it into. This new model rejects the ontology of music as a composition of a genotype in the form of a score or a recording that is rendered as phenotypes through performances or playbacks. Instead, music is here something more akin to an invention, intervention, and installation that necessarily takes place in space. Sound art is yet another example of music effacing or redefining itself, through new technological formats and cultural contexts.

Conclusion

Only forty years ago practices were simpler for musicians interested in the dissemination of their music beyond the local pub or town square: in general terms, they would compose a piece and hope to get it performed by a performer, an ensemble, or an orchestra. Or they would enter a studio, record some music to be released on vinyl, most likely followed by a concert tour to advertise the album. Today's musical practices have shattered into pieces of different media types, and related social and economic customs. However, considering the distribution formats of music, through channels so multidimensional and diverse, could it be that the most engaging, interactive, interpretative, ergodynamic, and personally fulfilling form of musical consumption might be to perform it ourselves? Perhaps the most profound way of listening to music is to play it ourselves, as such an engagement with the piece is slow and marked by discovery and experimentation (testing the piece, trying out different interpretations), until a version of the piece emerges that is uniquely ours, contrived for our own ears and body, and possibly the friends we play with. Music was always "supposed" to be performed, some seem to have forgotten after a whole century of recorded music. Let us remind ourselves of how sound recording was seen in the early days of phonographic technologies by quoting Philip G. Hubert, who, in 1893, wrote that "it is highly probable that, when the phonograph is found in every house, a phonographic version of every piece of music will accompany the printed sheet" (Hubert [1893] 2012: 42). That is, we might possess a professional recording that could help us understand the music better, supplementing our study of the printed score, aiding our practice, and augmenting our enjoyment of the piece. Not the other way around!

The circle closes upon itself: with interactive multimedia, we expect to participate in the works, to explore, navigate, and interpret them. It is perhaps a natural extension of this development of further and deeper interactivity, that artists who previously recorded their music in studios and released on fixed media have begun to explore the idea of publishing their music in the form of sheet music; it suffices to mention Beck's 2012 *Song Reader*, or Björk's 2016 collaboration with Jónas Sen for her *34 Scores for Piano, Organ, Harpsichord and Celeste*. This opens up for an exploration of a piece, to study the composer's intentions, and the possibilities of interpretation. The question here is not about the technology we use to make the work, whether acoustic or digital, but the possibilities given to us in playing the piece at a fundamental level of interpretation. The notated musical work – in whatever form of notation it might take, and we have explored a few in this book – is perhaps the ultimate musical gift to us listeners, as it makes itself available for interactivity, interpretation, and understanding at a level far deeper than we have ever had with the new formats of the electronic and digital age.

Part V

Conclusion

17

A Future of Music Tech

Code is the functional material of new digital technologies, but programming is never a design from scratch. In programming new instruments, we apply code libraries, compilers, synthesis and mapping algorithms, HCI guidelines, standards, and protocols such that we do not have to constantly reinvent the wheel. These elements constitute the particular technicity of new music technologies, but they also carry with them ergomimetic traits from older instruments and media[1] that are implemented in the new technology. In translating actions over to new designs, we reduce some features of the phenomenal world whilst amplifying others. Ergophors are necessarily abstractions. When writing established musical gestures into new technologies, we leave things behind and perhaps add novel features that are given by the nature of the new medium. Our question has been, like emigrants moving to a new continent: what should we bring with us from the old world, and what are the gifts of the new land? A subsequent and more important question then emerges of how we adapt and redefine ourselves in this new world: we are not the same people anymore.

Until now we could have divided digital music technologies into two main camps: *production* versus *performance* systems. Production systems include digital audio workstations, sound editors, audio plugins, virtual instruments, sequencers, notation software, and other tools whose primary purpose is to render a piece of music in the standard linear format to which we have become accustomed, that is, with notation and recording. Performance systems are music technologies that are made for real-time rendition, where the user plays, interacts, improvises, and composes through listening, with inventions such as instruments, software, live coding languages, apps, audio games, and virtual reality worlds. The former is primarily aimed at the stable media formats of the twentieth century (vinyl, CD, radio, TV, film, streaming) while the other represents a much longer and deeper musical tradition, one of liveness, performance, participation, study, exploration, and social cohesion. Current technological developments fuse these two modes of engaging with music, and in so doing, formally conflate the roles of the composer and the listener, or rather, the acts of composing and listening. This book has argued that music and its technologies are a much wider cultural, technological, and social domain than what the "music industry" has been occupied with during the past century. Indeed, it contends that the advent of recording, which framed music as a commercial product with related industry actors, is simply a short blip in a much longer musical history that predates humanity itself, and will continue serving as part of its existential grounding until its end.

Figure 17.1 A picture from a performance of Robert Henke's *Lumière II* in Italy, 2016. © Emmanuele Coltellacci.

The system of the crowd

Digital music technologies have been designed equally for the making and the listening of music, and on most digital platforms we are witnessing a change to music systems to which we delegate fine-grained action, with our steering becoming more that of general influence than direct control. This delegation of control applies equally to the way we create and listen to music. Alberto De Campo has described this shift as "lose control, gain influence" (De Campo 2014), and in one spelling it became "loose control, gain influence" which is an equally good description. We now tend to opt for higher-level control, and the automation written into software takes care of lower-level minuscule details. Of course, this is happening in areas beyond music: automation is being implemented in aviation, transport, cooking, economy, farming, education, and other fields. In many cases, automation extends not only our bodily movements, but also our thought processes and social patterns. Technology turns epistemic; it becomes an instrument of thought, equally augmenting and amplifying our mental and social capacities.

Automation in the field of music means that software will pull users' creative "data" onto central databases that apply machine learning of emerging musical styles, in turn feeding some fresh sounds and techniques back to the user. These new music clouds will apply subscription models, as we know from consumer services like Netflix, production software like Adobe, or computer game entertainment like Steam. We know that social relations shape machines, but also that machines shape society. Furthermore, as Langton Winner (1980) points out, technology has politics, and within it we find the whole of society inscribed – its structures, its values, and its dreams. Musical aesthetics are written into the functions of music software, and these programmed theories of music affect the way people conceive, ideate, create, produce, and master musical work. With the online big data harvesting of musical creativity, we begin to see the music producer increasingly tangled up in a larger techno-aesthetic musical cloud. This is nothing new – music has always been a system people partake in – but the difference is the informational and AI-controlled machine-based network in which the user operates. In the new technological paradigm both the instruments of the mind and the body will reside in the online network, where the creative behaviour of the crowd morphs with user activities to the degree that the origins of creative ideas are made obscure. The system becomes an extended mind, a creative prosthesis, a cyborg-like structure in which an individual's instrumental extension is directly linked into an online cloud of swarm creativity.

This book has conducted genealogical studies in music technology and demonstrated that many of the ideas and design tropes we apply in our new inventions are transduced ideas and actions from older technologies, a process we described as ergomimesis. The thrust has been that new music technologies are systems of potential action and thought – they are frameworks to compose and perform with – and yet this is nothing new: the principal difference is that the framework is increasingly epistemic and music theoretical, *written into* the technology itself. In the early Middle Ages, Boethius wrote that "there are three classes concerned with the art of music. One class has to do with instruments, another invents songs, a third judges the work of instruments and the song" (Strunk 1998b: 32). The form of this book has explored these classes, and with epistemic tools, computer software, and artificial intelligence, we are further developing Boethius's third class. Like Boethius, it is important to stress that the third class does not override the other two: instrumental practise and composition should be considered as paramount in reaching musical depth and establishing social relations, but with computer software we enter a stage where the body can be bypassed in the production of music, a situation where artificial intelligence will infiltrate Boethius's third class:

> The third is that which acquires the skill of judging, so that it weighs rhythms and melodies and the whole of song. And this class is rightly reckoned as musical because it relies entirely upon reason and speculation. And that person is a musician who possesses the faculty of judging – according to speculation and reason that is suitable to music. (Boethius in Strunk 1998b: 32)

Today's music technologies present themselves this way: they are devices of aesthetic sophistication, of weighing and judging, not necessarily devices for demonstrating performance prowess. Daniel Miller, music producer, DJ, and founder of Mute Records, recently argued to such effect: "One of the things I like about electronic music, in general – and some people might disagree with this – but you don't actually have to have any musical knowledge whatsoever to make the music. You just have to have great ideas, and a good ear. But you don't have to have any technical knowledge" (Miller 2018). Miller's statement is a fitting expression of how conceptual and disembodied musical creativity becomes with our epistemic tools. Jean-Michel Jarre conveys something similar in his foreword to the *Push, Turn, Move* book on interface design in electronic music, but here pointing to the aesthetic role of the technology:

> As electronic artists we are dependent on the evolution of synthesis and synthesizers. You have to find every opportunity to break your habits. This evolution of technology, the various fantastic new ways we can interact with electronic sounds and express ourselves, this is what can keep electronic artists and the music fresh. (Jarre in Bjørn 2017: 5)

The current developments in music technologies can be seen in the context of what has been termed the fourth industrial revolution. If the third industrial revolution brought about computers, the Internet, and new communication technologies, then the fourth is one of machine learning, robotics, nanotechnology, biotechnology, and diverse forms of automation. It is evident that music technology will be at the forefront of research in the current revolution, just as music technology has been of axial importance in the other technological revolutions (think instrument design, mathematics, writing, print, automata, machines, electronics, digital systems, symbolic artificial intelligence, and now deep learning). Just as developments in peer-to-peer file sharing, streaming, swarming file transfer, and compression responded to developments and demands in music, the domain of music will continue to occupy a position of prominence in applying robotics, machine learning, computational creativity and other forms of automation in creative work. There is no coincidence that some of the current cutting-edge research in deep learning is conducted by Google's Magenta Lab, whose *raison d'être* is to conduct music research. Indeed, the history of experimental music is one that is intrinsically connected to research in new technologies, whether those are developed in modest workshops and laboratories as in the examples of automata, player pianos, or early phonography, to the more formally funded contexts of radio labs (RTF in Paris, WDR in Cologne, RAI in Milan, the Polish Radio Experimental Studio in Warsaw, and the Radiophonic workshop in London), telecommunication labs (Bell Labs, Cahill, Gray), state-funded research centres (GRM, IRCAM, STEIM, EMS, etc.), university labs (CCRMA, CNMAT, IDMIL, CMC, C4DM, etc.), commercial companies (Steinberg, Apple, Native Instruments, Ableton), and now tech industry research labs (Google Magenta, DeepMind, etc.).

Industry developments: a case study

Cybernetic thinking has been with us for decades and has deeply influenced the arts of our time, but with online clouds, big data, and machine learning the potential increases for a more powerful collective network of musical intelligence. The music software manufacturer Native Instruments have done extensive research on the opportunities and are planning to move their production software to an online subscription model. They argue that music can become more democratic and user-friendly by removing the perceived barrier of entry that comes from the need for a thorough understanding of music theory and high-level instrumental skills that characterised previous musical epistemes. Daniel Haver, the CEO of the company, describes a survey the company did in the Western hemisphere in which they discovered that while 20 per cent of survey participants express themselves musically in some form or another, 50 per cent would like to do so. Music is an elementary and fundamental part of our psychology and culture, and Native Instruments want to tap into the desire to make music. With such a subscription model of music production technologies, the collective system of music (as in the diverse genres and affiliated theories) will be filtered through a particular technological node, and this will have aesthetic and ergodynamic repercussions. Haver predicts that their subscription platform, Sounds.com, will become "the world's hub for all the sound and all music tools there is"[2] and we can only begin to imagine the power this gives the company in terms of data analytics and potential control.[3] The plans do not end here: the company also want to combine the production and consumption platform of music such that financial transactions between a listener and a producer become easier; for example, when such optimisations are in place, a DJ might effortlessly use a stem (a packaged sound file format containing a beat, bass-line, melody, and vocal) or a sample from another producer without having to be concerned with the old-fashioned copyright problems of the past.

Native Instruments' plans are not necessarily ground-breaking in terms of IT: software as a service is a model already introduced in many areas of production and consumption. However, for musicians, this represents a drastic change, as they have typically owned their instruments and forged emotional bonds with them. Many instrumentalists would abhor the idea of a subscription service for their instruments. But herein lies one of the differences between the acoustic and the digital: software is immaterial, ephemeral, and we relate to it in a different way than to our acoustic instruments (or writing instruments like a pen for that matter). The music software from the early 2000s does not work on our current operating systems, so we might as well subscribe to software. The attraction with Native Instruments' model is how rich and deep their data sources will be. Not only will they be able to detect how people use their software, which sounds they download or which presets they use; they will also be able to detect emerging trends before the creators are conscious of them. The system will allow the company to observe creative methods, sketches, throwaways, and other steps in the creative process that are normally hidden, and thus gain hitherto unobtainable insights into creative work methods. They will be able to study hundreds of thousands of users, analyse their music, use of instruments, and extra-musical or sociological behaviour too.

Furthermore, when listeners begin to use the same software platform as music creators, producers can share music and information, interact with their listeners, and study their behaviour, applying further statistical algorithms in their creative process. User behaviour becomes an eminent parameter in the production of new content, as it has always been (any singer can detect how well their new song goes down with the audience), but the difference is that now it can be statistically modelled and compared, and improvements can be suggested by AI algorithms, based on global or narrow statistical trends.

The software industry is not just predicting a change in production methods; it is also anticipating emerging musical formats, in which generativity (here used as a general term signifying all non-linear formats) becomes a key element of future music consumption. This state of affairs has not gone unnoticed by Native Instruments who are implementing compositional and listening formats that go beyond the concept of the linear song and open up avenues for adaptive music. The Wotja and Bronze systems have already been discussed in this context, but a powerful company with a large market share will certainly be a weightier player. This development is indicative of a shift from conceiving of music as a linear rendering of digital samples to something more akin to a system (what we have called *invention*), with potential that can be expressed differently according to the time, place, use, people, and other parameters that might be directly controllable (interactive) or evolve with the piece as generative features.

Future music software will take a leap beyond being a simulation of instruments or recording studios. With networked computational machine learning devices, new software will connect people, frame the foundations for collaborations, and shape the emergence of music from multiple origins. This will happen through social platforms, where human and AI-agents can collaborate in the creative process, and new technical protocols, such as audio busses, shared interfaces, networked platforms, and polycontrol,[4] will be developed further for co-located or networked creative collaboration. In a recent interview, techno-producer Richie Hawtin described how he sees one role for intelligent software in the music production process: "I feel that those types of technologies will allow us to release more of our individuality; it is about throughput and pipelines and interfaces and allowing our movements to become data and then other data coming from other programs or other people or even from AI, to come and merge and collaborate together and come out into a wonderful experience."[5] Smart instruments, plugins, and tools based on collective intelligence will help us achieve masterful performance and sound. Examples include the *Noiiz* sample-library plugin, which uses machine learning technologies to search for sound samples that can be downloaded and used in a DAW. Here, users can describe sounds, search for similar sounds, combine sounds, and the system will suggest sounds from an online database with over one hundred and eighty thousand samples which can be quickly downloaded into the software (in one to two seconds). This is all based on machine learning audio descriptor features such as those explored in Chapter 12. The *Landr* plugin is another potential case study: a machine learning mastering tool that analyses how canonical pieces are mastered and achieves that technique through an online service where users can upload their tracks and get them back mastered, all without an involvement of the human ear.[6]

Shifting models, but *who* shifts?

It is fascinating to see how current developments in commercial music software and industry models echo centuries-old practices. The ideological and aesthetic repercussions are emanating: we are moving towards a mindset akin to one that existed before the advent of phonography, but perhaps also before the eighteenth-century establishment of the work-concept, and so more reminiscent of how music was practised during the Renaissance and Baroque periods. Practitioners of computer music have been experimenting with this since the invention of the computer, but it is only now that our technological infrastructure is ripe to support these ideas of shared, non-linear, emerging musical collaboration. Key elements include networked computing, database technologies, computational algorithms, artificial intelligence, machine listening, machine learning, blockchain technology for attributions and payments, changes in the cultural perception of copyright, and many more. In the age of networked digital media, a solution to the copyright issues that have plagued music in the late twentieth century is extremely important, as music has always been about referencing, borrowing, and sampling (think of Mozart borrowing from Haydn, or Beethoven from Handel), and everyone agrees that the current model is outdated.

There is a great quote, often misattributed to McLuhan, which says that "we shape our tools and thereafter our tools shape us." The problem is that if we take a more granular look, these generic "we" and "us" are not referring to the same entity. The people defining the ergodynamics of modern musical software are few and far between. A relatively small band of actors in the Bay Area, Berlin, and Stockholm are largely responsible for how our musical culture develops, with some further research taking place in Paris, London, and other cities. Although these industry players have a clear understanding of how music develops, and although they try to support diverse practices, they are bound to concentrate on the aesthetics of their primary customer group, which is often regarded as music producers of beat-based dance music. To many, the big data software aesthetics described above might seem like a nightmare: a machine learning data-parsing database of musical creativity! However, this is perhaps not so different from the group mentality we observe in early twentieth-century music where serialism reigned, or the less articulated, yet rigid, constraints of blues, rock, disco, or house. If anything, the late twentieth century opened up a platform where anything is possible – any music, any practice, using any instruments. We have broken out of the ideological constraints that often characterised music in the past, indeed experiencing the liberation of sound Edgard Varèse predicts and we discussed in the introduction of this book. The strength and diversity of DIY cultures, open source, audio programming languages, hacklabs, and other cultural practices mitigate the menace of aesthetic centralisation and enable a thriving musical culture where the experimental feeds into the mainstream. Online collaborative cultures have substantially improved the situation for learning and developing. The flow between experimentation and commercial platforms in both technological and artistic domains has increased.[7] In fact, as someone who teaches music, machine learning in music production is an exciting opportunity. It enables us, biological musicians, to attest to our humanity, to speak truth to power and show the machine how music can be unique and idiosyncratic, made with methods that the creative normalisation processes of the databases could never come up with. Machine creativity not only helps us to understand our creativity, but it also challenges it in fruitful ways.

Conclusion

In analysing current music technologies, I hesitate drawing a close parallel with Heidegger's description of the enframing (*Gestell*) power of technology (Heidegger 1977), but there is a sense in which big data and machine learning are becoming an engulfing framework that directs our thinking in ways that might efface our individuality. If the crowd-sourced software people use to make music suggests that the track is too fast, or would be better in a 4/4 time signature (because that's what the system has learned statistically), it is conceivable that some producers will follow the advice. The software might even be so sophisticated that such decisions will never have be taken, as the user might be led through the creative process so seamlessly and effortlessly that the stage of critical questioning and creative struggle is bypassed through a state of "creative flow."[8] Heidegger is critical of the enframing power of technology, but in the conclusion of his essay, he speculates that it might be *through* technology that we find a "saving power," and perhaps it is indeed when people begin to create their own deep learning networks, using open source software, collaborating on open data sample libraries, production methods, and music theoretical databases that new musical aesthetics and evolutionary sprouts will grow.

Rejecting machine-produced music, we use machines to produce *different* music, but this will require an understanding and control over its mechanisms. We need open machine learning libraries, where users can experiment with their artificial neural network design, train the systems their own way, and produce unique, strange, and idiosyncratic work that takes music further and critically engages with the society we are making. When our machines will make the music we hear on the radio's top charts (and arguably that is not too difficult), it will be something special to be able to display unique and individual human expression, and that is what future music will celebrate. Of course, these new musical inventions can come about just as much through someone playing the banjo in a barn as through someone applying deep learning networks in high-tech computer labs, but the focus here is on the technology that will be used by the music-making masses of the future.

18

Transformation of Tradition

New music technologies are moving beyond the historical weight of notation and recorded sound, with the design of new instruments equally and comprehensively rooted in older acoustic instruments, notational, and recording practices. These are the material, symbolic, and signal writing ancestral techniques of music that constitute the new practices. There can be no return to earlier conceptions of music, as the totality of musical actors has changed in our perception: the instruments, the score, the composer, the performer, the audience, etc. Such a change reflects a changed worldview, or as Eco states: "In every century, the way that artistic forms are structured reflects the way in which science or contemporary culture views reality" (Eco 1989: 13). Music is how we explore our human condition, our place in the world; it shapes our relationship with each other, with our technology, and now with our extended *other*: artificial intelligence. In Attali's political-economic theory, music *precedes* the way culture is organised. If we accept his argument that the age of recording is transcended by new performative and collaborative practices, this evidently does *not* mean that we will not notate or record music anymore. Composed music, either notated or recorded, is an established creative practice and the linear music track pressed on vinyl or streamed across the Internet will keep its status and relevance; consider the music of Hildegard von Bingen or Iannis Xenakis, Muddy Waters or Björk. Attali's paradigm shift merely signifies a new emphasis in musical practice, an addition to an already diverse flora that is not based on studio recording releases. This is an argument substantiated by the limited income musicians have from their recorded work, compared with playing concerts, composing for other media, and various merchandise.

Where is the new stuff?

Mark Fisher sometimes talked about how the music of the twenty-first century has ceased to be a signifier for the time: music is not revolutionary anymore, and we do not notice any drastic change or musical revolutions. For Fisher, there is a lack of cultural time or specificity in music, and we do not get the "future shock" any longer, the feeling of being stunned by the innovation and originality of music so strange that one has never heard anything like it before (Fisher 2014). This is a striking observation from an excellent critic, and might ring true to many of us if the focus is exclusively on music as a recorded product, released on vinyl, CD, or streaming. However, what Fisher may have overlooked is that new musical practices involve more than just recorded sound, and that the centre of musical creativity has shifted platform. Today we celebrate ingenious musical creativity that has become multimedia, interactive, interdisciplinary, high-tech, and involving collaboration and input from engineering and the sciences.

A related critique by Nicolas Collins, whose impactful musical career has been within new music, including writing a key book for new musical composition practice *Handmade Electronic Music: The Art of Hardware Hacking* (Collins 2006), reaches deeper into the core of the problem. Collins, who studied with Alvin Lucier and was engulfed by the experimental music of the 1960s and 1970s, praises the revolutionary spirit of those decades, then asserts that he has "detected no shift in the fundamental terrain of music that rivals the magnitude of the changes that took place in the 60s and 70s. I find this admission more than a little depressing" (Collins 2015: 1). Collins is not experiencing the shock of the new anymore, so he aptly asks his colleagues for advice – calling for a paradigm shift: "Show me a sign of a Next Big Thing" – and the answers are illuminating. Some point to the democratisation of music through new technologies, that everyone is an artist (Fluxus), and that now we operate with plural narratives, as opposed to the single-narrative past, and this explains the impossibility of a revolution (as when Cage shocked everyone with his work). David Toop even urges him away "from obsessing about movements and paradigm shifts" (Collins 2015: 4). Personally, I suspect that the problem here relates to what we call a revolution, and what kind of timescale we should apply to such a process. Furthermore, I suspect that Collins is unable to see the change taking place as he *epitomises* this very change.

This book has attempted to demonstrate how major media technological revolutions change music and musical practice: instruments, notation, printing press, recording, electronic, and digital systems. If the digital brought a potential for revolution, it is happening now – but this is a "long now." This revolution is constituted by a technical change, just as we experienced with the printing press or electricity, and on top of it surf new modes of making music, new genres and aesthetics. This revolution has seeds in 1950s mainframe computer music with Hiller and Isaacson's *Illiac Suite* and Max Mathews's first MUSIC language, and 1980s developments, such as the MIDI protocol and digital audio workstations software, but it is just now that the digital potential – the algorithmic nature in its variety – is being explored more widely. Digital music of the twentieth century was largely in the business of remediating and simulating previous media (instruments, notation, tape studio), and it is only now that we are seeing revolutionary practices, with generative music, game sound, live coding, artificial intelligence, and other networked and computational creativity work. A major media revolution does not happen in a few years: it takes decades.[1]

This is where we see the power of music: as cultural form that explores what the future might have in store. Alvin Lucier's brain interface mapped to sound was decades ahead of its time, and today we operate with these interfaces on a regular basis, still exploring what cognitive science and music might come up with through collaborative projects that benefit both fields. Furthermore, the ground-breaking innovations in new music are often characteristic of a strong creative engagement with the format itself. These works require critics to engage with instruments; circumnavigate installations;

use interactive software; rethink relations between sound and graphics in new audiovisual systems; study filmed documentation; engage with modular synths; scrutinise the code behind a piece; watch audiovisual clips; follow artists on social media; and so on. The revolutionary music of the future will not come to us in the form of a streaming stereo file. Today's music critic – and lovers of music in general – will have to navigate alien territories and consider the artistry and skill in the soldering of a new instrument, the programming of computers, and the use of artificial intelligence, and generally demonstrate knowledge of materials ranging from wood, strings, and reed to coding environments and deep learning. This same critic will have to be equipped with a multifaceted scientific understanding that underpins this new musical work, one that encompasses HCI, engineering, programming, acoustics, synthesis, design, and composition. Musicology can never be the same again. The field has exploded, and the term "music" may not cover the diversity of activities any longer. But this is where etymology comes to our aid: as the multitalented Muses demonstrate, *music* is the best of words to describe this networked multiplicity of activities!

Fisher's comment clearly focuses on recorded music and the industry as we knew it in the twentieth century, omitting to study the magnitude and diversity of emergent musical cultures. In the 1960s, music was popular, but the world of recorded and released music was small, defined, and stabilised. There weren't that many composers or bands around, and they practically all knew each other.[2] Today's music industry is fractured, spread into multiple platforms with diverse practices, making it harder to get a comprehensive overview of the fields. Where will we find the innovative music of the future when there are well over 135 million tracks on SoundCloud? When twelve hours of new content is uploaded onto the platform every minute?[3] Or when key musical work is composed for games and virtual reality? Clearly, music journalists and curators of music will continue what they have done for decades, but in light of the monumental volume of new music, more refined instruments for investigation are needed, and perhaps machine listening algorithms – the very technique that aids in the creation of more homogenous music – might also help us to discover new uncategorisable music, the music of the future that does not fit into existing boxes.

From consumption to production

New music is being produced on an unprecedented scale and level of availability: where before albums would have national releases and in very few cases be shipped to other countries, there is now global access. The music streaming platform SoundCloud has over 175 million unique *listeners* a month (not unique listens), for example. Considering how the role of the record label has receded as a curatorial filter, all this new available music cannot be of the highest quality, so our response might as well be, so what? Or, who cares? It is positive that so many people are making music, and that the criteria for what was released in the age of scarce media (LPs/CDs) simply do not apply anymore. It is a matter of celebration that school children make cover versions and a teenage band records their latest song, uploading their tracks onto SoundCloud,

YouTube, or iTunes. We are therefore operating with a network of networks, where patterns emerge, listener behaviours can be studied, and links and suggestions can be forged by simple algorithms ("people who listened to this also listened to that"), as well as by more complex AI recommendation systems. The current challenge for data scientists and musicologists is therefore how to engage with and understand this abundance of recorded music; how to explore emerging genres, styles, practices; how to cut through the mass and detect things in this cloud of big data that was never possible before. Here musicologists become digital humanities scholars (Berry 2012) that will resolve to the application of distant listening (Cook 2013) technologies in the study of music.

A related change in the overall picture in the industries of music is that the recording industry's global revenue has nearly halved since 2004 (from $33.6 billion in 2004 to $17.3 billion in 2017), while at the same time revenue in the musical instruments industry has boomed and is now around $16 billion (according to a 2015 NAMM report, though this report does not include music software sales, music apps, or games, so we can estimate that this number is much higher today). This change indicates that global populations are beginning to enjoy music in more participatory and engaging ways that remind us of pre-recording practices. A concrete example is the Rough Trade shop in Shoreditch, London, half of whose shop floor is dedicated to DIY electronics, books, and a coffee shop – occupying space that would previously have been devoted to vinyl and CDs. Music is more than the recording: it is also about making, thinking, and socialising. Furthermore, and perhaps as a result of the above development, the music industry is exploring visionary solutions that involve the coming together of the creation and consumption of music. This has not happened in commercial music technology before, though we do have the predecessors for this in open source software and in theories of generative music. Musicians and industry players are experimenting with new cryptocurrency as micropayment methods that might benefit the artist better and also solve the copyright problems from which we currently suffer when wanting to use a sample, a melody, or even an effect setting. This is one area that has not kept up with technological changes in our media and musical practice.

Composition and system

Generalising on practices of music in their global diversity is impossible, but with the sudden change in technology over the past two decades, we can detect clear changes in how people make, disseminate, and consume music. What characterises the new musical practices of the twenty-first century is a certain move from the linear format to the focus on the musical work as a system, an invention, assemblage or installation of sorts, whose materiality, spatiality, and situatedness separates it from the abstract notion of music in twentieth-century work, expressed as generalisable notes on the score, or "objectively" captured sounds designed for ubiquitous playback. The move is characterised by an understanding of *the system as a composition*, a concept we can trace back for two millennia, and, indeed, the two words – system and composition –

Figure 18.1 A picture from Marianthi Papalexandri-Alexandri's performance of *Untitled II*, at Festival Con Voce in Lucerne Switzerland in 2014. © Pe Lang.

mean the same thing from an etymological perspective.[4] The suggestion here is not to reject the usage "composing a work" in favour of "inventing a system"; it is merely to say that, from the perspective of new musical composition and performance, the latter might better describe the emerging musical practices of this new systematicity. Neither is this a rejection of the work-concept, but rather an observation of how its core and boundaries are being tested and tried, like an object on the scientific lab table, in order to understand what the work might mean in twenty-first-century music. Furthermore, the new approach clearly supports and even reinforces the role of the artistic signature, the composer's voice, which is not the same thing as the author function previously analysed as being tied up with the work-concept.

There is an abundance of new types of work that represent the system-invention approach and that trace their ergomimetic ancestry to the diverse practices of instrument building, notation, and recording. Examples include Tristan Perich with his *One-bit* symphony CD box, which contained a chip that produced music in real time, or his *Drift Multiply* piece for fifty one-bit chips and fifty violinists; Marianthi Papalexandri-Alexandri's instrument building and performance of sonic sculptures; Stelios Manousakis's (2016) *Hertzian Field* performance system that senses electromagnetic Wi-Fi waves in surround-sound space; Onyx Ashanti's *Sonocybernetics* project that includes an "exo-voice" cyborg DIY body-sensor kit; Claudia Molitor's aforementioned *Entangled* piece or her *Remember me . . .* performance where she archaeologically deconstructs a musical score by cutting through semi-transparent paper layers of new notes with a sharp knife; Paul Stapleton and Tom Davis's building of sculptural instruments in their *Ambiguous Devices* project; Toshimaru Nakamura with his no-input mixer system; Cathy van Eck's *Wings* speaker installation or amplification of music stands; Sarah Nicoll's *Inside-out* piano, an upright transportable instrument she made specifically for prepared piano alternative techniques; Simon Steen-Andersen's combination of orchestral music, installation, and performance art; Alejandro Montes de Oca's "Case-Specific Electroacoustic Instruments," which encompasses installations with speakers on water or four vibrating guitars; Larnie Fox's *Sonic Windmill* sculptures; the *Singing Ringing Tree* by Mike Tonkin and Anna Liu, which is a sculptural/musical composition installed on a hill in Lancashire; the modular patch inventions of Moor Mother; the extended instrumental cybernetic body of Ainara Legardon; Ray Lee's *Chorus* installation of rotating speakers; Ligeti's *Poème symphonique* for 100 metronomes; Pamela Z's technological extension of her voice, controlled through motion sensors; Patricia Alessandrini and Freida Abtan's *Orpheus Machines*, which extends early keyboard instruments with new technologies; the large array of systems and instruments made by Sergi Jordà and colleagues, such as the *Reactable*; Lou Reed's *Drones* guitar installation of feedback; Anton Mobin's *Autonomous Objects* amplified object inventions; Scott McLaughlin's dynamical and emergent systems; the conceptual music of Johannes Kreidler; Agostino Di Scipio's sonic ecosystems that apply feedback in installations; the transient paper instruments and *Vibrobot* installations of Kristina Andersen; Laurie Spiegel's *MusicMouse* software, which incorporates a music theory to explore; the generative music systems of Autechre or Farmers Manual; Llorenç Barber's bell pieces that transform towns into instruments by playing all the town's church bells; Holly Herndon's pieces that were written by creating software on her laptop which she equally considers to be her workshop and her instrument; Kat Hope's graphic scores; Ryan Ross Smith's animated score studies; Marco Donnarumma and Atau Tanaka's treatment of the body as a musical instrument through sensors and mapping; the diverse experimental activities of the SLÁTUR composers' collective; the theatrical and props-based music of Jessie Marino; Alex McLean or Renick Bell's live coding systems that afford some very bespoke manipulations of rhythmic (and polyrhythmic) patterns, Joana Chicau and Kate Siccio's application of live coding as choreographic notation; Enrique Tomás's tangible scores resulting in his definition of the instrument as a score; Yoko Ono's pieces of verbal instructions; John Zorn's game pieces for group improvisation; George Lewis's *Voyager* system of musical dialogue with a self-playing piano; Dick Raaijmakers's *Intona*, which turns microphones into instruments in an open compositional structure; the sonification work of Carla Scaletti; Peter Kirn's diverse projects and workshop activities, including his comprehensive *Create Digital Music* blog, and many, many more, as this paragraph could continue ad infinitum.

This book has not engaged in the analysis of musical pieces, but rather the study of how new technologies bring forth new compositional approaches that then become an inextricable part of the media's qualities. Although I have grouped the people above together in one very long paragraph as examples of musicians inventing systems, I would not dream of placing a label on their practices, denoting a style, genre, or

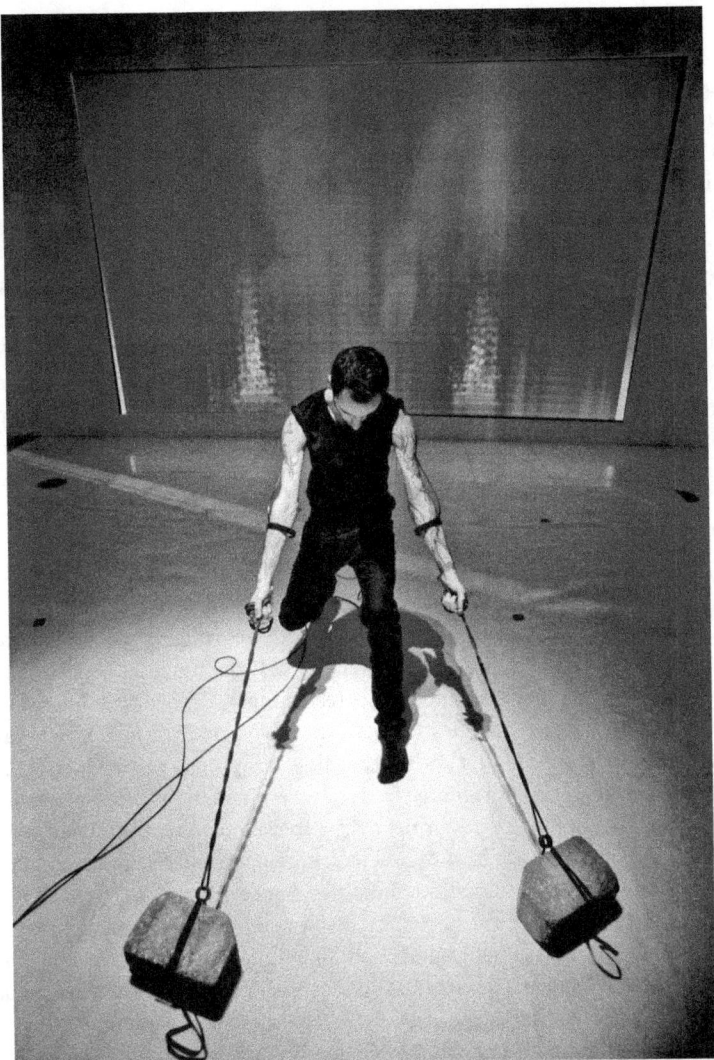

Figure 18.2 Marco Donnarumma performing *Hypo Chrysos* at Inspace, University of Edinburgh, 2011. © Chris Scott.

aesthetics. To the people mentioned, I apologise for having grouped them as part of a flock, and I am fairly convinced that they would reject any classification or denotation of an -ism that would presume to formalise their practices into an identifiable grouping or movement.[5] That said, a cluster of resemblances does distinguish these pieces: for example, the renewed focus on materiality and the technological conditions of musical works. In contradistinction to Aristotle's inert matter, we now operate within a framework of a new materialism (of people such as Manuel DeLanda or Karen Barad), where music cannot be understood as abstract theory. Another common thing that distinguishes these pieces is that a recording (audio or video) can only partially represent their character, which does not mean, however, that documentation of them will not be of any worth, nor that recorded music is "dead." While the recorded musical work will be an integral and important part of the new practices, continuing as it will to provide pleasure for listeners, and some financial income for some musicians, it will not occupy the same central position in our musical culture as it has done in the past century. We might ask if this focus on the system, the instrument, the software, or indeed this new materiality (Lange-Berndt 2015) is something that will pass, like "spectralism," "minimalism," or "new complexity" has in the past (Toop 2010: 89), but it is unlikely. We are not moving *towards* a solid concretised new practice that can be identified and spoken about at this stage: the feeling is rather that we are moving *away from* the hegemony of linear composition (the score and the recording), and multiple new worlds are heaving into view, waiting to be explored.

Conclusion

The argument in previous chapters has been that our new digital instruments combine three distinct levels of sonic writing in their bodies: the material (instruments), the symbolic (notation), and the signal (audio recordings). It does this equally through applying new material technologies in the form of sensors, hardware, and standards; notation, protocols, and code; and audio recording, sampling, analysing, and recreating. New music technologies do not emerge from nowhere, and the accumulated knowledge and conventions of past centuries in the above-mentioned forms of sonic writing are rigidly inscribed into them. Whilst digital instruments may lack the material depth of their acoustic counterparts, its unique qualities, haecceity, and haptic feedback, they can augment the embodied performance by various other means, such as motion capture analysis, machine learning, and auxiliary representation through visual or haptic means for example using screens and motors. The examples of musical pieces as systems given above demonstrate postdigital approaches to composition that are inspired by the building, programming, assembling, and unique creativity which go into every new piece of software. Understanding how the logic of music is maintained with new technologies will support future musical creativity, where we embrace new modes of making music and reflect upon our human condition through the mirror of music. In light of the changed music-technological landscape, the last chapter of this speculative conclusion will reflect upon the impact this new framework will have on music education.

19

New Education for New Music

This book has focused on new technologies and related changes in the way we make music. It is therefore inevitable that we have studied practices that are critical in spirit and experimental in approach. For this reason, the book has operated in the space of learned music, written music, preserved music, or music that is intrinsically linked to the theory and institutions of its time. Because of the ubiquity and permanence of writing, there is more historical material about the musical practice of a medieval monk than a medieval troubadour. Equally, because of documentation, it is easier to study how radio engineers in Paris or Cologne experimented with tape, compared with amateurs doing private experiments in garages around the world.[1] We know more about new synthesis techniques from academic papers written by their inventors and who use them in their music than from their use in popular music. Today people are exploring the use of AI in music all around the world, but it is the written documented accounts that will establish the history. Studios, labs, and R&D centres have always been places that document and publish their technologies. Indeed, we might, somewhat controversially, proclaim that Bell Labs have had more effect on music production than Apple, as the former published their research and practical projects (e.g., Max Mathews, James Tenney, and Laurie Spiegel), influencing the international development of new technologies, whereas the latter, even if they are the developer of one of the most popular digital audio workstations, Logic Pro, does not share its research.[2]

New terms for new practices

By engaging with a selective genealogy of music technologies, and not attempting to provide a historical narrative, this book has explored how inventions of the past support the ergodynamics of our current devices and practices. The emphasis has been on tracing the evolution of musical techniques, not on analysing technological objects or music. The geographical focus has mainly been on places and countries where technologies were developed and where critical research has been conducted. The protagonists of this book are the musicians who invent new music and new methods, through new technological means (instruments, notation, recording, coding). To these, I have applied terms such as composer, performer, producer, musician, interpreter, inventor, and artist, but none of these words are adequate in the context of new music, except perhaps "musician." Musician is a very broad word, which is its strength, but at times its weakness too. Yet, people in jazz or electronica do not tend to use terms such as "composers" and "interpreters," and contemporary music practitioners do not use terms such as "producer" for the creator of music. These terms largely fuse in new musical practices, where the composer becomes an engineer, the producer an instrument maker, and the interpreter a co-creator. The music of the twenty-first century will need a new language for describing these roles, but it is most likely that this will happen naturally "in the field," and not be suggested in an academic book.

This book has presented a picture of changing musical practice: one where the traditional constellations of instrument-maker-composer-performer, or composer-score-performer, or performer-instrument-audience are breaking down. What replaces these cannot be clearly described as one thing or gestalt, as these are a multiplicity of practices where lines that might previously have been clear are fading. The instrument merges with a sound installation, the performer live codes their instrument, the composer becomes an instrument maker, the audience participate in the composition of a piece with their networked devices, the instruments learn from their players and exchange information, the music changes depending on social and spatiotemporal parameters, the distinction between music and visual elements blurs in new immersive media environments, playing an instrument and playing a game are the same thing, and so on.

Music education in the future

In light of the drastic changes brought forth by computational media, it is pertinent to ask how musicians will learn and practice their skills in the future. It is not easy to answer the question of what a musical curriculum should comprise to be progressive, creative, and culturally relevant, as opposed to solely serving a conservative element of the museum culture (Boulez 1977). Clearly, we will practice acoustic instruments and compose for them: they are technologies of infinite mystery and depth, which will be rediscovered in any cultural period. There is no need to burn the instruments, as the Italian Futurists and Spanish Anarchists demanded at the start of the twentieth century, but rather rethink them in the modern context. However, young people often enter the world of music through digital technologies (e.g., apps and games), and their natural instruments for musical expression become digital systems of some sorts. This prompted Berkelee College of Music in 2018 to define the "Electronic Digital Instrument" as one of the principle instruments students can focus on whilst at the institution.[3]

What kind of education might be offered to a new generation of musicians who need to understand electronics in order to create an interface, computer programming to define the sound and music of their instrument, and music theory to design the compositional aspects of the instrument? They will also need performance skills in order to play their piece, general research skills to be able to ask relevant questions, and knowledge of how to apply apposite research methodologies that might include cross-disciplinary collaborations. How will future musicians subsequently share this knowledge, not only as performed music or technological objects, but also in the form of description and explanation, sketches and code, and audiovisual documentation, such that musicologists and critics will be able to engage with the work at a level that it deserves? How will they communicate their works with other performers or transmit knowledge to the next generation of learners? Will such education find a place in the music conservatories and university music departments? Will this education be interdisciplinary, flexible, and modular, or will we continue to maintain disciplinary silos where composition, performance, product design, computer science, film, production, and computer games are all separate fields? Higher education should be able to respond to changing circumstances and create flexible environments for research tailored to this world of new musical practices. However, what hinders progress is the slow pace of secondary education (up to eighteen years old), where regulations are stricter and the areas of music and music technology are defined as two separate fields (where the former focuses on musical performance and the latter on music production) – definitions that separate these "disciplines" as they might have appeared in the early 1990s, but not in the current heterogeneous interdisciplinary practice of new music.[4]

The entrenched ideas we find in educational establishments of composers writing for virtuosic performers, in "music," on the one hand, and producers recording rock music or making new electronic music tracks, in "music technology," on the other, are becoming slightly dated. The world of practice with sound and music met by graduating students does not reflect this binary division of disciplines. Music is a highly collaborative social form that focuses on cohesion, not competition (like sports), and the modern technological context presents opportunities for highly interdisciplinary work methods. The model of the "virtuoso" performer can no longer be the purpose of music education. In future music education, it might not be for everyone to dedicate time to reach an expert level on an instrument (although playing one is essential for everyone, I would argue). When computers can spew out thousands of notes per minute or per second, should we wish, the idea of reaching Franz Liszt's piano skills is transformed: while it is still a noble goal, our musical languages have surpassed the need for such skill. Indeed, we might rather shift our educational goals from that of "virtuoso" to "amateur." Any study into the concept of the *amateur* will demonstrate that it stands for the lover of the arts, someone who is serious about their dedication to their practice. It can be argued that it is the capitalist professionalisation of our labour that stratifies work against pleasure, generation of monetary value versus enjoyment (Merrifield 2017). New music technologies and the practices involved, such as DJing, conflate these terms, and we become virtuoso amateurs of music, not necessarily through our physical motoric skills, but the understanding of the technology, how to apply it, navigate aesthetic taste, and the dedication to music as an interdisciplinary intermedia art form.

The argument here is not that one thing is replaced by another. Past practices will continue, but the emphasis will clearly shift: that is the nature of cultural evolution. Therefore, if the focus of musical education shifts from the traditional conception of the virtuosi as people with extreme motor skills and interpretative ingenuity, to support system designers, programmers, makers, and multi-instrumentalists, we might indeed be better off with a word like "inventor." The new inventors of music will be interdisciplinary makers and creative thinkers whose research, engineering and creative skills will have been trained through music.[5] The twenty-first-century musician will be able to connect sensors to the Internet of things, read specs, solder wires, program analogue-to-digital converter chips, interpret numerical data, write code, design synthesis engines or deep learning networks, map data to sound, conjure up compositional structures that are interesting to play, and generally make their instrument engaging and understandable for both performers and audience. This is no small skill set, but these are some of the compositional, instrument-making, and performance skills that are currently of value and should be part of musical education. Coincidentally, from the perspective of education and transferable skills, in addition to the unquestionable value of music itself, these are also the skills that are required for any good job in technology, game design, or the new media industries.

How to teach invention?

The new music pedagogy will not be an engineering education: it is simply that digital technologies become the means through which new musical works are expressed, so technology design has become part of our compositional language. However tempting it is to compare this to a novelist who does not think much about the instruments of writing, the epistemic dimension of music technologies means that the equipment is impregnated with music, which is another reason for establishing a critical relationship with it, one based on knowledge and practical skill. Instrumental learning on digital instruments will be different from the embodied non-symbolic means we are used to with acoustic instruments, as learning digital instruments is somewhat more textual (with manuals, new vocabulary learning, menus, symbols, icons, and endless signifiers in software and digital instruments). We might indeed consider this a different education altogether, one of information and explicit learning (as opposed to the tacit learning on the acoustic instruments), one where the technology is in continual flux and where the understanding of problems is emphasised, not the exact solutions to them. In this new music, every piece and every instrument will pose its own problems, so it is up to the student of music to come up with solutions. Instead of the mimetic teaching of performance or instrument building, this new education will teach the *methods of research and knowledge acquisition*, the procedural know-how research knowledge of the new luthier. The classroom then becomes a research laboratory, but in that lab there are terminals that connect to the world's YouTubes and MOOCs, where the student can seek the information required for the particular problem at hand. Students now find examples of code, diagrams, instructions, in online fora where questions can be asked and answers provided that outperform the knowledge of the tutor. The new MOOC platforms (massive open online courses) such as Kadenze, Coursera, edX, or Udacity also contain courses for people in the creative industries. These developments are amenable to students' needs in that when music becomes such a dispersed creative domain, they will need to be able to tailor their education around their agendas and goals. Most importantly, a solid music education of the twenty-first century must always centre on the core concepts of musical meaning: of expression, collaboration, inventing, sharing, and partaking. These are not easily teachable through online learning, videos, or textbooks, and it is here that the old conceptions workshops and research labs substantiate their educational worth.

In his introduction to Stiegler's book *The Neganthropocene*, Daniel Ross articulates the incompleteness of the human being, in response to which we typically establish a threefold operation: "1) to *produce* artificial organs; 2) to *learn to practise* these artificial organs; 3) to *institute*, for the purposes of such learning and such practices, social organizations that articulate the relations between the generations, metastabilizing the forms of knowledge that *are* these practices and these cares" (Ross in Stiegler 2018: 19). Stiegler's book is political: it transcends the nihilism that results in the denial of climate change and post-factuality, and explores how we can move beyond the Anthropocene. The direness of the current situation is such that we cannot engage in resistance any longer, as Derrida, Deleuze, or Lyotard suggest, but we need to actually *invent* (Stiegler 2018: 254). Music is a paradigmatic example of how we produce artificial organs, practise them, and form social institutions around their use. It is an ideal example of invention and innovation, of collaboration and communication beyond language, across generations, cultures, geography, and historical periods. It consolidates patterns of attention, care, and love for something that is everyone's at all times, something that is immaterial and free, yet with the potential for immense wealth generation and consumerist practices (examples are concert houses, festivals, instrument collectors and music collectors, including the $2 million paid for the only copy of a Wu Tang Clan album). As such, music is a laboratory for experimentation in human–machine relations, and there is a qualitative difference in our new learning machines that separates them from the algorithmic mechanisms of our documented history. Music now serves as a demystification instrument for experiments in automation, artificial intelligence, crowd intelligence, cryptocurrencies, intellectual property, and machine-assisted creativity.

Conclusion

Music education should embrace, reflect, and develop further the historical role of music, both in terms of theory and practice, in the world's diverse cultures. Music plays a fundamental role in all societies, exemplified by the important role of music in childhood development and child rearing. We were all musicians once! Music is an activity in which we can investigate issues of what it means to be human, which today involves embodied relations with technology, the application of artificial intelligence in our lives, new intellectual property and financial models, and networked global collaboration. Teaching and training compositional and performance practices from the perspective of aesthetics, history, and theory is therefore crucial. From such a background, a basis is given to ask the following questions: What is the role of the musical instrument? How is it played? How does it work as part of an ensemble with other instruments? How does the performer relate to other performers? Or to the audience? How much theory is inscribed in the system and how much freedom does the performer have? Which other technologies are enrolled? Where do they originate from? Who makes the decisions that are written into our new systems? How does listening to music change? How do the new inventions transform the way we make and distribute music? Why?

At the dawn of writing, Plato, in his dialogue on etymology called *Cratylus*, describes how the origins of the term "music" derive from the Muses' search for truth: "The name of the Muses and of music would seem to be derived from their making philosophical inquiries (μῶσθαι)" (*Cratylus* 406a). This is a statement that echoes throughout the ages with authors such as St Augustine and Isidore of Seville. The idea that music is a method by which we carry out research on our physical and social worlds has indeed been a subtheme in this book: that musical practice is about developing theory and its instruments, exploring mind and body, logic and harmony, self and others. It is about the instrument as an organ that extends our body outwards as well as inspects our psyche inwards. Music is a search, a probe, a *ricercare*, an invention. It is at the core of what makes us human, and we witness the joy of music that children possess before they attain language, as well as in old people who have lost both memory and language; we see, in other words, that music is a fundamental, primary mode of human existence. It is precisely the fathomable nature of current music technologies that accentuate the nature of creative expression: how do we express our individuality when our tools draw from the statistical means of the crowd? Yet again, music becomes the Petri dish for the study of human kind. Music will therefore continue to be one of the key sciences, as it has always been, at least since historical records began; and it might be through musical expression that we can best investigate how we might use the emergent intelligent technologies, yet preserve our individual humanity and soul.

Notes

Preface

1. Indeed, this choice of a medium is not about the materiality necessarily, although paper books do have different qualities than ebooks, but the process that lies behind the book. No matter how wonderful our new technologies are, existing as they do in the digital material domain, they are additions to our existing media forms, not replacements. That said, there is a website (www.sonicwriting.org) which contains information about the book and relevant audiovisual materials. This site will be updated, for example with the interviews I conducted while researching the book.

Introduction

1. Odin's (or Óðinn's) method of learning the runes was through sound ("æpandi nam" = yelling, I picked them up), and his name is etymologically related to *óður*, or poesy (cf. also the English word "ode" deriving from Greek *oide*). Óðinn is, of course, the god of poetry and magic. Magic in Nordic is *galdur*, related to the verb *gala*, which means to chant or shout.
2. This is a theory further supported by the work of cognitive philosophers, such as Andy Clark and David Chalmers with their notion of the extended mind.
3. We often change our instruments during performance: we retune string instruments, change effect settings in electronics, and the *whole point* of live coding is to create and redefine the instrument during play.
4. Although the words *organ* and *ergos* have the same root, the possible concept of "organdynamics" would not be of interest here, as we are not really talking about the qualities of the instrument itself, but the quality of our performance or work (*ergon*) on the instrument, with affiliated exploration, learning, and teaching.
5. For example, the fiddle and the violin are the same physical objects, but they have different ergotypes.

Chapter 1

1. See http://imaginaryinstruments.org. These instruments could be called "ideophones" if that concept wasn't so close to the already existing "idiophones," a category defined by the above-mentioned von Hornbostel and his colleague Sachs, and which we will discuss in Chapter 4.
2. Aristotle makes this clear in *Politics* 1341a "the flute happens to possess the additional property telling against its use in education that playing it prevents the employment of speech."

3   There are differing opinions on when the monochord was invented, who invented it, and so on, but that discussion is only remotely relevant here as the theory of ratios can be done with any string instrument, such as the kithara. See Creese (2010) for further information. See also Rehding (2016).
4   The three key systems of belief in Japan can all mix in the same person – it is often said that a Japanese person might be born into the Shinto religion, get married as a Christian, and die as a Buddhist (Kisala 2006: 3).
5   The latter project is a physical violin with the neck only (no resonating body), but applying new convolution digital signal processing techniques to simulate any other violin in the world. This is achieved by recording the impulse response from historical violins (including Stradivari and Guarneri), so the player can choose the sound from classical instruments when playing.

Chapter 2

1   For a more serious engagement with the idea of speculative aesthetics (of the perception of alien objects), see N. Katherine Hayles's response to how aesthetics are treated in object-oriented ontology (Hayles 2014).
2   We are reminded of the old farmer who had a favourite shovel that had lasted him for decades, although he had exchanged the blade three times, and the handle three times too.
3   For further information, visit www.steim.nl
4   For further study of commercial NIME controllers, Garth Paine has written ably about these in his 2015 NIME paper and McPherson et al. have recently conducted a study in new marketing techniques of new musical instruments, comparing the distinct approaches in NIME, HCI and crowdfunded approaches (McPherson et al. 2019).
5   Further information here: http://karlax.tommays.fr/
6   Composer Andrew Stewart has composed and performed extensively with the Karlax instrument (www.dandrewstewart.ca), and so has Francis Faber, who is part of the Ensemble Fabrique Nomade. See video here: https://vimeo.com/67049071
7   See www.liveinterfaces.org/2016/#workshops for example.
8   Indeed, Google's Ngram Viewer shows that there is hardly any use of the word before 1920, and that it slowly increases up to the 1960s, where it suddenly starts to boom exponentially. This is clearly related to the design discourse in electronic and digital technology.

Chapter 3

1   A Moog engineer, Jim Scott, explains: "It was something like vacuum tubes, in that the circuitry would not suddenly go into clipping, it would distort gracefully ... Also, the circuitry was inherently wide band ... It passed frequencies beyond the audio range ... And we're getting into guesswork here, but the feeling is that there were things that happened up in the ultrasonic range that can cause inner modulation and distortions, [this] reflects back and can be heard in the audible range" (Pinch and Trocco 2002: 235).
2   Sergi Jordà, a pioneer in new musical instrument design and leader of the *Reactable* team, wrote his PhD thesis on the "digital luthier" (Jordà 2006).
3   Epistemic tools are related to Norman's cognitive artefacts (1991: 17) and Clark's cognitive technology (2001). It also relates to Mithen's (1998, 2000) and Sterilny's (2004) material and epistemic artefacts.

Chapter 4

1. Belinda Barnet conducts an illuminating interview with Eldredge in the *Fiberculture Journal* that further contextualise his ideas and experiments: http://three.fibreculturejournal.org/fcj-017-material-cultural-evolution-an-interview-with-niles-eldredge/.
2. This has clearly been changing in recent years. Georgina Born's (2014) work is relevant here, and so is Gary Tomlinson's, Ian Cross', Christopher Small's, Aniruddh Patel's, and Emily Dolan's, amongst others.
3. Such as Hornbostel-Sachs's system focusing on material and acoustic nature, Dräger's system emphasising appearance and playing technique, or Heyde's natural bottom-up classification. See Kvifte (2007).
4. The notion of "resist" is quite particular here. It could have been written as "re-sist" (*re-* "against", *-sistare* "stand" in Latin), closely related to the word "exist," or to "stand against something," like an object or *Gegenstand* (re = "against", sistere = "stand firm") – see note 2 in Chapter 13.

Chapter 5

1. Cage's well-known piece 4'33" also derived from Tudor's interest in performing Cage's vague ideas of a "silent piece" (Kotz 2007: 15).
2. See https://philandgayleneuman.com.
3. These mathematical arts dealt with numbers in their different modalities – that is, where numbers were static in arithmetic, static and spatial in geometry, in motion in music, and in motion and spatial in astronomy.
4. A recent discovery by Varelli (2013) of a manuscript in the British Library containing polyphonic notation named "cithara notation," based on the arrangement of intervals represented by the lines that follows the tuning of the six-string Greek *cithara*, a system that brings together what is often called "diastematic-syllabic notation", replaced by the graphical symbols of "-" and "o" designating two voices, written on top of horizontal lines.
5. The term com-position, illustrates well how this works, as it is about putting (-ponere/-positus), stuff together (com-), or placing things next to each other.

Chapter 6

1. Such circular scores later appeared in the renewed interest in musical scores in the mid-twentieth century, for example the pieces *A Rose is a Rose is a Round* by James Tenney from 1970, or *Makrokosmos* by Georg Crumb from 1972.
2. This history has been told from the economic perspective by Jacques Attali in his book *Noise: The Political Economy of Music* (Attali 1985). Attali's periods of representing, repetition, and composition map roughly onto the epistemes defined in this book as those of acoustic, electric, and digital instruments.
3. For an exploration of the Roman de Fauvel manuscript with translation, see the Digital Fauvel project here: www.digitalfauvel.org.

4   Chapter 8 discusses combinatorics, a mathematical discipline introduced to Europeans via Arabic scholarship in the thirteenth century, with prominent practitioners being Llull, Fibonacci, and de Nemore, and further developed in the seventeenth century by Leibniz, Pascal, and others.
5   For a more practical engagement with this topic see also *Improvisation and Inventio in the Performance of Medieval Music: A Practical Approach* by Angela Mariani (2018).

Chapter 7

1   We should note that Goehr's book was first published in 1992, so although her description is absolutely correct, it would have been written slightly differently today (or in 2007, which is the publication year of the book I reference).
2   The politics of technology in art is beyond the scope of this book, but there are innumerable factors that play into this situation, such as computers being seen as military technologies in the 1970s or the use of art in the culture war, for example where modernist art and media art was used in Cold War territorial agendas (Beck and Bishop 2016).
3   Eno has been an active promoter of generative music ever since, often describing how, on seeing the Game of Life in execution (a cellular automata artificial life system based on three simple rules but exhibiting weirdly complex behaviour), he began to think of musical composition as the invention of rules that can be executed differently every time they are followed.
4   Before Christmas 2010, due to an orchestrated effort by music lovers in the UK, the piece nearly topped one of Simon Cowell's *X Factor* pop productions in the music charts.
5   I say "computer" here, but this need not be in digital format necessarily. Rather thinking in postdigital terms, these are musical practices that are either digital or define themselves in relation to the digital – either as complementary or critical.
6   Conceived of this way, we see parallels in how the tape and computer music studios were only accessible to the privileged few in the 1970s (but now we have all that equipment in our laptops) and today the cutting-edge of new music has again moved to research labs, operating behind closed walls. This will be explored in later chapters.

Chapter 8

1   A phrase often used in the marketing of early synthesizers and the title of a fine book on the topic by Paul Théberge (1997).
2   See Nick Collins's chapter in an excellent compendium called the *The Oxford Handbook of Algorithmic Music* (McLean and Dean 2018).
3   In later chapters we will see how the development of current digital music technologies echo this statement.

Chapter 9

1   This desire relates to the dream of an objective recording of an actual sound, written down, as it appears in a natural context. Studio engineers know that this is a pipe

dream, since every microphone type and placement will affect the sound recorded, rendering sound recording just as much a framed representation as photography.

2. From Bacon's *New Atlantis*: "We have also sound-houses, where we practice and demonstrate all sounds and their generation. We have harmonies, which you have not, of quarter-sounds and lesser slides of sounds. Divers instruments of music likewise to you unknown, some sweeter than any you have, together with bells and rings that are dainty and sweet. We represent small sounds as great and deep, likewise great sounds extenuate and sharp; we make divers tremblings and warblings of sounds, which in their original are entire. We represent and imitate all articulate sounds and letters, and the voices and notes of beasts and birds. We have certain helps which set to the ear do further the hearing greatly. We also have divers strange and artificial echoes, reflecting the voice many times, and as it were tossing it, and some that give back the voice louder than it came, some shriller and some deeper; yea, some rendering the voice differing in the letters or articulate sound from that they receive. We have also means to convey sounds in trunks and pipes, in strange lines and distances." This paragraph was apparently hung upon one of the walls in the BBC Radiophonic Workshop.

3. Chladni demonstrated his system widely, including to Napoleon and Goethe, both of whom were very impressed (Stockmann 2007).

4. See www.phonozoic.net/n0130.htm and www.phonozoic.net/n0131.htm.

5. The technique of streaming music via cable would later be implemented by General George Owen Squier, who founded the Wire Inc. company in 1922 with the aim of "employ[ing] electric power lines to transmit news programs, music, lectures, general entertainment and advertising directly into private homes" (Vanel 2013: 46). Squier came up with the term "muzak" to describe his invention of piped music aimed to be light background sound for factories, restaurants and supermarkets. The predecessor of Spotify can therefore be found equally in the 1880s and the 1920s, although decades would pass before the technology became perfectly ripe for such innovations.

6. This fantastic story is beyond the scope of this book, but the work of Sterne (2003), Schmidt Horning (2013), and Katz (2010) gives further information. See also the compilation of historical texts on the advent of phonography by Taylor, Katz, and Grajeda (2012).

Chapter 10

1. Christopher Redgate's twenty-first-century oboe is an excellent example of how an instrument can be redesigned and worked even after the standardisation point of the instrument, offering new expressive potential to composers and players: www.21stcenturyoboe.com.

2. Satinder P. Gill (2015) warns against the belief in explicit knowledge, and suggests that Polanyi's tacit knowledge needs to be part of our interface design in the future. True knowledge should be able to deal with uncertainty and ambiguity, but this fits easily to the symbolic requirements of language or technological states. It is here that music trains us and opens up a channel for non-linguistic communication, pre-linguistic, based on intuition, and indeed incorporating equally the *know-what* (explicit knowledge), *know-how* (implicit knowledge), but also, importantly, *know-when* (of rhythm and music).

3 Tom Richards, a researcher at Goldmiths College, London, has recently created a "mini" version of the Oramics Machine and performed with it widely. See http://minioramics.blogspot.co.uk.
4 Miller S. Puckette wrote Max at IRCAM in the 1980s, with the intention of providing a friendly programming environment for composers to work on symbolic processing of note values, as this was before computers were fast enough to deal with real-time sound.
5 Interestingly, and supporting this argument, the GRM tools (audio plugins available for most digital audio workstations) have been very popular with composers working with field recordings and abstract sound, which can be contrasted with the symbolic nature of Max and its roots in written, algorithmic compositional practices.
6 Incidentally, the maximum playing time of a CD (the data it can potentially store) is seventy-four minutes, and the story goes that this was in order to be able to fit Beethoven's Ninth onto it, in its longest recording which lasted seventy-four minutes. According to a popular narrative, Karajan had been consulted by Sony regarding this matter, but the actual facts on this matter are naturally a little bit more complex (Immink 2007).
7 I do not want to go deeply into this distinction of oral and aural culture here, but the former would involve be mouth-to-mouth or gestural ergomimetic processes of transmission, whereas the latter is a musical culture in which transmission is not written in symbols, but often recorded and shared through diverse musical media.
8 Arguably, DJing is one way of interpreting recorded music in context, but this is often done through mixing with other music, and therefore has a different idea of authorship. We could also speculate about the use of sampling and the Stem file format, but here we are stopped in our tracks by copyright law which stalls the creative process.

Chapter 11

1 Joseph Hyde and colleagues at Bath Spa University have for a number of years celebrated visual music through their excellent *Seeing Sound* symposia. It is a biennial event that focuses on audiovisual works and their website contains a collection of interesting productions. See www.seeingsound.co.uk.
2 This software was written as part of a research project, "New Multimedia Tools for Electroacoustic Music Analysis," in electroacoustic music listening, directed by Simon Emmerson and Leigh Landy.
3 "Spectromorphology is concerned with perceiving and thinking regarding spectral energies and shapes in space, their behaviour, their motion and growth processes, and their relative functions in a musical context" (Smalley 1997: 124).
4 See IRiMaS website: https://research.hud.ac.uk/institutes-centres/irimas/.

Chapter 12

1 In that sense, a tree falling in the forest will definitively emit sound.
2 The longer the sample window is, the better frequency resolution you get (more bands), but clearly the time resolution suffers. With shorter windows, more frames are analysed, but at the cost that it results in fewer frequency bands.

3. Other tools include: Essentia 2.0, MIRtool-box, jAudio, jMIR, LibXtract, AuditoryToolbox, SCMIR, librosa, and CLAM Music Annotator.
4. The reason I use the term "extra-audio signals" and not "extra-musical" is that I do think human gestures, relationships, and other such communication in live performance are intrinsically musical.

Chapter 13

1. In reality, things are clearly not as simple as this provocative abstraction portrays. There are hackers who dabble with electronics and code, not understanding fully what they are doing but are obtaining interesting results, and there are musicians who make great music without necessarily knowing much about theory or even playing an instrument.
2. The Latin roots of the word "object" signify that it is something which is thrown in front of us (*ob-* "in front of," *-iacere*, "to throw"); something we are faced with and have to relate to. The German word *Gegenstand* has a similar meaning, namely something that stands against us, facing us. See note 4 in Chapter 4.
3. In an interview with the virtuosic cellist Pablo Casals, George Carlin asked him why he was still practising three hours a day at the age of 93. Casals answered: "I'm beginning to show some improvement." https://masbury.wordpress.com/2009/04/04/casals-at-93-im-beginning-to-show-some-improvement/.
4. Jimi Hendrix was not the first person to experience feedback between his guitar and amplifier, but he was able to conceive of feedback as an ergodynamic quality of the instrument, thus adding dimensions of expressivity to people's understanding of its function.
5. Clearly, if the string breaks we might have a breakdown situation, but there are other cases where a broken instrument becomes an ergodynamic to explore, as we find, for example in Keith Jarret's Köln Concert (see Bin 2018: 225).
6. Indeed, one of the reasons why physical systems, like the acoustic or electric, are often more interesting to play than digital simulation systems, is the exploration of their non-linearity and chaos. This is evident in the playing of so many performers: think Miles or Hendrix. Tom Mudd has studied extensively the potential for imbuing digital systems with simulated non-linear behaviour (Mudd et al. 2015).
7. See also the http://logosfoundation.org website, where Raes's work is described; for example, his collaboration with Aphex Twin on musical drum robots.
8. It warrants a dedicated research project to study how social media and video sharing websites have changed the way we learn musical instruments today.
9. See the discussion on Stiegler in the introduction. Here the ergophor is that of hammering. Likewise, the new "sweep" interaction pattern that we have become so accustomed to on mobile device screens traces its roots to how we flick a page on a printed book (an experience people in the future might become unfamiliar with).
10. This conversational engagement with technology has been explored for decades, for example cybernetician Gordon Pask when developing his Conversation Theory as a result of developing cybernetic systems consisting of electronic technologies and human actors, or the human–machine interaction as a form of conversation in which the relevant elements evolve together. This idea of a dialogue with an object is

expressed in diverse ways in the literature; for example, Newton Armstrong has written about this as "realisational interfaces" (Armstrong 2007).

Chapter 14

1. We could say that we discover things in algorithms, code libraries, hardware limitations, and so on, but these are designed affordances or limitations of the materials we work with, not of nature, to temporarily set up a silicon vs. carbon dichotomy.
2. For example, I have personally started creating something as a piece for live coded multichannel spectral composition (the Threnoscope), which later became an instrument and a compositional framework.
3. It is quite remarkable that we can attribute so many of the key techniques in gestural instruction or conducting, musical notation, and algorithmic music to this same person from the eleventh century.
4. The presentations by participants can be viewed on www.sonicwriting.org/ircam.html.
5. For further information on Angliss's website, see www.sarahangliss.com.

Chapter 15

1. Note that we are talking about music composition here, not the human expression/interpretation of that music. A true human–machine comparison, a musical Turing test, would have to be played by machines only or humans only for a fair trial.
2. For interested readers, see the book *Computers and Creativity* (McCormack and d'Inverno 2012).
3. For early examples of such systems, see, for example, Al Biles's *GenJam* (Biles 1994), George Lewis's *Voyager* system (Lewis 2000), or François Pachet's *Continuator* (Pachet 2003).
4. The project is still supported in a new incarnation as Wotja: https://intermorphic.com/wotja/. For historical information about the Koan project, see this page: https://intermorphic.com/sseyo/koan/.
5. We might note that if you ask a composer or a painter how they made their piece, and why they made it this way, you are not likely to get many answers, as these work processes are implicit, involving tacit knowledge, something that artists rarely rationalise, unless they are writing a textbook, teaching, or writing software. What musicologists and art historians often rely upon is historical data, biographical information, interviews, letters, and compositional sketches and versions of the piece.
6. The experience of time is not important for a machine, so it could listen to Beethoven's Ninth Symphony in a couple of seconds.
7. This is essentially a question of interpretation, and if we are to compare music composed by human and machines, we need the same type of rendering of the music. It is interesting to observe here that the features to which we pay most attention when analysing a player's interpretation of a notated musical work are ones that are not specified so rigidly in the traditional musical score itself: dynamics and timing.
8. Some musicians have already begun using these services, and as an example, pop singer Taryn Southern (www.youtube.com/channel/UCK5YnMYmiU_i6cF8plPlTBQ)

proudly presents her music as "Composed with AI" and references the online cloud-based Amper "artificial intelligence composer, performer and producer" service (www.ampermusic.com).
9   See https://nsynthsuper.withgoogle.com for further information about that project.
10 I should declare my partiality in this research as an investigator in the recently started MIMIC (Musically Intelligent Machines Interacting Creatively) research project, involving researchers from Goldsmiths, Durham, and Sussex Universities.

Chapter 16

1   Indeed, such exclusive focus on sound is most possibly *an effect of* our twentieth-century recording media. Björn Heile, writing about the instrumental theatre of Mauricio Kagel, states that the spatial, embodied, visual and kinetic nature of performance has always been a key element of music: "integral elements of music in most cultures; no other culture conceives of music as disembodied pure sound, and in Western music, too, this conception is relatively new" (Heile 2006: 37).
2   It is possible to map these stages to what I have discussed as material, symbolic, and signal inscriptions of sonic writing.
3   We explored the concept of invention as the metaphor for modern musical practices in Chapter 14.
4   These projects have substantial information and websites, see Howse's project, http://econtact.ca/16_4/howse_gallery.html and at http://1010.co.uk/org/ErdWormOnlinewhitemanual.pdf; Owen Green's Cardboard Cutout: http://owengreen.net/portfolio/cardboard-cutout/; John Bowers's Victorian Synthesizer: www.jmbowers.net/works/victorian.html; and John Richard's Dirty Electronics project www.dirtyelectronics.org.
5   See, for example, https://musiclab.chromeexperiments.com.
6   It is perhaps interesting in this context, considering that Schnell's work is being conducted at IRCAM, that Boulez, the institute's founder and director, was less keen on audience participation, rejecting a commissioned work by David Bedford in 1971 on the grounds that audience participation (they were supposed to play the kazoo) would have the effect that people would not take the work seriously. Or was it perhaps the kazoo itself that had this effect? For a performance of Bedford's work, see www.goo.gl/UoCkpt.

Chapter 17

1   Trained actions such as hit, press, turn, pluck, stroke, shake, flick, etc., but also logical operations such as we see in the evolution of the pin-cylinder through the punchcard and piano rolls to the DAW.
2   The talk can be seen at www.youtube.com/watch?v=UTCeluaYa7Y (accessed 20 February 2018).
3   This does not suggest that the company has anything sinister in mind. On the contrary, their explicit statements of sharing knowledge and democratising musicianship are

admirable. However, looking at companies such as Facebook and Google, we also note their original noble intentions.
4 Practically, the violin has space for only one person playing, and the sound is typically in one stream. Digital instruments are multi-stream and can have multiple interfaces to the same instrument or system.
5 See https://youtu.be/iBRNkz2Hg5U?t=12m48s.
6 See https://blog.landr.com/better-with-lydian/.
7 An example might be how the Google Magenta team has open sourced their TensorFlow machine learning system, now used in the work of music experimentalists all around the world. Another example between cutting-edge investigation and commercial culture would be to point out that there is just one degree of separation between Stockhausen and Madonna, or Laetitia Sonami and Ariana Grande – with Björk and Imogen Heap connecting them, respectively.
8 A word the software industry likes to use, based on Csikszentmihalyi's (1996) psychology of creativity.

Chapter 18

1 We might also ask, which Collins hints at, whether this might be a problem of his age, not *the* age. Who is to judge whether new complexity, post-minimalism, or punk, disco, house, hip-hop, drum'n'bass, and grime are not as interesting and revolutionary as the music of the 1960s? Has the music of Autechre not been as revolutionary as that of Lucier? Impossible to answer, of course, but whilst we might agree with many of Collins's observations, and greatly sympathise with his argument, the natural lens of age could be operative here. One that I am, of course, conditioned by too.
2 Any interview with an old rocker describing the 1960s suffices to demonstrate this fact; here is Ronnie Wood discussing how practically everyone knew each other in the popular music of the 1960s, a "small world . . . of comradery": www.youtube.com/watch?v=VqjEIRIONgo.
3 See Peter Kirn's interview with Eric Wahlforss (CTO, SoundCloud) and Daniel Haver (CEO, Native Instruments) here: www.youtube.com/watch?v=xdLqPlMMc7A.
4 Greek: *syn+histanai*, Latin *com+ponere*, or to set up or place (*histanai/ponere*) something together (*syn/com*). As already mentioned, *systema* (σύστημα) was also the word the ancient Greeks used for musical scales.
5 A good example of how such an -ism can cause unnecessary problems is the example of Nyman's (1974) denotation of some composers as minimalists, a label rejected by most of the composers discussed.

Chapter 19

1 As an example, Schaeffer's work with tape, albeit pioneering, is not the first case of musicians working with tape as a compositional method, suffice to point to the work of Halim El-Dabh in Cairo, who began composing tape music four years earlier than Schaeffer (Young 2007).

2 I am talking about music *production* technologies here and their inclusion in music technology history. It is unquestionable how much effect Apple has had on the *consumption* of music, equally with their iPod/iPhone players and the iTunes store.
3 See https://www.youtube.com/watch?v=vJOhWKL_6FU&.
4 See my blog post "Critical Music Technologies" written after spending a whole morning studying Ofcom policies in secondary music education: www.sonicwriting.org/blog/musictechnology. This outdated division between "music" and "music technology" also has an unfortunate effect on gender distribution within these two "fields." See Born and Devine (2015) to learn more about this problematic.
5 A good example of the changes taking place are the interdisciplinary educational programmes in musical interface design that are being established all over the world, or even "Music Design" as a recent Master's programme has been named at Hochschule Furtwangen University.

Bibliography

Aaron, Sam and Blackwell, Alan (2013), "From sonic Pi to overtone: creative musical experiences with domain-specific and functional languages," in *FARM '13 Proceedings of the First ACM SIGPLAN Workshop on Functional Art, Music, Modeling and Design*: 35–46.

Abbado, Adriano (2018), *Visual Music Masters: Abstract Explorations: History and Contemporary Research*. Milan: Skira.

Acland, Charles C. (Ed.) (2007), *Residual Media*. Minneapolis and London: University of Minnesota Press.

Adorno, Theodor W. (1990), "The Form of the Phonograph Record," translated by Thomas Y. Levin, in *October*, 55 (Winter): 56–61. First published as "Die Form der Schallplatte," in *23: Eine Wiener Musikzeitschrift* 17–19 (15 December 1934).

Adorno, Theodor W. (2006), *Towards a Theory of Musical Reproduction: Notes, a Draft and Two Schemata*. Cambridge: Polity Press.

Aho, Marko (2016), *The Tangible in Music: The Tactile Learning of a Musical Instrument*. London: Taylor and Francis.

Akrich, Madeleine (1992), "The De-scription of Technical Object," in W. Bijker and J. Law (Eds.), *Shaping Technology/Building Society*. Cambridge: MIT Press.

Ames, Charles (1987), "Automated Composition in Retrospect: 1956–1986," in *Leonardo*, 20 (2): 169–185.

Andersen, Christian Ulrik and Pold, Søren Bro (2018), *The Metainterface: The Art of Platforms, Cities, and Clouds*. Cambridge, MA: MIT Press.

Aristoxenus (1902), *The Harmonics of Aristonexus*. Trans. Henry Stewart Macran. Oxford: Clarendon Press.

Armstrong, Newton (2007), *An Enactive Approach to Digital Musical Instrument Design*. VDM Verlag Dr. Mueller e.K.

Attali, Jacques (1985), *Noise: The Political Economy of Music*. Minneapolis: University of Minnesota Press.

St Augustine (2007), *On Order: De Ordine*. South Bend, IN: St Augustine's Press.

Ausoni, Alberto (2009), *Music in Art*. Los Angeles: Getty Publications.

Ayrey, Craig (2005), "Pousseur's Scambi (1957), and the new problematics of the open work," in *Pousseur Seminar*. London: Middlesex University.

Bachelard, Gaston (1984), *The New Scientific Spirit*. Boston: Beacon Press.

Bacon, Francis (1670), *Sylva Sylvarum*. London: Published by J.R. For William Lee.

Bacon, Francis (2000), *Francis Bacon: The New Organon*. Cambridge: Cambridge University Press.

Bacon, Francis ([1905] 2011), *The Philosophical Works of Francis Bacon*. Edited by John M. Robertson. New York: Routledge.

Bader, Rolf (Ed.) (2018), *Springer Handbook of Systematic Musicology*. Berlin and Heidelberg: Springer-Verlag.

Bailer-Jones, Daniela M. (2009), *Scientific Models in Philosophy of Science*. Pittsburgh: University of Pittsburgh Press.

Bailey, Derek (1992), *Improvisation: Its Nature and Practice in Music*. London: British Library National Sound Archive.
Baily, John (1977), "Movement Patterns in Playing the Herati Dutār," in John Blacking (Ed.), *The Anthropology of the Body*. London: Academic Press, 275–330.
Baird, David (2004), *Thing Knowledge: A Philosophy of Scientific Instruments*. Berkeley: University of California Press.
Bannan, Nicholas (2012), *Music, Language, and Human Evolution*. Oxford: Oxford University Press.
Barad, Karen (2003), "Posthumanist Performativity: Toward an Understanding of How Matter Comes to Matter," in *Signs*, 28 (3), *Gender and Science: New Issues* (Spring): 801–831.
Barad, Karen (2007), *Meeting the Universe Halfway: Quantum Physics and the Entanglement of Matter and Meaning*. Durham, NC: Duke University Press.
Barbosa, Álvaro (2003), "Displaced Soundscapes: A Survey of Network Systems for Music and Sonic Art Creation," in *Leonardo Music Journal*, 13: 53–59.
Barker, Andrew (1990), *Greek Musical Writings. Volume 2, Harmonic and Acoustic Theory*. Cambridge: Cambridge University Press.
Barker, Andrew (2007), *The Science of Harmonics in Classical Greece*. Cambridge: Cambridge University Press.
Barney, Stephen (2005), *The Etymologies of Isidore of Seville*. Cambridge: Cambridge University Press.
Barthes, Roland (1977), "The Death of the Author," in *Image-Music-Text*. New York: Hill and Wang, 142–148.
Bartók, Béla and Lord, Albert B. (1951), *Serbo-Croatian Folk Songs*. New York: Columbia University Press.
Bates, Eliot (2012), "The Social Life of Musical Instruments," in *Ethnomusicology*, 56: 363–395.
Beament, James (1997), *The Violin Explained: Components, Mechanism, and Sound*. Oxford: Oxford University Press.
Beck, John and Bishop, Ryan (Eds.) (2016), *Cold War Legacies: Systems, Theory, Aesthetics*. Edinburgh: Edinburgh University Press.
Beer, David (2017), "The Social Power of Algorithms," in *Information, Communication and Society*, 20 (1): 1–13.
Bélis, Annie (2001), "Aulos," in *Grove Music Online*. www.oxfordmusiconline.com/grovemusic/view/10.1093/gmo/9781561592630.001.0001/omo-9781561592630-e-0000001532 [accessed 2 May 2018].
Benjamin, Walter (1969), "The Work of Art in the Age of Mechanical Reproduction," in *Illuminations*. Edited by Hannah Arendt. New York: Schocken Books, 217–251.
Berger, Anna Maria Busse (2005), *Medieval Music and the Art of Memory*. Berkeley: University of California Press.
Bergerac, Cyrano de (1889), *A Voyage to the Moon*. Edited by Curtis Hidden Page, translated by Archibald Lovell. New York: Doubleday and McClure Co. (Originally published in 1657 as "The Other World: Comical History of the States and Empires of the Moon.")
Bergström, John A. (1903), "A New Type of Ergograph, with a Discussion of Ergographic Experimentation," in *The American Journal of Psychology*, 14 (3/4) (July–October): 246–276.
Berlioz, Hector (1948), *Treatise on Instrumentation*. New York: Edwin F. Kalmus.

Bernardini, Nicola and Vidolin, Alvisi (2005), "Sustainable live electro-acoustic music," in *Proceedings of the International Sound and Music Computing Conference*. Salerno, Italy.

Bernstein, Nicolai A. (1996), "On Dexterity and its Development," in M. Latash and M.T. Turvey (Eds.), *Dexterity and its Development*, Mahwah, NJ: Lawrence Erlbaum Associates, 3–244.

Berry, David (2011), *The Philosophy of Software: Code and Mediation in the Digital Age*. Basingstoke: Palgrave Macmillan.

Berry, David (Ed.) (2012), *Understanding Digital Humanities*. London: Palgrave MacMillan.

Berry, David M. and Dieter, Michael (2015), *Postdigital Aesthetics: Art, Computation and Design*. Basingstoke: Palgrave Macmillan.

Bertin-Mahieux, Thierry; Ellis, Daniel P.W.; Whitman, Brian; and Lamere, Paul (2011), "The million song dataset," in *Proceedings of the 12th International Conference on Music Information Retrieval* (ISMIR 2011).

Beyer, Robert T. (1999), *Sounds of Our Times: Two Hundred Years of Acoustics*. New York: Springer Verlag.

Bhagwati, Sandeep; Giordano, Marcello; Berzowska, Joanna; Bachmayr, Alex; Stein, Julian; Browne, Joseph; Del Tredici, Felix; Egloff, Deborah; Sullivan, John; Wanderley, Marcelo; and Cossette, Isabelle (2016), "Musicking the Body Electric. The 'body:suit:score' as a polyvalent score interface for situational scores," in *Proceedings of the TENOR Conference*, Anglia Ruskin University, Cambridge, UK.

Bharucha, Jamshed (1993), "MUSCAT: A Connectionist Model of Musical Harmony," in S. Schwanauer and D. Levitt (Eds.), *Machine Models of Music*. Cambridge, MA: MIT Press, 497–509.

Biles, John A. (1994), "GENJAM: A Genetic Algorithm for Generating Jazz Solos," in *Proceedings of International Computer Music Conference*. San Francisco: ICMA.

Bin, Astrid (2018), *The Show Must Go Wrong: Towards an Understanding of Audience Perception of Error in Digital Musical Instrument Performance*. PhD thesis. Queen Mary University of London.

Bjørn, Kim (2018), *Push, Turn, Move: Interface Design in Electronic Music*. Copenhagen: Push Turn Move.

Blackburn, Manuella (2011), "The Visual Sound-Shapes of Spectromorphology: An Illustrative Guide to Composition," in *Organised Sound*, 16 (1): 5–13.

Boden, Margaret A. (1990), *The Creative Mind: Myths and Mechanisms*. London: Wiedenfield and Nicholson.

Boden, Margaret A. (2013), *Creativity and Art: Three Roads to Surprise*. Oxford: Oxford University Press.

Bogost, Ian (2012), *Alien Phenomenology*. Ann Arbor, MI: Open Humanities Press.

Bolter, Jay David and Grusin, Richard (1999), *Remediation: Understanding New Media*. Cambridge: MIT Press.

Bonner, Anthony (2007), *The Art and Logic of Ramon Llull*. Leiden: Brill Academic Pub.

Born, Georgina (1995), *Rationalizing Culture: IRCAM, Boulez, and the Institutionalization of the Musical Avant-Garde*. Berkeley and Los Angeles: University of Los Angeles Press.

Born, Georgina (2011), "Music and the Materialization of Identities," in *Journal of Material Culture*, 16 (4): 376–388.

Born, Georgina and Devine, Kyle (2015), "Music Technology, Gender, and Class: Digitization, Educational and Social Change in Britain," in *Twentieth-Century Music*, 12 (2): 135–172.

Boulez, Pierre (1976), *Pierre Boulez: Conversations with Celestin Deliege*. London: Eulenburg Books.

Boulez, Pierre (1977), "Technology and the Composer," in *The Times Literary Supplement*, 6 May 1977.

Bourriaud, Nicolas (2002), *Relational Aesthetics*. Paris: Les Presses du réel.

Bovermann, Till; de Campo, Alberto; Egermann, Hauke; Hardjowirogo, Sarah-Indriyati; and Weinzierl, Stefan (Eds.) (2017), *Musical Instruments in the 21st Century*. Singapore: Springer.

Bowers, John and Haas, Annika (2014), "Hybrid Resonant Assemblages: Rethinking Instruments, Touch and Performance in New Interfaces for Musical Expression," in *NIME 2014: New Interfaces for Musical Expression*. London: Goldsmiths, University of London.

Bowker, Geoffrey C. and Star, Susan Leigh (2000), *Sorting Things Out: Classification and its Consequences*. Cambridge: MIT Press.

Bretan, Mason; Oore, Sageev; Engel, Jesse; Eck, Douglas; and Heck, Larry (2017), "Deep Music: Towards Musical Dialogue," in *Proceedings of the Thirty-First AAAI Conference on Artificial Intelligence* (AAAI-17).

Brougher, Kerry and Mattis, Olivia (2005), *Visual Music: Synaesthesia in Art and Music Since 1900*. New York: Thames and Hudson.

Brown, Earle (2002), *Interview*. Interviewed by Cornelius Duffalo and Gregg Bendian on 5 May 2002, http://musicmavericks.publicradio.org/programs/program7.html [accessed 28 October 2017].

Bull, Michael (2000), *Sounding Out the City: Personal Stereos and the Management of Everyday Life*. Oxford: Berg.

Bullock, Jamie and Coccioli, Lamberto (2006), "Modernising Musical Works Involving Yamaha DX-based Synthesis: A Case Study," in *Organised Sound*, 11 (3): 221–227.

Burroughs, William S. (1962), *The Ticket That Exploded*. New York: Grove Press.

Button, Graham and Dourish, Paul (1996), "Technomethodology: Paradoxes and Possibilities," in *Proceedings of the ACM Conference on Human Factors in Computing Systems CHI'96*. New York: ACM, 19–26.

Buxton, William (1977), "A Composer's Introduction to Computer Music," in *Interface 6*: 57–72.

Cadoz, Claude (2009), "Supra-instrumental Interactions and Gestures," in *Journal of New Music Research*, 38 (3): 215–230.

Cage, John (1939), *Silence: Lectures and Writings*. Middletown, CO: Wesleyan University Press.

Cage, John (1969), *Notations*. New York: Something Else Press.

Campbell, Murray; Greated, Clive; and Myers, Arnold. (2004), *Musical Instruments: History, Technology, and Performance of Instruments of Western Music*. Oxford: Oxford University Press.

Cardew, Cornelius (1967), *Treatise*. New York: Gallery Upstairs Press.

Casey, Michael (2009), "Soundspotting: A New Kind of Process?" in *The Oxford Handbook of Computer Music*. Oxford: Oxford University Press, 421–456.

Castells, Manuel (1996), *The Rise of the Network Society: The Information Age: Economy, Society and Culture*. Oxford: Blackwell Publishers.

Chomsky, Noam (1957), *Syntactic Structures*. The Hague: Mouton and Co.

Chion, Michel (1994), *Audio-Vision: Sound on Screen*. New York: Columbia University Press.
Ciani, Suzanne (1976), *Report to the National Endowment of the Arts. RE: Composer Grant*.
Clark, Andy (1996), *Being There: Putting Brain, Body, and World Together Again*. Cambridge, MA: MIT Press.
Clark, Andy (2001), "Reasons, Robots and the Extended Mind (Rationality for the New Millenium)," in *Mind and Language*, 16 (2): 121–145.
Clark, Andy (2016), *Surfing Uncertainty: Prediction, Action, and the Embodied Mind*. New York: Oxford University Press.
Clark, Andy and Chalmers, David (1998), "The Extended Mind," in *Analysis*, 58 (1): 7–19.
Clarke, Michael; Dufeu, Frédéric; and Manning, Peter (2016), "Using Software Emulation to Explore the Creative and Technical Processes in Computer Music: John Chowning's Stria, a Case Study from the TaCEM project," in *Proceedings ICMC 2016: Is the sky the limit?* HKU University of the Arts, 218–223.
Clerc, Miguelángel (2013), "The In(visible) Sound," in *Sound and Score: Essays on Sound, Score and Notation*. Leuven: Orpheus Institute Series, Leuven University Press.
Cohen, Dalia and Katz, Ruth (1960), "Explorations in the Music of the Samaritans: An Illustration of the Utility of Graphic Notation," in *Ethnomusicology*, 4: 67–73.
Cole, Hugo (1974), *Sounds and Signs: Aspects of Musical Notation*. Oxford: Oxford University Press.
Collins, Nicolas (2006), *Handmade Electronic Music: The Art of Hardware Hacking*. New York: Routledge.
Collins, Nicolas (2015), "Quicksand," www.nicolascollins.com/texts/quicksand.pdf [accessed May 2018]
Collins, Nick (2008), "The Analysis of Generative Music Programs," in *Organised Sound*, 13 (3): 237–248.
Collins, Nick (2010), *Introduction to Computer Music*. Chichester: John Wiley.
Collins, Nick (2016), "Towards Machine Musicians who have Listened to more Music than us: Audio Database-led Algorithmic Criticism for Automatic Composition and Live Concert Systems," in *Computers in Entertainment*, 14 (3): 1–14.
Collins, Nick (2016), "A Funny Thing Happened on the Way to the Formula: Algorithmic Composition for Musical Theatre," in *Computer Music Journal*, 40 (3): 41–57.
Collins, Nick (2017), "Corposing a History of Electronic Music," in *Leonardo Music Journal*, 27: 47–48.
Collins, Nick (2018), "Origins of Algorithmic Thinking in Music," in Alex McLean and Roger Dean (Eds.), *The Oxford Handbook of Algorithmic Composition*. New York: Oxford University Press.
Collins, Nick; McLean, Alex; Rohrhuber, Julian; and Ward, Adrian (2003), "Live Coding in Laptop Performance," in *Organised Sound*, 8 (3): 321–330.
Collins, Nick and McLean, Alex (2014), "Algorave: Live Performance of Algorithmic Electronic Dance Music," in *Proceedings of 2014 NIME Conference*. London: Goldsmiths.
Collins, Nick and Sturm, Bob (2011), "Sound Cross-synthesis and Morphing Using Dictionary-based Methods," in *Proceedings of ICMC2011, International Computer Music Conference*, Huddersfield.
Colton, Simon (2008), "Creativity Versus the Perception of Creativity in Computational Systems," in *Proceedings of the AAAI Spring Symposium on Creative Intelligent Systems*.
Cook, Nicholas (2013), *Beyond the Score: Music as Performance*. Oxford: Oxford University Press.

Cook, Perry (2001), "Principles for Designing Computer Music Controllers," in *Proceedings of the NIME Conference*. Also published in: Cook, Perry (2017), "2001: Principles for Designing Computer Music Controllers," in A. Jensenius and M. Lyons (Eds.), *A NIME Reader. Current Research in Systematic Musicology*, Vol. 3. Cham: Springer.

Cope, David (1996), *Experiments in Musical Intelligence*. Madison, WI: A-R Editions.

Corbella, M. and Windisch, A. (2013), "Sound Synthesis, Representation and Narrative Cinema in the Transition to Sound (1926–1935)," in *Cinémas*, 24 (1): 59–81.

Costello, Diamuid (2018), *On Photography: A Philosophical Inquiry*. Oxford: Routledge.

Couprie, Pierre (2016), "EAnalysis: Developing a Sound-based Music Analytical Tool," in Leigh Landy and Simon Emmerson (Eds.), *Expanding the Horizon of Electroacoustic Music Analysis*. Cambridge: Cambridge University Press, 170–194.

Couture, Nadine; Bottecchia, S.; Chaumette, S.; Cecconello, M.; Rekalde, J.; and Desainte-Catherine, M. (2018), "Using the Soundpainting Language to Fly a Swarm of Drones," in Chen (Ed.), *Advances in Human Factors in Robots and Unmanned Systems. AHFE 2017. Advances in Intelligent Systems and Computing*, 595. Cham: Springer.

Covach, John (1997), "We Won't Get Fooled Again: Rock Music and Musical Analysis," in David Schwarz et al. (Eds.), *Keeping Score: Music, Disciplinarity, Culture*. Charlottesville, VA: University of Virginia Press, 75–89.

Cowell, Henry ([1952] 1981), "Current Chronicle," in *Musical Quarterly* (January); reprinted in *John Cage: An Anthology*. Edited by Richard Kostelanetz. New York: Da Capo Press, 97.

Cox, Geoff (2012), *Speaking Code: Coding as Aesthetic and Political Expression*. Cambridge, MA: MIT Press.

Cox, Geoff; McLean, Alex; and Ward, Adrian (2000), "The Aesthetics of Generative Code," in *International Conference on Generative Art*. Milan.

Craenen, Paul (2014), *Composing under the Skin: The Music-making Body at the Composer's Desk*. Leuven: Leuven University Press.

Cranmore, Jeff and Tunks, Jeanne (2015), "Brain Research on the Study of Music and Mathematics: A Meta-Synthesis," in *Journal of Mathematics Education*, 8 (2): 139–157.

Creese, David (2010), *The Monochord in Ancient Greek Harmonic Science*. Cambridge: Cambridge University Press.

Cross, Ian (2001), "Music, Mind and Evolution," in *Psychology of Music*, 29 (1): 95–102.

Csikszentmihalyi, Mihaly (1996), *Creativity: Flow and the Psychology of Discovery and Invention*. New York: Harper Collins.

Cypess, Rebecca (2016), *Curious and Modern Inventions: Instrumental Music as Discovery in Galileo's Italy*. Chicago: The University of Chicago Press.

Dack, John (2005), "The 'open' form – literature and music," in *The "Scambi Symposium."* London: Goldsmiths College.

Dahlback, Karl (1958), *New Methods in Vocal Folk Music Research*. Oslo: Oslo University Press.

Davies, Stephen (2001), *Musical Works and Performances: A Philosophical Exploration*. Oxford: Clarendon Press.

Davis, Tom (2017), "The Feral Cello: A Philosophically Informed Approach to an Actuated Instrument," in *Proceedings of the NIME Conference*. Copenhagen: Aalborg University.

Davis, Whitney (1996), *Replications: Archaeology, Art History, Psychoanalysis*. University Park: Pennsylvania State University Press.

Dawkins, Richard (1976), *The Selfish Gene*. Oxford: Oxford University Press.

Dawkins, Richard (1989), *The Selfish Gene* (2nd edition). Oxford: Oxford University Press.

De Campo, Alberto (2014), "Lose control, gain influence – concepts for metacontrol," in *Proceedings of the International Computer Music Conference*. Athens, Greece.

De Souza, Jonathan (2017), *Music at Hand: Instruments, Bodies, and Cognition*. Oxford: Oxford University Press.

Deleuze, Gilles and Guattari, Felix (1987), *A Thousand Plateaus: Capitalism and Schizophrenia*. London: Continuum.

Derrida, Jacques (1997), *Of Grammatology*. Baltimore and London: Johns Hopkins University Press.

Devlin, L. (2002), "Book of the month: Athanius Kircher – Musurgia Universalis." https://special.lib.gla.ac.uk/exhibns/month/nov2002.html [accessed January 2017].

d'Errico, Francesco; Villa, Paola; Llona, Ana C. Pinto; and Idarraga, Rosa Ruiz (1998), "A Middle Palaeolithic origin of music? Using cave-bear bone accumulations to assess the Divje Babe I bone 'flute'", in *Antiquity*, 72 (275): 65–79.

d'Errico, Francesco; Henshilwood, Christopher; Lawson, Graeme; Vanhaeren, Marian; Tillier, Anne-Marie; Soressi, Marie; Bresson, Frédérique; Maureille, Bruno; Nowell, April; Lakarra, Joseba; Backwell, Lucinda; and Julien, Michèle (2003), "Archaeological Evidence for the Emergence of Language, Symbolism, and Music – An Alternative Multidisciplinary Perspective," in *Journal of World Prehistory*, 17 (1) (March): 1–70.

Dimpker, Christian (2013), *Extended Notation: The Depiction of the Unconventional*. Berlin: Lit Verlag Dr. W. Hopf.

Dolan, Emily I. (2012), "Toward a Musicology of Interfaces," in *Keyboard Perspectives*, 5: 1–13.

Dolan, Emily I. (2015), "Musicology in the Garden," in *Representations*, 132 (1): 88–94.

Doornbusch, Paul (2009), "A Chronology of Computer Music and Related Events," in *The Oxford Handbook of Computer Music*. Oxford: Oxford University Press, 557.

Dreyfus, Laurence (1996), *Bach and the Patterns of Invention*. Cambridge: Harvard University Press.

Dwyer, Terence (1971), *Composing with Tape Recorders: Musique Concrète for Beginners*. London: Oxford University Press.

Eacott, John (2001), *Morpheus: Emergent Music*. London.

Ebcioğlu, Kemal. (1988), "An Expert System for Harmonizing Four-part Chorales," in *Computer Music Journal*, 12 (3): 43–51.

Eck, Cathy van (2017), *Between Air and Electricity: Microphones and Loudspeakers as Musical Instruments*. New York: Bloomsbury.

Eco, Umberto (1989), *The Open Work*. Cambridge, MA: Harvard University Press.

Edison, Thomas (2012), "The Phonograph and its Future," in Timothy D. Taylor, Mark Katz, and Tony Grajeda (Eds.), *Music, Sound, and Technology in America*, Durham and London: Duke University Press.

Eigenfeldt, Arne (2014), "Generative Music for Live Performance: Experiences with Real-time Notation," in *Organised Sound*, 19: 276–285.

Eimert, Herbert (1958), "What is Electronic Music," in *Die Reihe*. Pennsylvania: Theodore Presser Co.

Eldredge, Niles and Gould, Stephen Jay (1972), "Punctuated Equilibria: An Alternative to Phyletic Gradualism," in T.J.M. Schopf (Ed.), *Models in Paleobiology*. San Francisco: Freeman Cooper, 82–115.

Eldridge, Alice; Casey, Michael; Moscoso, Paola; and Peck, Mika (2016), "A New Method for Ecoacoustics? Toward the Extraction and Evaluation of Ecologically-meaningful Soundscape Components Using Sparse Coding Methods," in *PeerJ*, 4: e210.

Eldridge, Alice and Kiefer, Chris (2017), "The self-resonating feedback cello: interfacing gestural and generative processes in improvised performance," in *The Proceedings of 2017 Conference on New Interfaces for Music Expression*. Copenhagen: Aalborg University.

Engel, Jesse; Resnick, Cinjon; Roberts, Adam; Sander, Dieleman; Eck, Douglas; Simonyan, Karen; and Norouzi, Mohammad (2018), "Neural Audio Synthesis of Musical Notes with WaveNet Autoencoders." https://arxiv.org/abs/1704.01279 [accessed May 2018].

Engström, A. and Stjerna, Å. (2009), "Sound Art or Klangkunst? A Reading of the German and English Literature on Sound Art," in *Organised Sound*, 14 (1): 11–18.

Eno, Brian (1983), "The Recording Studio as a Compositional Tool," in *Downbeat*, July: 56–57.

Erickson, Robert (1975), *Sound Structure in Music*. Berkeley and Los Angeles: University of California Press.

Estrada, Ernesto (2011), *The Structure of Complex Networks: Theory and Applications*. Oxford: Oxford University Press.

Feaster, Patrick (2011), "'A Compass of Extraordinary Range': The Forgotten Origins of Phonomanipulation," in *ARSC Journal*, 42 (2) (Fall): 163–203.

Feaster, Patrick (2012), *Pictures of Sound: One Thousand Years of Educed Audio: 980–1980*. Atlanta, GA: Dust-to-digital.

Fels, Sidney (2004), "Designing for intimacy: creating new interfaces for musical expression," in *Proceedings of the IEEE*, 92 (4).

Fels, Sidney and Hinton, Geoffrey E. (1993), "Glove-talk: A neural network interface between a data-glove and a speech synthesizer," in *IEEE Trans. Neural Networks*, 4: 2–8.

Ferrari G. (2006), "Note on the Historical Rotation Seismographs," in R. Teisseyre, E. Majewski, and M. Takeo (Eds.) *Earthquake Source Asymmetry, Structural Media and Rotation Effects*. Berlin and Heidelberg: Springer.

Fiebrink, R. (2011), "Real-time human interaction with supervised learning algorithms for music composition and performance." PhD thesis, Princeton University.

Finberg, Joshua (2000), "Guide to the Basic Concepts and Techniques of Spectral Music," in *Contemporary Music Review*, 19 (2): 81–113.

Fisher, John A. (1998), "Rock'n' Recording: The Ontological Complexity of Rock Music," in Philip Alperson (Ed.), *Musical Worlds: New Directions in the Philosophy of Music*. Pennsylvania: Pennsylvania State University Press.

Fisher, Mark (2014), *Ghosts of my Life: Writings on Depression, Hauntology and Lost Futures*. Winchester and Washington: Zero Books.

Foley, John Miles (1988), *The Theory of Oral Composition: History and Methodology*. Bloomington: Indiana University Press.

Fuller, Matthew (2006), *Softness: Interrogability; General Intellect; Art Methodologies in Software*. Aarhus: Center for Digital Æstetik-forskning.

Foucault, Michel (1977), "What is an Author?" trans. Donald F. Bouchard and Sherry Simon, in Donald F. Bouchard (Ed.), *Language, Counter-Memory, Practice*. Ithaca, New York: Cornell University Press, 124–127.

Foucault, Michel (1984), "What is an Author," in Paul Rabinow (Ed.), *The Foucault Reader*. New York: Pantheon Books, 101–121.

Fowler, C.B. (1967), "The Museum of Music: A History of Mechanical Instruments," in *Music Educators Journal*, 54 (2): 45–49.

François, Jean-Charles (2015), "Improvisation, Orality, and Writing Revisited," in *Perspectives of New Music*, 53 (2): 67–144.

Friedman, Ken and Smith, Owen F. (Eds.) (2006), "Fluxus and Legacy," Special Issue of *Visible Language*, 40 (1).

Fritz, Claudia; Curtin, Joseph; Poitevineau, Jacques; Morrel-Samuels, Palmer; and Tao, Fan-Chia (2012), "Player preferences among new and old violins," in *Proceedings of the National Academy of Sciences*, 109 (3): 760–763.

Gagen, Justin and Cook, Nicholas (2016), "Performing in Second Life," in *The Oxford Handbook of Music and Virtuality*. Oxford: Oxford University Press, 191–210.

Galanter, Philip (2009), "Thoughts on Computational Creativity," in *Proceedings of the Computational Creativity: An Interdisciplinary Approach*. Dagstuhl, Germany: Schloss Dagstuhl – Leibniz-Zentrum fürr Informatik.

Galileo Galilei (1914), *Dialogues Concerning Two New Sciences*. New York: The Macmillan Company.

Galloway, Alex (2012), *The Interface Effect*. Cambridge, UK: Polity Press.

Gauvin, Jean-François (2006), "Artisans, Machines, and Descartes's Organon," in *History of Science*, 44 (2): 187–216.

Gauvin, Jean-François (2011), "Instruments of Knowledge," in Desmond M. Clarke and Catherine Wilson (Eds.), *The Oxford Handbook of Philosophy in Early Modern Europe*. Oxford: Oxford University Press, 315–337.

Geslin, Yann and Lefevre, Adrien (2004), "Sound and musical representation: the Acousmographe software," in *Proceedings of 2004 International Computer Music Conference*. Miami: University of Miami.

Getsy, David J. (2011), *From Diversion to Subversion: Games, Play, and Twentieth-Century Art*. Pennsylvania: The Pennsylvania State University Press.

Godlovitch, Stan (1998), *Musical Performance: A Philosophical Study*. London and New York: Routledge.

Goehr, Lydia (2007), *The Imaginary Museum of Musical Works: An Essay in the Philosophy of Music*. Oxford: Oxford University Press.

Gill, Satinder P. (2015), *Tacit Engagement: Beyond Interaction*. London: Springer.

Golden, Ean (2007), "Music Maneuvers: Discover the Digital Turntablism Concept, 'Controllerism,' Compliments of Moldover," in *Remix Magazine*, October.

Goodman, Nelson (1968), *Languages of Art: An Approach to a Theory of Symbols*. Oxford: Oxford University Press.

Gouk, Penelope (1980), "The Role of Acoustics and Music Theory in the Scientific Work of Robert Hooke," in *Annals of Science*, 37 (5): 573–605.

Gouk, Penelope (1999), *Music, Science, and Natural Magic in Seventeenth-Century England*. New Haven: Yale University Press.

Gracyk, Theodore (1996), *Rhythm and Noise: An Aesthetics of Rock*. Durham: Duke University Press.

Greie-Ripatti, Antye and Bovermann, Till (2017), "Instrumentality in Sonic Wild{er}ness," in T. Bovermann et al. (Eds.), *Musical Instruments in the 21st Century*. Singapore: Springer.

Grierson, Mick and Boon, Tim (2013), "The Oramics Machine: The Lost Legacy of British Electronic and Computer Music," in Frode Weium and Tim Boon (Eds.), *Material Culture and Electronic Sound* (Vol. 8, Artefacts: Studies in the History of Science and Technology). Lanham, MD: Rowman and Littlefield Publishers.

Hadjeres, Gaëtan; Pachet, François; and Nielsen, Frank (2017), "DeepBach: A Steerable Model for Bach Chorales Generation," in *Proceedings of the 34th International Conference on Machine Learning*.

Hagel, Stefan (2009), *Ancient Greek Music: A New Technical History*. Cambridge: Cambridge University Press.

Hajdu, Georg (2007), "Playing Performers," in *Proceedings of the Music in the Global Village Conference*. Budapest, Hungary, 41–42.
Hankins, Thomas L. and Silverman, Robert J. (1995), *Instruments and the Imagination*. Princeton, NJ: Princeton University Press.
Harman, Graham (2002), *Tool-Being: Heidegger and the Metaphysics of Objects*. Peru, IL: Open Court.
Harris, Yolande (2013), "Score as Relationship," in *Sound and Score: Essays on Sound, Score and Notation*. Leuven: Orpheus Institute Series, Leuven University Press.
Harvey, Jonathan (1975), *The Music of Stockhausen: An Introduction*. Berkeley and Los Angeles: University of California Press.
Hayles, N. Katherine (2014), "Speculative Aesthetics and Object-Oriented Inquiry (OOI)," in *Speculations: A Journal of Speculative Realism*, 5.
Heidegger, Martin (1962), *Being and Time*. Oxford: Blackwell Publishers.
Heidegger, Martin (1977), *The Question Concerning Technology and Other Essays*. New York: Harper and Row.
Heile, Björn (2006), *The Music of Mauricio Kagel*. Farnham: Ashgate.
Helmholtz, Hermann (1875), *On the Sensations of Tone: As a Psychological Basis for the Theory of Music*. London: Longmans, Greens, and Co.
Herremans, Dorien; Chuan, Ching-Hua; and Chew, Elaine (2017), "A Functional Taxonomy of Music Generation Systems," in *ACM Comput. Surv.*, 50 (5): Article 69.
Heselwood, Barry (2008), "Features of tablature notation in the current International Phonetic Alphabet chart," in *Leeds Working Papers in Linguistics and Phonetics*, 13.
Heyde, Herbert (2001), "Methods of Organology and Proportions in Brass Wind Instrument Making," in *Historic Brass Society Journal*, 13.
Hill, Donald R. (1974), *The Book of Knowledge of Ingenious Mechanical Devices*. Dordrecht and Boston: D. Reidel Publishing Company.
Hoffmann, Peter (2009), "Music Out of Nothing?: A Rigorous Approach to Algorithmic Composition by Iannis Xenakis." PhD thesis, Berlin: Technischen Universität Berlin.
Hood, Mantle (1971), *The Ethnomusicologist*. New York, NY: McGraw-Hill.
Hope, Cat (2017), "Electronic Scores for Music: The Possibilities of Animated Notation," in *Computer Music Journal*, 41 (3): 21–35.
Hornbostel, Erich von (1933), "The Ethnology of African Sound Instruments," in *Africa 6*, 129: 277–311.
Hornbostel, Erich M. von and Sachs, Curt (1961), "Classification of Musical Instruments," in *The Galpin Society Journal*, 14: 3–29. Originally published in 1914 as "Systematik der Musikinstrumente. Ein Versuch," *Zeitschrift für Ethnologie*, xlvi.
Horner, Andrew and Goldberg, David (1991), "Genetic Algorithms and Computer-Assisted Composition," in *Proceedings of the Fourth International Conference on Genetic Algorithms*. San Diego, CA: Morgan Kaufman.
Horwood, Wally (1983), *Adolphe Sax 1814–1894: His Life and Legacy*. Baldock: Egon Publishers.
Hubert, Philip G. ([1893] 2012), "What the Phonograph Will do for Music and Music-Lovers," in Timothy D. Taylor, Mark Katz, and Tony Grajeda (Eds.), *Music, Sound, and Technology in America*. Durham and London: Duke University Press.
Ihde, Don (1979), *Technics and Praxis*. Dordrecht, Holland: D. Reidel Publishing Company.
Ihde, Don (1990), *Technology and the Lifeworld: From Garden to Earth*. Bloomington: Indiana University Press.

Immink, Kees A. Schouhamer (2007), "Shannon, Beethoven, and the Compact Disc," in *IEEE Information Theory Newsletter*, December.

Impett, Jonathan (2000), "Situating the *Invention* in Interactive Music," in *Organised Sound*, 5 (1): 27–34.

Ingold, Tim (2001), "Beyond Art and Technology: The Anthropology of Skill," in M.B. Schiffer (Ed.), *Anthropological Perspectives on Technology*. Albuquerque: University of New Mexico Press, 17–31.

Isaacson, Walter (2017), *Leonardo Da Vinci: The Biography*. New York: Simon and Schuster.

Jack, Robert H.; Stockman, Tony; and McPherson, Andrew (2017), "Rich gesture, reduced control: the influence of constrained mappings on performance technique," in Kiona Niehaus (Ed.), *Proceedings of the 4th International Conference on Movement Computing (MOCO '17)*. New York: ACM.

Jander, Owen (2001), "Virtuoso," in *Grove Music Online*. www.oxfordmusiconline.com/grovemusic/view/10.1093/gmo/9781561592630.001.0001/omo-9781561592630-e-0000029502 [accessed 28 March 2018].

Jensenius, Alexander Refsum; Wanderley, Marcelo M.; Godøy, Rolf Inge; and Leman, Marc (2009), "Musical Gestures: Concepts and Methods in Research," in Rolf Inge Godøy and Marc Leman (Eds.), *Musical Gestures: Sound, Movement, and Meaning*. Oxon: Routledge, 12–35.

Jeserich, Philipp (2013), *Musica Naturalis: Speculative Music Theory and Poetics, from Saint Augustine to the Late Middle Ages in France*. Baltimore: John Hopkins University Press.

Jin, Zeyu; Mysore, Gautham J.; Diverdi, Stephen; Lu, Jingwan; and Finkelstein, Adam (2017), "VoCo: Text-based Insertion and Replacement in Audio Narration," in *ACM Transactions on Graphics*, 36 (4): Article 96.

Johnston, Andrew; Candy, Linda; and Edmonds, Ernest (2008), "Designing and Evaluating Virtual Musical Instruments: Facilitating Conversational User Interaction," in *Design Studies*, 29 (6): 556–571.

Johnston, Andrew; Candy, Linda; and Edmonds, Ernest (2009), "Designing for conversational interaction," in *Proceedings of New Interfaces for Musical Expression*. Carnegie Mellon University.

Jordà, Sergi (2005), "Digital Lutherie: Crafting musical computers for new musics' performance and improvisation." PhD Thesis, University of Pompeu Fabra, Barcelona.

Kahn, Douglas D. (2014), "Revelation of Hidden Noises: Improbable Musical Instruments and Sound Devices in Early Twentieth Century Literature," in G. Celant (Ed.), *Art or Sound*. Milan: Progetto Prada Arte.

Kania, Andrew (2006), "Making Tracks: The Ontology of Rock Music," in *The Journal of Aesthetics and Art Criticism*, 64 (4): 401–414.

Kartomi, Margaret. J. (1990), *On Concepts and Classifications of Musical Instruments*. Chicago, IL: University of Chicago Press.

Kassler, Jamie C. and Oldroyd, David R. (1983), "Robert Hooke's Trinity College 'Musick Scripts', His Music Theory and the Role of Music in his Cosmology," in *Annals of Science*, 40 (6): 559–595.

Katz, Mark (2010), *Capturing Sound: How Technology Has Changed Music*. Berkeley and Los Angeles: University of California Press.

Katz, Mark (2012), "Introduction," in Timothy D. Taylor et al. (Eds.), *Music, Sound, and Technology in America*. Durham and London: Duke University Press.

Kim-Boyle, David (2014), "Visual Design of Real-Time Screen Scores," in *Organised Sound*, 19: 286–294.

Kim-Boyle, David (2018), "Reframing the Listening Experience through the Projected Score," in *Tempo* 72 (284): 37–50.

Kircher, Athanasius (1650), *Musurgia universalis sive ars magna consoni et dissoni in X. libros digesta*. Rome: Grignandi Corbelletti.

Kirsh, David (2009), "Knowledge, Explicit vs Implicit," in T. Bayne, A. Cleeremans, and P. Wilken (Eds.), *The Oxford Companion to Consciousness*. Cambridge: Oxford University Press, 397–402.

Kirsh, David (2013), "Embodied Cognition and the Magical Future of Interaction Design," in *ACM Transactions on Computer-Human Interaction*, 20 (1): Article 3.

Kirsh, David and Maglio, Paul (1994), "On Distinguishing Epistemic from Pragmatic Actions," in *Cognitive Science*, 18: 513–549.

Kirchmeyer, Helmut (1962), "On the Historical Construction of Rationalistic Music," in *Die Reihe* 8: 11–29.

Kisala, Robert (2006), "Japanese Religions," in P.L. Swanson and C. Chilson (Eds.), *Nanzan Guide to Japanese Religions*. Honolulu: University of Hawaii, 1–13.

Kitano, Naho (2007), "Animism, Rinri, Modernization; the Base of Japanese Robotics," in *2007 IEEE International Conference on Robotics and Automation*. Rome, Italy.

Kittler, Friedrich (1999), *Gramophone, Film, Typewriter*. Stanford: Stanford University Press.

Knobloch, Eberhard (1979), "Musurgia universalis: Unknown Combinatorial Studies in the Age of Baroque Absolutism," in *History of Science*, 17: 258–275.

Knobloch, Eberhard (2005), "Mathematics and the Divine: Athanasius Kircher," in Teun Koetsier and Luc Bergmans (Eds.), *Mathematics and the Divine: A Historical Study*. Amsterdam: Elsevier.

Knobloch, Eberhard (2013), "Renaissance Combinatorics," in R. Wilson and J. Watkins (Eds.), *Combinatorics: Ancient and Modern*. Oxford Scholarship Online.

Koetsier, Teun (2001), "On the Prehistory of Programmable Machines: Musical Automata, Looms, Calculators," in *Mechanism and Machine Theory*, 36: 589–603.

Kotz, Liz (2007), *Words to Be Looked At: Language in 1960s Art*. Cambridge, MA: The MIT Press.

Kubik, Gerhard (1962), "The Phenomenon of Inherent Rhythms in East and Central African Instrumental Music," in *African Music*, 3 (1): 33–42.

Kvifte, Tellef (2007), *Instruments and the Electronic Age*. Oslo: Taragot Sounds.

Kvifte, Tellef (2011), "Musical Instrument User Interfaces: The Digital Background of the Analogue Revolution," in *Proceedings of New Interfaces for Musical Expression*. Oslo.

Landy, Leigh (2009), "Sound-Based Music 4 All," in Roger T. Dean (Ed.), *The Oxford Handbook of Computer Music*. Oxford: Oxford University Press, 518–535.

Lange-Berndt, Petra (2015), *Materiality*. Cambridge, MA: MIT Press.

Latour, Bruno (1987), *Science in Action: How to Follow Scientists and Engineers through Society*. Cambridge: Harvard University Press.

Latour, Bruno (1991), "Technology is Society Made Durable application/pdf icon," in J. Law (Ed.), *A Sociology of Monsters Essays on Power, Technology and Domination, Sociological Review Monograph*, 38: 103–132.

Latour, Bruno (1992), "Where are the Missing Masses? The Sociology of a Few Mundane Artifacts," in W. Bijker and J. Law (Eds.), *Shaping Technology/Building Society: Studies in Sociotechnical Change*. Cambridge, MA: MIT Press.

Latour, Bruno (1999), *Pandora's Hope. Essays on the Reality of Science Studies*. Cambridge, MA: Harvard University Press.

Latour, Bruno (2005), *Reassembling the Social: An Introduction to Actor-Network-Theory*. New York: Oxford University Press.

Law, John (1991), *A Sociology of Monsters: Essays on Power, Technology, and Domination*. New York: Routledge.

Leibniz, Gottfried ([1680] 1951), "Precepts for Advancing the Sciences and Arts," in Philip Wiener (Ed.), *Leibniz Selections*. New York: Scribner's.

Lely, John and Saunders, James (2012), *Word Events: Perspectives of Verbal Notation*. New York: Continuum.

Leman, Marc (2008), *Embodied Music Cognition and Mediated Technology*. Massachusetts: MIT Press.

Leroi-Gourhan, André (1993), *Gesture and Speech*. Cambridge, MA, and London: MIT Press.

Levin, Thomas Y. (2003), "Tones from out of Nowhere: Rudolph Pfenninger and the Archaeology of Synthetic Sound," in *Grey Room*, 12: 32–79.

Lewis, George (1996), "Improvised Music after 1950: Afrological and Eurological Perspectives," in *Black Music Research Journal*, 16 (1): 91–122.

Lewis, George (2000), "Too Many Notes: Computers, Complexity and Culture in Voyager," in *Leonardo Music Journal*, 10: 33–39.

Liley, Thomas (1998), "Invention and Development," in R. Ingham (Ed.). *The Cambridge Companion to the Saxophone*. Cambridge: Cambridge University Press.

Loy, Gareth (2006), *Musimathics: The Mathematical Foundations of Music*. Cambridge, MA: The MIT Press.

MacCurdy, Edward (1938), *Notebooks of Leonardo da Vinci*. London: Jonathan Cape.

Mackenzie, Adrian (2002), *Transductions: Bodies and Machines at Speed*. London: Continuum.

Maes, Laura; Raes, Godfried-Willem; and Roger, Troy (2011), "The Man and Machine Robot Orchestra at Logos," in *Computer Music Journal*, 35 (4): 28–48.

Maes, Pieter-Jan; Leman, Marc; Lesaffre, Micheline; Demey, Michiel; and Moelants, Dirk (2010), "From Expressive Gesture to Sound: The Development of an Embodied Mapping Trajectory Inside a Musical Interface," in *Multimodal User Interfaces*, 3 (67).

Magnusson, Thor (2009), "Of Epistemic Tools: Musical Instruments as Cognitive Extensions," in *Organised Sound*, 14: 168–176.

Magnusson, Thor (2010), "Designing Constraints: Composing and Performing with Digital Musical Systems," in *Computer Music Journal*, 34(4): 62–73.

Magnusson, Thor (2011), "ixi lang: a SuperCollider parasite for live coding," in *Proceedings of the International Computer Music Conference*, University of Huddersfield, UK, 31 July–5 August: 503–506.

Magnusson, Thor (2014), "Herding Cats: Observing Live Coding in the Wild," in *Computer Music Journal*, 38 (1): 8–16.

Magnusson, Thor (2017), "Musical Organics: A Heterarchical Approach to Digital Organology," in *Journal of New Music Research*, 46 (3): 286–303.

Mann, Steve (2007), "Natural Interfaces for Musical Expression: Physiphones and a Physics-based Organology," in *Proceedings of the NIME Conference*. New York: New York University.

Mariani, Angela (2018), *Improvisation and Inventio in the Performance of Medieval Music: A Practical Approach*. Oxford: Oxford University Press.

Mathiesen, Thomas J. (1999), *Apollo's Lyre: Greek Music and Music Theory in Antiquity and the Middle Ages*. Lincoln: University of Nebraska Press.

Matyja, Jakub R. (2015), "The Next Step: Mirror Neurons, Music, and Mechanistic Explanation," in *Frontiers in Psychology*, 14.

Mayer, Leonard B. (1989), *Style and Music: Theory, History, and Ideology*. Chicago: The University of Chicago Press.

Mays, Tom and Faber, Francis (2014), "A Notation System for the Karlax Controller," in *Proceedings of the NIME Conference*. London: Goldsmiths, University of London.

McCormack, Jon and d'Inverno, Mark (2012), *Computers and Creativity*. London: Springer.

McCullough, Malcolm (1996), *Abstracting Craft: The Practiced Digital Hand*. Cambridge: MIT Press.

McLean, Alex (2014), "Making Programming Languages to Dance to: Live Coding with Tidal," in *Proceedings of the 2nd ACM SIGPLAN International Workshop on Functional Art, Music, Modelling and Design*.

McLuhan, Marshall (1964), *Understanding Media*. Cambridge, MA: MIT Press.

McPherson, Andrew (2012), "Techniques and Circuits for Electromagnetic Instrument Actuation," in *Proceedings of New Interfaces for Musical Expression*. Ann Arbor, MI, USA.

McPherson, Andrew; Berdahl, Edgar; Lyons, Michael J.; Jensenius, Alexander Refsum; Bukvic, Ivica Ico; and Knudsen, Arve (2016), "NIMEhub: Toward a repository for sharing and archiving instrument designs," in *Proceedings of the NIME Conference*. Brisbane: Griffith University.

Medeiros, Carolina Brum and Wanderley, Marcelo M. (2014), "A Comprehensive Review of Sensors and Instrumentation Methods in Devices for Musical Expression," in *Sensors*, 14 (8): 13556–13591.

Melley, Timothy (2000), *Empire of Conspiracy: The Culture of Paranoia in Postwar America*. Ithaca: Cornell University.

Merrifield, Andy (2017), *The Amateur: The Pleasures of Doing What You Love*. London: Verso.

Metfessel, Milton (1928), *Phonophotography in Folk Music: American Negro Songs in New Notation*. Chapel Hill: University of North Carolina Press.

Metfessel, Milton (1928), "The Collecting of Folk Songs by Phonophotography," in *Science* 67 (13 January): 28.

Miller, Daniel (2018), "The Future of Music" (Mate Galic, Daniel Haver, Daniel Miller, Sebastian Kuss), DLD 18. www.youtube.com/watch?v=UTCeluaYa7Y.

Miller, Dayton Clarence (1916), *The Science of Musical Sounds*. New York: The Macmillan Company.

Miranda, Eduardo Reck (1993), "Cellular Automata Music: An Interdisciplinary Project," in *Interface/Journal of New Music Research*, 22 (1): 3–21.

Mithen, Steven (1998), "A Creative Explosion? Theory of the Mind, Language and the Disembodied Mind of the Upper Palaeolithic," in Steve Mithen (Ed.), *Creativity in Human Evolution and Prehistory*. New York: Routledge.

Mithen, Steven (2000), "Mind, Brain and Material Culture: An Archaeological Perspective," in P. Carruthers and A. Chamberlain (Eds.), *Evolution and the Human Mind: Modularity, Language and Metacognition*. Cambridge: Cambridge University Press.

Mithen, Stephen (2005), *The Singing Neanderthals: The Origins of Music, Language, Mind and Body*. London: Weidenfeld and Nicholson.

Moholy-Nagy, László (1983), "New Form in Music: Potentialities of the Phonograph," in Krisztina Passuth (Ed.), *Moholy-Nagy*. London: Thames and Hudson.

Moholy-Nagy, László (2017), "Production – Reproduction," in Christoph Cox and Daniel Warner (Eds.), *Audio Culture, Revised Edition: Readings in Modern Music*. London: Bloomsbury.

Molnar-Szakacs, Istvan and Overy, Katie (2006), "Music and Mirror Neurons: From motion to 'e'motion," in *Social Cognitive and Affective Neuroscience*, 1 (3): 235–241.

Mooney, James (2017), "The Hugh Davies Collection: Live Electronic Music and Self-built Electro-acoustic Musical Instruments, 1967–75," in *Science Museum Group Journal*, 7.

Moretti, Franco (2013), *Distant Reading*. London: Verso.

Morton, Timothy (2013), *Realist Magic: Objects, Ontology, Causality*. Ann Arbor: Open Humanities Press.

Moseley, Roger (2016), *Keys to Play: Music as Ludic Medium from Apollo to Nintendo*. Oakland: University of California Press.

Mudd, Tom; Holland, Simon; Mulholland, Paul; and Dalton, Nick (2015), "Investigating the effects of introducing nonlinear dynamical processes into digital musical interfaces," in *Proceedings of Sound and Music Computing Conference*. Sound and Music Computing Network.

Murch, Walter (1995), "Sound Design: The Dancing Shadow," in *Projections 4*. London: Faber and Faber.

Navarro, J.; Sendra, J.J.; and Muñoz, S. (2009), "The Western Latin Church as a Place for Music and Preaching: An Acoustic Assessment," in *Applied Acoustics*, 70: 781e9.

Nettl, Bruno (2015), *The Study of Ethnomusicology: Thirty-Three Discussions*. Urbana: University of Illinois Press.

Newville, Leslie J. (2009), "Development of the Phonograph at Alexander Graham Bell's Volta Laboratory Contributions from the Museum of History and Technology," in *United States National Museum Bulletin*, 218 (5): 69–79.

Niehaus, Gerhard (2009), *Algorithmic Composition: Paradigms of Automated Music Generation*. Vienna: Springer Verlag.

Nilson, Click (2016), *Collected Rewritings: Live Coding Thoughts, 1968–2015*. Burntwood: Verbose.

Nolan, Catherine (2000), "On Musical Space and Combinatorics: Historical and Conceptual Perspectives in Music Theory," in *Bridges: Mathematical Connections in Art, Music, and Science*. Winfield: Kansas, 201–208.

Norman, Donald (1991), "Cognitive Artifacts," in *Designing Interaction*. New York: Cambridge University Press, 17–38.

Nuovo, Victor (2017), *John Locke: The Philosopher as Christian Virtuoso*. Oxford: Oxford University Press.

Nyman, Michael (1974), *Experimental Music: Cage and Beyond*. Cambridge: Cambridge University Press.

O'Callaghan, Casey (2007), *Sounds: A Philosophical Theory*. Oxford: Oxford University Press.

O'Hara, Mick (2015), "Deconstruction and Aesthetics Extract from an Interview with Bernard Stiegler," in *InPrint*, 3 (1). http://arrow.dit.ie/inp/vol3/iss1/5.

O'Modhrain, Sile (2011), "A Framework for the Evaluation of Digital Musical Instruments," in *Computer Music Journal*, 35 (1): 28–42.

Oram, Daphne (1972), *An Individual Note on Music, Sound and Electronics*. London: Galliard Ltd.

O'Riordan, Kate (2017), *Unreal Objects: Digital Materialities, Technoscientific Projects and Political Realities*. Digital Barricades. London: Pluto Press.

Oliveros, Pauline (1984), "Tape Delay Techniques for Electronic Music Composers," in *Software for People: Collected Essays 1962–1981*. Baltimore, MD: Smith Publications.

Olsen, Dale (1980), "Note on a 'Corpophone'," in *Newsletter of the Society for Ethnomusicology*, 4 (5).

Ono, Yoko (1964), *Grapefruit*. Tokyo: Wunternaum Press.
Orning, Tanja (2013), "*Pression* Revised: Anatomy of Sound, Notated Energy, and Performance Practice," in *Sound and Score: Essays on Sound, Score and Notation*. Leuven: Orpheus Institute Series, Leuven University Press.
Oyama, Susan (1985), *The Ontogeny of Information: Development Systems and Evolution*. Cambridge: Cambridge University Press.
Pachet, François (2003), "The Continuator: Musical Interaction with Style," in *Journal of New Music Research*, 32 (3): 333–341.
Paine, Garth (2010), "Towards a Taxonomy of Realtime Interfaces for Electronic Music Performance," in *Proceedings of New Interfaces for Musical Expression*. Sydney: University of Technology.
Paine, Garth (2015), "Interaction as Material: The Techno-somatic Dimension," in *Organised Sound*, 20: 82–89.
Palisca, Claude V. (2006), *Music and Ideas in the Sixteenth and Seventeenth Centuries*. Illinois: University of Illinois Press.
Palisca, Claude V. and Bent, Ian D. (2001), "Theory, theorists," in *Grove Music Online*. www.oxfordmusiconline.com/grovemusic/view/10.1093/gmo/9781561592630.001.0001/omo-9781561592630-e-0000044944?rskey=LdiYz2andresult=1.
Papadopoulos, Alexandre; Roy, Pierre; and Pachet, François (2014), "Avoiding plagiarism in markov sequence generation," in *Proceedings of the Innovative Applications of Artificial Intelligence Conference (AAAI'14)*. Quebec.
Papadopoulos, George and Wiggins, Geraint (1999), "AI methods for algorithmic composition: A survey, a critical view and future prospects," in *Proceedings of the AISB Symposium on Musical Creativity*. Edinburgh, UK, 110–117.
Parch, Harry (1979), *Genesis of a Music: An Account of a Creative Work, Its Roots, and Its Fulfillments*. Second Edition. New York: Da Capo Press.
Parikka, Jussi (2012), *What is Media Archaeology?* Cambridge, UK: Polity Press.
Patel, Aniruddh D. (2008), *Music, Language, and the Brain*. New York: Oxford University Press.
Patterson, Benjamin (2012), *Born in the State of FLUX/us*. Exhibition catalogue. Houston: Contemporary Arts Museum.
Patteson, Thomas (2016), *Instruments for New Music: Sound, Technology, and Modernism*. Oakland: University of California Press.
Peacock, Kenneth (1988), "Instruments to Perform Color-Music: Two Centuries of Technological Experimentation," in *LEONARDO*, 21 (4): 397–406.
Pearce, Celia (2006), "Prologue: Portrait of the Artist as a Young Gamer," in *Visible Language*, 40 (1): 66–89.
Peirce, Charles S. (1868), "On a New List of Categories," in *Proceedings of the American Academy of Arts and Sciences*, 7: 287–298.
Pescatello, Ann M. (1992), *Charles Seeger: A Life in American Music*. Pittsburgh: University of Pittsburgh Press.
Petersen, Sonja (2013), "Craftsmen-Turned-Scientists? The Circulation of Explicit and Working Knowledge in Musical-Instrument Making, 1880–1960," in *Osiris*, 28 (1): 212–231.
Phillips, Tom (1997), *Music in Art: Through the Ages*. Munich: Prestel.
Pickering, Andrew (1995), *The Mangle of Practice: Time, Agency, and Science*. Chicago, IL: University of Chicago Press.
Pinch, Trevor (2007), "Between Technology and Music: Distributed Creativity and Liminal Spaces in the making and selling of synthesizers," in *SHOT Workshop on the Animating Passions of the History of Technology*, 18 October.

Pinch, Trevor and Trocco, Frank (2002), *Analog Days: The Invention and Impact of the Moog Synthesizer*. Cambridge, MA: Harvard University Press.

Pinch, Trevor; Bijker, Wiebe E.; and Hughes, Thomas P. (1987), *The Social Construction of Technological Systems: New Directions in the Sociology and History of Technology*. Cambridge, MA: MIT Press.

Pinch, Trevor and Bijsterveld, Karin (2004), "Sound Studies: New Technologies and Music," in *Social Studies of Science*, 34 (5): 635–648.

Polanyi, Michael (1966), *The Tacit Dimension*. Garden City, NY: Doubleday and Company.

Polgár, Tamás (2008), *Freax Volume 1: The Brief History of the Computer Demoscene*. Winnenden, Germany: CSW Vorlag.

Polimeneas-Liontiris, Thanos (2018), "Low Frequency Feedback Drones: A non-invasive augmentation of the double bass," in *Proceedings of 2018 Conference of New Interfaces for Musical Expression*. Blackburg, VA: Virginia Tech.

Pye, David (1968), *The Nature and Art of Workmanship*. London: Cambium Press.

Pöhlmann, Egert, and West, Martin L. (Eds.) (2001), *Documents of Ancient Greek Music. The Extant Melodies and Fragments Edited and Transcribed with Commentary*. Oxford: Clarendon Press.

Read, Gardner (1979), *Music Notation: A Manual of Modern Practice*. 2nd edn. Miami, FL: Taplinger Publishing Company.

Rebelo, Pedro (2010), "Notating the Unpredictable," in *Contemporary Music Review*, 29 (1): 17–27.

Rehding, Alexander (2016), "Instruments of Music Theory," in *Music Theory Online*, 22. http://mtosmt.org/issues/mto.16.22.4/mto.16.22.4.rehding.html.

Reich, Steve (2002), "Music as a Gradual Process," in *Writings on Music, 1965–2000*. Oxford: Oxford University Press.

Restle, Conny (2008), "Organology: The Study of Musical Instruments in the 17th Century," in J. Lazardzig, L. Schwarte, and H. Schramm (Eds.), *Theatrum Scientiarum – English Edition, Volume 2, Instruments in Art and Science: On the Architectonics of Cultural Boundaries in the 17th Century*. Berlin: De Gruyter, 257–268.

Rheinberger, Hans-Jörg (2016), "Afterword: Instruments as Media, Media as Instruments," in *Studies in History and Philosophy of Biological and Biomedical Science*, 57: 161–162.

Richards, John (2011), "Lead and Schemas," in *Institute of Contemporary Arts: Roland Magazine*, 9: 23–25.

Rietmüller, Albrecht (1994), "The Matter of Music is Sound and Body Motion," in Hans Ulrich Gumbrecht and K. Ludwig Pfeiffer (Eds.), *Materialities of Communication*. Stanford, CA: Stanford University Press, 147–156.

Risset, Jean-Claude (1992), "The Computer as an Interface: Interlacing Instruments and Computer Sounds; Real-time and Delayed Synthesis; Digital Synthesis and Processing; Composition and Performance," in *Journal of New Music Research*, 21 (1): 9–19.

Ritchie, Graeme (2006), "The Transformational Creativity Hypothesis," in *New Generation Computing*, 24: 241. https://doi.org/10.1007/BF03037334.

Roads, Curtis (1996), *The Computer Music Tutorial*. Cambridge, MA: The MIT Press.

Roads, Curtis (2001), *Microsound*. Cambridge, MA: The MIT Press.

Roberts, Adam; Engel, Jesse; Raffel, Colin; Hawthorne, Curtis; and Eck, Douglas (2018), "A Hierarchical Latent Vector Model for Learning Long-Term Structure in Music." https://arxiv.org/abs/1803.05428.

Roberts, Charlie and Kuchera-Morin, JoAnn (2012), "Gibber: Live Coding Audio in the Browser," in *Proceedings of the International Computer Music Conference*: 64–69.

Rohrhuber, Julian; de Campo, Alberto; Wieser, Renate; Van Kampen, Jan-Kees; Echo, Ho; and Hölzl, Hannes (2007), "Purloined Letters and Distributed Persons," in *Music in the Global Village Conference*, Budapest.

Ronell, Avital (1989), *The Telephone Book: Technology, Schizophrenia, Electric Speech*. Lincoln: University of Nebraska.

Rosen, Jody (2008), "Researchers Play Tune Recorded Before Edison," in *The New York Times*, 27 March. www.nytimes.com/2008/03/27/arts/27soun.html.

Sachs, Curt ([1940] 2006), *The History of Musical Instruments*. Mineola, NY: Dover Publications.

Salen, Katie and Zimmerman, Eric (2003), *Rules of Play: Game Design Fundamentals*. Cambridge, MA: MIT Press.

Sallis, Friedmann (2015), *Music Sketches*. Cambridge: Cambridge University Press.

Sauer, Theresa (2009), *Notations 21*. New York: Mark Batty Publisher.

Saunders, Rob (2012), "Towards Autonomous Creative Systems: A Computational Approach," in *Cognitive Computation*, 4: 216. https://doi.org/10.1007/s12559-012-9131-x.

Saussure, Ferdinand de (1986), *Course in General Linguistics*. 3rd edition. Chicago: Open Court Publishing Company.

Schaeffer, John (1996), "Who is La Monte Young?" in W. Duckworth and R. Fleming (Eds.), *Sound and Light La Monte Young and Marian Zazeela*. Lewisburg: Bucknell University Press.

Schaeffer, Pierre (2017), *Treatise on Musical Objects: An Essay Across Disciplines*. Berkeley and Los Angeles: University of California Press.

Schedel, Margaret (2018), "Colour is the Keyboard: Case Studies in Transcoding Visual to Sonic," in *The Oxford Handbook of Algorithmic Music*. Oxford: Oxford University Press.

Schmidt Horning, Susan (2013), *Chasing Sound: Technology, Culture, and the Art of Studio Recording from Edison to the LP*. Baltimore: The John Hopkins University Press.

Schnell, Norbert; Robaszkiewicz, Sébastien; Bevilacqua, Frederic; and Schwarz, Diemo (2015), "Collective Sound Checks – Exploring Intertwined Sonic and Social Affordances of Mobile Web Applications," in *Proceedings of the Ninth International Conference on Tangible, Embedded, and Embodied Interaction*. New York, 685–690.

Schoenherr, Steven E. (1999), "Recording Technology History." www.aes-media.org/historical/html/recording.technology.history/notes.html [accessed March 2018].

Schulenberg, David (2001), *Music of the Baroque*. New York: Oxford University Press.

Schwab, Michael (2013), "Introduction," in Michael Schwab (Ed.), *Experimental Systems: Future Knowledge in Artistic Research*. Leuven: Leuven University Press.

Schwarz, Diemo (2000), "A System for Data-Driven Concatenative Sound Synthesis," in *Digital Audio Effects (DAFx)*. Verona, Italy.

Schwarz, Diemo; Cahen, Roland; and Britton, Roland (2008), "Principles and Applications of Interactive CorpusBased Concatenative Synthesis," in *Journées d'Informatique Musicale (JIM)*. Albi, France.

Scott, Édouard-Léon ([1878] 2010), *The Phonautographic Manuscripts of Édouard-Léon Scott de Martinville*. Edited and translated by Patrick Feaster. FirstSounds.org, published December 2009; Edition 1.1 (March 2010). www.Phonozoic.Net/Fs/Phonautographic-Manuscripts.Pdf [accessed January 2018].

Scripture, Edward Wheeler (1906), *Researches in Experimental Phonetics. The Study of Speech Curves*. Washington, DC: Carnegie Institution of Washington.

Seeger, Charles (1951), "An Instantaneous Music Notator," in *Journal of the International Folk Music Council*, 3: 103–106.

Seeger, Charles (1958), "Prescriptive and Descriptive Music Writing," in *The Musical Quarterly*, 44 (2): 184–195.

Serafin, Stefania; Erkut, Cumhur; Kojs, Juraj; Nilsson, Niels C.; and Nordahl, Rolf (2016), "Virtual Reality Musical Instruments: State of the Art, Design Principles, and Future Directions," in *Computer Music Journal*, 40 (3): 22–40.

Serra, Xavier (2007), "State of the Art and Future Directions in Musical Sound Synthesis," in *International Workshop on Multimedia Signal Processing*.

Simondon, Gilbert (1958), *Du Mode d'Existence des Objets Techniques*. Paris, France: Aubier.

Simondon, Gilbert (2007), "Technical Individualization," trans. Karen Ocana with Brian Massumi, in Joke Brouwer and Arjen Mulder (Eds.), *Interact or Die!* Rotterdam: V2_/NAi, 206–213.

Simondon, Gilbert (2017), *On the Mode of Existence of Technical Objects*. Minneapolis: University of Minnesota Press.

Small, Christopher (1998), *Musicking: The Meanings of Performing and Listening*. Hanover: University Press of New England.

Smalley, Dennis (1997), "Spectromorphology: Explaining Sound-shapes," in *Organised Sound*, 2 (2): 107–126.

Smirnov, Andrey (2013), *Sound in Z: Experiments in Sound and Electronic Music in Early 20th-Century Russia*. London: Koenig Books.

Smith, Sylvia (1984), *Scribing Sound: An Exhibition of Music Notations (1952-1984)*. Hartford, CT: New Music America Festival.

Sonnenfeld, Alexander and Hansen, Kjetil Falkenberg (2016), "S-notation: A complete musical notation system for scratching and sample music derived from 'Theory of Motions,'" in *Proceedings of the TENOR Conference*. Oxford.

Star, Susan Leigh and Griesemer, James R. (1989), "Institutional Ecology, 'Translations' and Boundary Objects: Amateurs and Professionals in Berkeley's Museum of Vertebrate Zoology, 1907-39," in *Social Studies of Science*, 19 (3): 387–420.

Stein, Leon (1979), *Structure and Style: The Study and Analysis of Musical Forms*. Miami: Summy-Birchard Music.

Sterilny, Kim (2004), "Externalism, Epistemic Artefacts and the Extended Mind," in Richard Schantz (Ed.), *The Externalist Challenge*. New York: Walter de Gruyter.

Sterne, Jonathan (2003), *The Audible Past: Cultural Origins of Sound Reproduction*. Durham, NC: Duke University Press.

Sterne, Jonathan (2007), "Media or Instruments? Yes," in *OFFSCREEN*, 11 (8–9).

Sterne, Jonathan (2014), "'What Do We Want?' 'Materiality!' 'When Do We Want It?' 'Now!'" in Tarleton Gillespie, Pablo J. Boczkowski, and Kirsten A. Foot (Eds.), *Media Technologies: Essays on Communication, Materiality, and Society*. Cambridge, MA: MIT Press.

Sterne, Jonathan and Akiyama, Mitchell (2011), "'The Recording that Never Wanted to be Heard' and Other Stories of Sonification," in Trevor Pinch and Karin Bijsterveld (Eds.), *The Oxford Handbook of Sound Studies*. New York: Oxford University Press, 544–560.

Stevens, John; Butterfield, Ardis; and Karp, Theodore (2001), "Troubadours, trouvères," in *Grove Music Online*. www.oxfordmusiconline.com/grovemusic/view/10.1093/gmo/9781561592630.001.0001/omo-9781561592630-e-0000028468 [accessed 5 May 2018].

Stiegler, Bernard (1998), *Technics and Time, 1: The Fault of Epimetheus*. Stanford: Stanford University Press.

Stiegler, Bernard (2010a), *For a New Critique of Political Economy*. Cambridge: Polity Press.

Stiegler, Bernard (2010b), "Memory," in W.J.T. Mitchell and Mark B.N. Hansen (Eds.), *Critical Terms in Media Studies*. Chicago: University of Chicago Press.

Stiegler, Bernard (2018), *The Neganthropocene*. Edited, translated, and with an introduction by Daniel Ross. London: Open Humanities Press.

Stiegler, Bernard and Hughes, Robert (2014), "Programs of the Improbable, Short Circuits of the Unheard-of," in *Diacritics*, 42 (1): 70–108.

Stingelin, Martin (1994), "Comments on a Ball: Nietzsche's Play on the Typewriter," in Hans Ulrich Gumbrecht and K. Ludwig Pfeiffer (Eds.), *Materialities of Communication*. Stanford: Stanford University Press, 70–82.

Stockhausen, Karlheinz (2004), "Electronic and Instrumental Music," in Christoph Cox and Daniel Warner (Eds.), *Audio Culture: Readings in Modern Music*. New York: Continuum Books.

Stockmann, H.-J. (2007), "Chladni meets Napoleon," in *European Physical Journal Special Topics*, 145: 15–23.

Stroppa, Marco (1984), "The Analysis of Electronic Music," in *Contemporary Music Review*, 1: 175–180.

Strunk, W.; Treitler, L.; and McKinnon, J. (1998a), *Source Readings in Music History: Greek Views of Music*. Rev. edn. New York: Norton.

Strunk, W.; Treitler, L.; and McKinnon, J. (1998b), *Source Readings in Music History: The Early Christian Period and the Latin Middle Ages: Early Christian Period and the Latin Middle Age*. Rev. edn. New York: Norton.

Sueur, Jérôme and Farina, Almo (2015), "Ecoacoustics: The Ecological Investigation and Interpretation of Environmental Sound," in *Biosemiotics*, 8 (3): 493–502.

Théberge, Paul (1997), *Any Sound You can Imagine: Making Music/Consuming Technology*. Hanover, NH: Wesleyan University Press.

Théberge, Paul (2017), "Musical Instruments as Assemblage," in T. Bovermann et al. (Eds.), *Musical Instruments in the 21st Century*. Singapore: Springer.

Thompson, Walter (2006), *Soundpainting: The Art of Live Composition*. New York: Author.

Todd, Peter (1989), "A Connectionist Approach to Algorithmic Composition," in *Computer Music Journal*, 13 (4): 27–43.

Toffler, Alvin (1980), *The Third Wave: The Classic Study of Tomorrow*. New York: Bantam.

Tomás, Enrique (2016), "Politics of Musical Interfaces: Ideologies and Digital Disenchantment," in *Proceedings of the First International Conference on Interface Politics*. Barcelona.

Tomás, Enrique (2016), "Musical Instruments as Scores: A Hybrid Approach," in *Proceedings of TENOR 2016, International Conference on Technologies for Music Notation and Representation*. Cambridge.

Tomlinson, Gary (2015), *A Million Years of Music: The Emergence of Human Modernity*. New York: Zone Books.

Toop, Richard (2010), "Against a Theory of Musical (New) Complexity," in Max Paddison and Irène Deliège (Eds.), *Contemporary Music: Theoretical and Philosophical Perspectives*. Farnham: Ashgate, 89–97.

Treitler, Leo (1992), "The 'Unwritten' and 'Written Transmission', of Medieval Chant and the Start-Up of Musical Notation," in *The Journal of Musicology*, 10 (2): 131–191.

Truitt, E.R. (2015), *Medieval Robots: Mechanism, Magic, Nature, and Art*. Philadelphia: University of Pennsylvania Press.

Tzanetakis, George; Kapur, Ajay; Schloss, W. Andrew; and Wright, Matthew (2007), "Computational Ethnomusicology," in *Journal of Interdisciplinary Music Studies*, 1 (2): 1–24.

Ulfarsson, Halldor (2018), "The Halldorophone: The Ongoing Innovation of a Cello-like Drone Instrument," in *Proceedings of New Interfaces for Musical Expression*. Blackburg, VA: Virginia Tech.

Vanel, Herve (2013), *Triple Entendre: Furniture Music, Muzak, Muzak-Plus*. Illinois: University of Illinois Press.

Varela, Francisco; Thompson, Evan; and Rosch, Eleanor (1991), *The Embodied Mind*. Cambridge, MA: MIT Press.

Varelli, Giovanni (2013), "Two Newly Discovered Tenth-Century Organa," in *Early Music History*, 32: 277–315.

Varèse, Edgard and Wen-chung, C. (1966), "The Liberation of Sound," in *Perspectives of New Music*, 5 (1): 11–19.

Vasquez Gomez, Juan; Tahirolu, Koray; and Kildal, Johan (2017), "Idiomatic Composition Practices for New Musical Instruments: Context, Background and Current Applications," in *Proceedings of the international conference on new interfaces for musical expression*.

Velardo, Valerio; Elmsley, Andrew; and Groves, Ryan (2018), "How Music Can Boost User Engagement by 40%. Melodrive White Paper." http://melodrive.com/material/how-music-can-boost-user-engagement-by-40.pdf [accessed May 2018].

Vergo, Peter (2005), *That Divine Order: Music and the Visual Arts from Antiquity to the Eighteenth Century*. London: Phaidon Press.

Verrando, Giovanni and AA, VV. (2014), *New Lutherie: Orchestration, Grammar, Aesthetics*. Milan: Edizioni Suvini Zerboni.

Waghmare, Kalyani C. and Sonkamble, Balwant A. (2017), "Raga Identification Techniques for Classifying Indian Classical Music: A Survey," in *International Journal of Signal Processing Systems*, 5 (4).

Walker, Gabrielle (2003), "The Collector," in *New Scientist*, 179 (2405): 38.

Walters, John L. (1997), "Sound, Code, Image," in *Eye*, 7 (26): 24–33.

Wanderley, Marcelo M. (2000), "Gestural Control of Music." http://recherche.ircam.fr/equipes/analyse-synthese/wanderle/Gestes/Externe/kassel.pdf [accessed June 2018].

Wanderley, Marcelo M.; Vines, Bradley W.; Middleton, Neil; McKay, Cory; and Hatch, Wesley (2007), "The Musical Significance of Clarinetists' Ancillary Gestures: An Exploration of the Field," in *Journal of New Music Research*, 34 (1).

Wang, Ge (2014), "Ocarina: Designing the iPhone's Magic Flute," in *Computer Music Journal*, 38 (2): 8–21.

Wang, Ge (2018), *Artful Design: Technology in Search of the Sublime, A MusiComic Manifesto*. Stanford: Stanford University Press.

Waters, Simon (2007), "Performance Ecosystems: Ecological Approaches to Musical Interaction," in *EMS: Electroacoustic Music Studies Network*: 1–20.

Waters, Simon (2017), "Entanglements with Instruments," in *The Life-Cycle of Musical Instruments*, MIRN keynote address. https://mirnblog.files.wordpress.com/2016/10/programme-summary.pdf

Weisser, Stéphanie and Quanten, Marteen (2011), "Rethinking Musical Instrument Classification: Towards a Modular Approach to the Hornbostel–Sachs System," in *Yearbook for Traditional Music*, 43: 122–146.

Wessel, David and Wright, Matthew (2001), "Problems and Prospects for Intimate Musical Control of Computers," in *Computer Music Journal*, 26 (3): 11–22.

Westgeest, Hans (1986), "Ghiselin Danckerts' 'Ave Maris Stella': The Riddle Canon Solved," in *Tijdschrift van de Vereniging voor Nederlandse Muziekgeschiedenis*, 36: 66–79.

Whitelaw, Mitchell (2004), *Metacreation: Art and Artificial Life*. Cambridge, MA: The MIT Press.

Wiggins, Geraint (2003), "Categorising creative systems," in *Proceedings of the IJCAI'03 Workshop on Creative Systems*.

Wiggins, Geraint (2006), "A Preliminary Framework for Description, Analysis and Comparison of Creative Systems," in *Knowledge-Based Systems*, 19 (7): 449–458.

Wiggins, Geraint (2008), "Computer Models of Musical Creativity: A Review of ComputerModels of Musical Creativity by David Cope," in *Literary and Linguistic Computing*, 23 (1).

Williams, Raymond (1977), *Marxism and Literature*. New York: Oxford University Press.

Wilson, Frank R. (1999), *The Hand: How its Use Shapes the Brain, Language, and Human Culture*. New York: Vintage Books.

Wing, Jeannette M. (2006), "Computational Thinking," in *Communications of the ACM*, 49 (3): 33–35.

Winkler, Peter (1997), "Writing Ghost Notes: The Poetics and Politics of Transcription," in D. Schwartz, A. Kassabian, and L. Siegel (Eds.). *Keeping Score: Music, Disciplinarity, Culture*. Charlottesville, VA: University of Virginia Press.

Winner, Langton (1980), "Do Artifacts Have Politics?" in *Dædalus: Journal of the American Academy of Arts and Sciences*, 109 (1).

Wishart, Trevor (1996), *On Sonic Art*. Amsterdam: Hardwood Academic Publishers.

Wittgenstein, Ludwig (1968), *Philosophical Investigations*. Oxford: Blackwell Publishers.

Wyse, Lonce (2017), "Audio Spectrogram Representations for Processing with Convolutional Neural Networks," in *Proceedings of the First International Workshop on Deep Learning and Music joint with IJCNN*, 1 (1): 37–41.

Young, Rob (2007), "Once Upon at Time in Cairo," in *The Wire*, March (Issue 277).

Zecher, Carla (2007), *Sounding Objects: Musical Instruments, Poetry, and Art in Renaissance France*. Toronto: University of Toronto Press.

Zils, Aymeric and Pachet, François (2001), "Musical Mosaicing," in *Proceedings of the COST G-6 Conference on Digital Audio Effects (DAFX-01)*, Limerick, Ireland, 6–8 December.

Index

Ableton, 41, 44, 226
Abtan, Freida, 236
Adamčiak, Milan, 100–1
Adobe, 162, 204, 224
Adorno, Theodor, 128, 136, 138–9
aesthetics
    of audiovisual, 138–9, 156
    and computational creativity, 114, 205
    of instruments xi, 10, 12, 35 n.1, 45, 59, 209
    and mimesis, 89
    music, 57, 92, 97–9, 105, 107, 113, 142, 154, 158, 168, 183, 201, 230, 232, 236, 241, 243
    of musical score, 101, 187
    of new media, 141
    and process, 103
    of software 51, 224, 229
    speculative, 35 n.1
    of systems, 208
    technolgy 52–3, 67, 104, 177, 181, 225–8
African music, 80, 98, 104, 147
agency
    delegation of, 175
    human, 137, 209
    instrumental, 24, 31, 51, 60, 71, 209
    material, 24, 68
    musical, 23, 182, 189
    technological, 52
Aho, Marko, 17
AIVA (Artificial Intelligence Virtual Artist), 203
Alessandrini, Patricia, 236
Al-Farabi, 22, 76
algorave, 190
Al-Kindi, 76
Alypius, 74
amateur, 137, 146, 204, 210, 239, 241
Amper Music, 203
Andersen, Kristina, 236
Angliss, Sarah, 191, 191 n.5

ANT (actor-network theory), 8, 51, 56, 133, 173
Aphex Twin, 172 n.7
Apollinaire, Guillaume, 131
Apollo, 2, 20–1, 25
app (mobile phone), 12, 41, 43, 86, 91, 107, 175, 179, 182, 191, 199, 207 213–15, 223–4, 240
Apple, 214–15, 226, 239, 239 n.2
Arabic, 22, 25, 67–8, 76, 78, 89, 91 n.4, 97–8, 104–5, 111, 114, 196
archaeology, 9, 172, 235
    media, 2, 141
Arduino, 38
Aristotle, 9, 18, 21, 21 n.2, 48–9, 73–4, 147, 237
Aristoxenus, 73–4, 76
Armstrong, Newton, 175 n.10
ars memorativa, 80
artificial
    intelligence (AI), 1, 7, 24, 34, 65, 96, 107, 160, 163, 177, 200, 203–4, 204 n.8, 216, 225–9, 231–4, 243, 239
    life (A-Life), 104 n.3, 115, 198, 200, 208
Ashanti, Onyx, 235
Athena, 20–1
Attaignant, Pierre, 86
Attali, Jacques, 210, 210 n.2, 231
audio
    games, 183, 213, 223
    programming languages, 56, 183, 229
audiovisual, 100, 138, 140, 147, 147 n.1, 156, 180, 189, 191–2, 207–8, 211–12, 215, 233, 241
augmented reality, 187, 215–16 (*see also* virtual reality)
Augustine of Hippo, St, 26, 74–5, 244
aulos, 20–1, 25, 47, 73
Autechre, 232 n.1, 236
authorship, 93, 100, 146 n.8, 179, 181, 188–9, 202
automata, 1, 7, 24, 30, 97, 104 n.3, 109, 112, 114, 143, 182, 196–8, 211, 226

autonomy, x, 52, 93, 98, 114, 236
avant-garde, ix, 13, 140, 195–6
Avraamov, Arseny, 140

Babbage, Charles, 114, 197
Bach, Johann Sebastian, 2, 25, 91, 99, 112–13, 118, 151, 182–3, 197, 211
Bachelard, Gaston, 62
Bacon, Francis, 47–9, 126, 126 n.2, 174
bagpipe, 18, 25
Bailey, Derek, 170
Baird, Davis, 8, 9, 48, 53–5, 57, 167
Banū Mūsā, 97
Barad, Karen, 55, 135, 237
Barbara, Joan Le, 60
Barber, Llorenç, 236
Bargeld, Blixa, 60
Baroque, 50, 91–2, 182, 200, 229
Barrett, Natasha, 155
Bartók, Béla, 113, 135, 146
Bauhaus, 97, 137
Beatles, The, 203
Beck, 219
Beckett, Samuel, xi, 104, 148
Bedford, David, 214 n.6
Beer, David, 163
Beethoven, Ludwig van, 25, 97, 113, 144, 144 n.6, 202 n.6, 229
Bela, 38
Bell, Alexander Graham, 130–3, 138, 148
Bell Labs, 114, 149, 226, 239
Bell, Renick, 236
Bent Leatherband, 211
Berger, Anna Maria Busse, 80–1
Bergerac, Cyrano de, 131–2, 206
Berio, Luciano, 106
Berlioz, Hector, 28–9, 39
Bernardini, Nicola, 109
Berry, David, 118, 208, 234
Betts, Tom, 199, 212–13
Bevilacqua, Frédéric, 191
Biles, Al, 198 n.3
Bingen, Hildegard von, 78, 231
Björk, 38, 199, 211, 229 n.7, 214–15, 219, 231
black box, 35, 51, 59, 63, 199
Blackburn, Manuella, 155
blues, 136, 151, 163, 172, 195, 229
Boden, Margaret, 196, 198

body, 225
  anatomy, 55, 218
  cyborg, 235–6
  extended, 38, 48, 51, 60, 140, 170, 244
  as instrument, 3, 20, 45, 235
  instrumental, 5, 17, 29 n.5, 34, 42, 57, 59, 66, 126
  of knowledge, 76
  language, 180
  and mind, 8, 22, 171, 225, 244
  performer, 6, 21, 171
  prosthetics, 31, 65, 72
  schema, 11, 170
  scientific, xii
  suit, 191
  technics, 4, 29
  of work, 36
Boethius, Anicius Manlius Severinus, 22–5, 75–6, 225
Bogost, Ian, 8, 34–5
Bolter, Jay David, 10, 133, 169, 215
Borgeat, Patrick, 199
Born, Georgina, 62 n.2, 241 n.4
bottleneck concept, 173
Boulez, Pierre, 66, 100, 106, 142, 214 n.6, 240
boundary objects, 45, 65
Bowers, John, 170, 209, 212, 212 n.4
Bowker, Geoffrey, 51–2
brain interface, 17, 186, 232
Brecht, George, 103
Bronze player, 199, 208, 228
Brown, Earle, 100
Buchla synthesizer, 118–19
Bull, Michael, 218
Burroughs, William, 100, 104
Bussotti, Sylvano, 99
Butcher, John, 60
Butler, Ken, 209

Cage, John, 17, 98, 100–1, 103, 105–6, 112, 140, 195, 207 n.1, 216, 232
canon, 22, 67–8, 84, 91, 98, 144, 155, 197, 228
Cardenas, Alexandra, 190
Cardew, Cornelius, 98, 100–1, 188
Cardiff, Janet, 217
Casey, Michael, 161–2
Cassiodorus, Flavius Magnus Aurelius, 75–6
categorising, 5, 18, 51, 233
Cavatorta, Andy, 211

CD (compact disk), 8, 133, 143, 144 n.6, 145–6, 180, 198–200, 207, 209, 210, 223, 231, 233–5
Cecilia, St, 25–6, 47
Chalmers, David, 4 n.2
Chen, Audrey, 60
CHI (Human Factors in Computing Systems conference), 30, 181
Chicau, Joana, 236
Chion, Michael, 158–9
Chladni, Ernst, 127–9, 127 n.3
Chomsky, Noam, 114
Chowning, John, 155
ChucK, 190
Church, 20–2, 24–5, 36, 44, 48, 78, 87, 236
Ciani, Susanne, 118–19
Cicero, Marcus Tullius, 75, 91, 183
clarinet, 19, 28, 39, 73, 135, 172
Clark, Andy, 4 n.2
Clarke, Michael, 155
Classical music, 50, 57, 92, 113, 191
Clerc, Miguelángel, 105
Coldcut, 212
Coldplay, 38
Collins, Nick, 110 n.2, 114–15, 153, 159, 189–90, 196, 201–2, 204
Collins, Nicolas, 232, 232 n.1
Colton, Simon, 198
combinatorics, 13, 50, 54, 91, 91 n.4, 110–15, 118, 183, 196–7, 206
Componium, 113–14
computational creativity, 107, 109–10, 114, 157, 179, 195–8, 201, 203–4, 226, 232
computer games, 45, 103, 168, 188, 212, 241
Conrad, Tony, 30, 209
constructivism, technological, xi
controllerism, 44
controllers, 19, 37, 39 n.4, 40–1, 44–6, 131, 161, 175, 181, 207–8, 212
conversational interfaces, 52, 175, 175 n.10, 206
Cook, Perry, 36–7, 40, 181, 234
Cope, David, 197
corposition, 202
Couprie, Pierre, 154–5
Couture, Nadine, 188
Cowell, Henri, 103
Crocker, Richard, 72
Cross, Ian, 62 n.2, 72
Crumb, George, 84 n.1

CSound, 67, 114
cybernetics, 62, 91, 102, 116, 161, 170, 175 n.10, 207–8, 226, 235–6
Cypess, Rebecca, 25, 91–2

dance, 1–2, 17, 19, 37, 44, 93, 101, 103, 147, 161, 171, 189–91, 216, 229
Danckerts, Ghiselin, 84–5
Dante Alighieri, 148
Dart, Thurston, 98
database, 160–2, 177, 202, 205–6, 224, 228–30
Davies, Angharad, 60
Davies, Stephen, 145
Davis, Hugh, 36
Davis, Miles, 60, 145, 170 n.6
Davis, Tom, 43, 236
Davis, Whitney, 5
Davorin-Jagodić, Martin, 100
Darwin, Charles, 1
DAW (Digital Audio Workstation), 33–4, 145, 223 n.1, 228
Dawkins, Richard, 61
death metal, 200
Debussy, Claude, 39, 147, 196
De Campo, Alberto, 224
deep learning, 1, 30, 46, 110, 157, 159, 162, 168, 200, 203–6, 208, 226, 230, 233, 242
DeepMind, 204–5, 226
d'Errico, Francesco, 3
D'Escrivan, Julio, 188
DeLanda, Manuel, 237
Deleuze, Gilles, xiv, 9, 52, 99, 117, 243
Derrida, Jacques, 4, 72, 243
Descartes, René, 90
determinism, 62
  technological, xi
dexterity, 171
didgeridoo, 151
digital humanities, 65, 157, 162, 202, 234
Dionysus, 20–1, 25
Dirty Electronics Ensemble, 209, 212, 212 n.4
disco music, 232 n.1
distant listening, 202, 234
DIWO (Do It With Others), 65, 211
DIY (Do It Yourself), 65, 118, 146 n.8, 168, 181, 211, 214, 229, 234–5
DJ, 141, 225, 227, 241

DMI (Digital Musical Instrument), 33, 39, 44
Dolan, Emily I., 8, 12, 62 n.2
Donnarumma, Marco, 236–7
Dörner, Axel, 60
drum'n'bass, 232 n.1
Duchamp, Marcel, 103, 195, 216
Dunning, Graham, 211
Dury, Remi, 39, 41

Eacott, John, 198–9
Ebcioğlu, Kemal, 197
Eck, Cathy van, 7, 236
Eco, Umberto, 92, 98–9, 231
Edison, Thomas, 130–3, 148
Eimert, Herbert, 111
Einarsson, Einar Torfi, 191
Einstein, Albert, 13
Einstürzende Neubauten, 30
El-Dabh, Halim, 239 n.1
Eldredge, Niles, 61, 61 n.1, 65
Eldridge, Alice, 42–3, 158, 174–5
electroacoustic music, 67, 109, 146, 151,
    153–5, 154 n.2, 201–2, 236
electronica, 67, 98, 199, 214, 240
Elektronische Musik, 142
Emmerson, Simon, 154 n.2
enactivism, 11, 55, 65, 168–9
enframing (Gestell), 230
Enlightenment, 19, 22, 28, 126
Eno, Brian, 62, 104, 104 n.3, 117, 146, 198–9,
    209, 214–15
epiphilogenesis, 10, 62
episteme, xii, 6, 6–7, 9, 13, 24, 30, 47, 55, 73–4,
    86 n.2, 142, 185, 227
epistemic, 13, 46, 57, 224–5
    artefacts, 4, 49, 56 n.3
    dimension, 242
    gap, 131
    model, 167
    nature, 65, 135, 163, 167
    objects, 68
    shift, 179
    signifiers, 155
    tools, xii, 4, 9, 13, 28, 31, 47–8, 53, 56,
        56 n.3, 68, 90, 111, 167, 175, 180, 225
Epplay, Vincent, 198
Erek, Cevdet, 217
ergodynamics, 10–12, 33, 35, 42–6, 59, 60, 63,
    68, 72, 102, 107, 118–19, 137, 157,
    167–70, 169 n.4, 170 n.5, 172–4, 177,
    181, 183–5, 192, 208, 211, 215, 218,
    227, 229, 239
ergography, 11–12
ergomimesis, 10–12, 36, 43, 46, 133, 144 n.7,
    172, 175–6, 208, 211, 223, 225, 235
ergonomics, 10, 25, 45, 55–6, 59, 62–3, 171,
    184, 209
ergophor, 11–12, 62, 104, 168, 172 n.9, 208, 223
ergotype, 12, 12 n.5, 167
Erickson, Robert, 7, 30
ethnography, 33, 150
ethnomethodology, 33, 53
ethnomusicology, 8, 19, 95, 98, 104, 123–4,
    135, 148–51, 161
evolution
    cultural, 241
    human, 3, 5, 9, 71
    of ideas, 9
    of instruments, 9, 13, 19, 25, 29, 55, 60–2,
        95, 109
    musical, 72, 239
    natural, 13
    of new media, 169, 210
    of notation, 13
    of objects, 12
    science of, ix
    technical, 66, 223 n.1, 226
expressionism, 95
extempore, 91, 99, 190
extended mind, 4 n.2, 92, 225
externalisation, 4–5, 53

Faber, Francis, 39–40, 40 n.6
Falkenstein, Jan T.v., 214
Farmers Manual, 236
Feaster, Patrick, 129, 140
Feldman, Morton, 100
Ferneyhough, Brian, 106
Ferrari, Luc, 140
FFT (Fast Fourier Transform), 46, 149, 152–5,
    157, 159, 204
fiddle, 12 n.5
Filmstro, 203
Fischinger, Oscar, 100, 102, 138, 140, 147, 188
Fisher, Mark, 206, 231, 233
Flamenco, 105
Flash, Grandmaster, 141
Flow Machines, 203

flute, 1, 3–6, 10, 17, 17 n.1, 20–1, 25, 28, 43, 47, 57, 59, 106, 163, 181
Fluxus, 102–3, 186, 232
folk, 25, 48, 98, 105, 144, 148–9, 151, 182, 202
Foucault, Michel, 24, 99, 182–3
Fox, Larnie, 236
Fripp, Robert, 62
frippertronics, 62
fugue, 80, 163, 195
Futurist, 14, 131, 185, 240
   manifesto, ix

Galanter, Philip, 114–15, 117
Galas, Diamanda, 60
Galileo, 18, 48, 74, 125–6
gamelan, 2, 200
gameplay, 11, 168, 175, 192, 212–13
Gauvin, Jean-François, 49, 90
genealogy, xi, xii, 6, 110, 142, 167, 225, 239
genetic algorithms, 197, 208
Gershwin, George, 196
gestalt, 240
Gestalt theory, 184
Gibber, 190
Gibson, William, 131, 215
Gille, Bertrand, 62
Gilmore, Joe, 199
Goehr, Lydia, 89–90, 92–3, 96–7, 96 n.1, 107, 120
Goodman, Nelson, 98, 106
Gouk, Penelope, 49–50, 126
Gould, Glenn, 145
grammatisation, 3–5, 9, 65, 68, 83, 92, 93
grammatology, 4
Grande, Ariana, 38, 229 n.7
graphophone, 132
Grateful Dead, 144
Greek
   ancient, xii, 2, 6, 19–25, 31, 50, 73, 76–7, 89, 97, 180, 235 n.4
   mousike, xii, 2
   music, 50, 74–5, 89, 142
   mythology, 2, 19–20, 31, 76
Green, Owen, 212, 212 n.4
Gregory the Great, 22
Grierson, Mick, 140
grime, 232 n.1
Grisey, Gérard, 30, 153–4

GRM (Groupe de Recherches Musicales), 141–2, 142 n.5, 153, 226
Grusin, Richard, 10, 133, 169, 215
Guarneri, 29, 29 n.5
Guattari, Felix, xiv, 52, 99
Guðnadóttir, Hildur, 60
Guido of Arezzo, 78–80, 96, 100, 110–11, 188, 197, 252 n.3
Guidonian Hand, 188
guitar, 10, 17–19, 30, 39, 41–2, 79, 89, 125, 162, 169 n.4, 172, 176, 184, 188, 236
Gutenberg press, 73, 86–7
Gysin, Brian, 104

haecceity, 60, 238
halldorophone, 42, 61, 174, 208
Hancock, Herbie, 33
Handel, George Frideric, 97, 229
Hankins, Thomas, 48
Hansson, Úlfur, 211
hardware, 38, 44, 50, 54, 59–60, 63, 67, 114, 120, 180, 181 n.1, 186, 210–11, 232, 238
harp, 1, 18, 20–1, 25, 28, 48, 57, 72, 74, 79, 84, 211
   aeolian, 196
Harris, Yolande, 94, 98, 181, 193
Harrison, Michael, 181
Harvey, Jonathan, 30, 106, 154
Haubenstock-Ramati, Roman, 100
Haver, Daniel, 227, 233 n.3
Haydn, Joseph, 113, 197, 229
Hayles, N. Katherine, 35 n.1
Hays, Sorrel, 100
Hawtin, Richie, 228
HCI (human–computer interaction), 8, 12, 36, 161 n.4, 168, 170, 172, 223, 233
Heap, Imogen, 38, 229 n.7
Heidegger, Martin, 6, 8, 54, 90, 170, 172, 209, 230
Heile, Björn, 207 n.1
Helmholtz, Hermann von, 56, 129–30, 136, 152
Hendrix, Jimi, 60, 169 n.4, 170 n.6, 208
Henke, Robert, 217, 224
Henry, Pierre, 140, 154
Heraclitus, 9
hermeneutics, 48, 51, 55, 148
Herndon, Holly, 236
Herron, Molly, 211
Hiller, Lejaren, 114, 197–8, 232

Index

hip-hop, 172, 232 n.1
Holzer, Derek, 211
Honegger, Arthur, 97
Hood, Mantle, 123, 135, 149–50, 153
Hooke, Robert, 126
Hope, Cat, 99, 188, 236
Hornbostel, Eric von, 17, 17 n.1, 63, 64 n.3
Howse, Martin, 212, 212 n.4
Hubert, Philip G, 218
Hucbald of St Amand, 77
Humtap, 203
Husserl, Edmund, 142
Hutchins, Carleen, 29

Ibn Rushd (Averroes), 76
ICMC (International Computer Music Conference), 181
identity, 5, 39, 95, 174, 183
Ihde, Don, 8, 48, 51, 57, 65, 169
Ikeda, Rioji, 217
Impett, Jonathan, 183, 211
impressionism, 95
improvisation, 1, 13, 71, 78, 91–2, 91 n.5, 104–7, 115, 118–19, 129, 144–5, 159, 180–2, 188, 192, 236
Indian music, 63, 78, 98, 104–5, 114, 159
individuation, 39, 60, 81, 139, 144, 173, 208
  (*see also* transidividuation)
interface
  affordances, 11, 43, 173–5
  brain, 17, 232
  critique, 176
  graphical, 189
  instrumental, 35, 44
  musical, 11, 34, 36–7, 39, 45, 59, 64, 109, 184–6, 191, 209, 226
  physical, 53, 55, 63, 135 n.2, 240, 242 n.5
Ingold, Tim, 171
innovation, 11, 13, 21, 28, 34, 38–9, 41, 45, 60–2, 72, 81, 83, 102, 107, 133, 133 n.6, 137, 154, 174, 179–80, 182, 186, 199–200, 203, 231–2, 243
inscription, 1–2, 4–6, 13, 51, 107, 147–8, 155, 157, 167, 179, 210 n.2
installations, 14, 62, 67, 91, 102, 191, 207–8, 211, 216–18, 232, 234, 236, 240
instruments (*see also individual instruments*)
  acoustic, 6–7, 19, 30, 34–6, 39, 42, 44–6, 50, 55, 57, 59–60, 62, 65–6, 109, 146, 167–8, 173, 176, 184–5, 227, 231, 240, 242
  conceptual, 17
  digital, xi, 7, 10, 13–14, 30, 33, 35–9, 44, 50–1, 53, 55–7, 59–61, 63–5, 68, 79, 86 n.2, 95, 118, 168, 175–6, 180–1, 184–6, 209, 228 n.4, 238, 240, 242
  economic, 18
  electronic, ix, 7, 30, 34–6, 55–6, 63, 184–5
  etymology of, 18
  mechanical, x, 10, 139, 211
  musicology of, 12
  scientific, 18, 28, 47–8, 62, 68, 89–90, 124–5, 128–9
Internet of Things (IoT), 168, 208, 242
invention, 5, 10, 13, 25, 41, 55, 62, 73, 78–9, 91, 95, 104 n.3, 115, 118, 128–33, 132 n.5, 136, 141, 147, 149, 182–3, 185, 200, 203, 206, 209, 211–12, 211 n.3, 218, 223, 225, 228–30, 234–6, 239, 242–4
IRCAM, 65, 67, 105, 142, 142 n.4, 153, 191, 214, 214 n.6, 226
Irish folk, 105, 135, 151
Isaacson, Leonard, 114, 197–8, 232
Ishiyanagi, Toshi, 102
Isidore of Seville, 75, 84, 91, 123, 244
Iwai, Toshio, 212
ixi lang, 190

Jarre, Jean-Michel, 225–6
Jarret, Keith, 170 n.5
Javanese music, 105
jazz, 8, 98, 104–5, 136, 143–5, 151, 240
Jóhannsson, Hans, 29
John, Elton, 33
Joplin, Scott, 196
Jordà, Sergi, 56 n.2, 236
Jukedeck, 203
Justka, Ewa, 209, 211

Kahn, Douglas, 17, 131, 218
Kandinsky, Wassily, 147
Kania, Andrew, 143, 145
Kapur, Ajay, 211
Karlax, 39–41, 40 n.6
Kenchington, Sarah, 211
keyboard
  musical, 29–30, 79, 86, 89, 100, 131, 140, 236
  type, 7, 57

Kiefer, Chris, 42–3, 174–5
Kim-Boyle, David, 188
Kim-Cohen, Seth, 218
Kinect, 38, 188
Kircher, Athanasius, 25, 27, 112–13
Kirn, Peter, 236
kithara, 20, 22, 22 n.3, 71 n.4
Kitundu, Walter, 211
Klee, Paul, 147
Knobloch, Eberhard, 50, 112–13
Knotts, Shelly, 190
Knowles, Allison, 101
Kraftwerk, 98
Kreidler, Johannes, 236
Kubish, Christina, 216
Kupka, Frank, 147
Kurenniemi, Erkki, 57
Kurzweil, Rey, 203
Kvifte, Tellef, 7, 30, 63 n.3
Kyma, 38, 185

Labelle, Brandon, 218
laboratory, 43, 47, 49, 62, 90, 119, 132, 167, 172, 176, 203, 210, 217, 226, 235, 242–3
Lachenmann, Helmut, 66, 99, 106, 209
Landy, Leigh, 30, 154 n.2
Latour, Bruno, 8, 51–2, 56, 133, 208
Lee, Ray, 236
Legardon, Ainara, 236
Leibniz, Gottfried, 91 n.4, 111–14
Leitner, Bernhard, 216
Leman, Marc, 36, 170
Leroi-Gourhan, André, 4, 5
Levin, Thomas Y., 138
Lewis, George, 104–5, 144, 198 n.3, 236
Ligeti, György Sándor, 153, 236
Linn, Roger, 40–1
LinnStrument, 39–41
Listenius, Nicolaus, 90, 93
Liszt, Franz, 241
Liu, Anna, 236
live coding, xiv, 7, 9 n.3, 17, 57, 67, 99, 107, 117, 139, 168, 180, 182–3, 188–90, 223, 232, 236
Llull, Ramon, 91 n.4, 111–13
logos, 5, 21, 47, 171
Loughridge, Deirdre, 17
Lovelace, Ada, 114, 197
Lucier, Alvin, 17, 128, 216, 232, 232 n.1
lute, 25, 47, 59, 79, 86, 89, 181
luthier, 29, 44, 55–6, 56 n.2, 60, 66, 169, 181, 242
Lyotard, Jean-François, 243
lyre, 3, 20–1, 25, 72, 84, 163

Macero, Teo, 145
machine learning, 30, 60–1, 64, 159–60, 168, 176–7, 202–5, 210, 224, 226, 228–30, 229 n.7, 238
machine listening, 14, 110, 124, 154–5, 157–9, 162–3, 168, 175, 179, 189, 202–3, 205, 229, 233
Mackenzie, Adrian, 10, 34
McLaren, Norman, 102, 140, 147, 188
McLaughlin, Scott, 236
McLean, Alex, 115, 190, 236
McLuhan, Marshall, 133, 169, 210–11, 229
McPherson, Andrew, 39 n.4, 42–3, 63, 172–4
Madonna, 229 n.7
Magenta, 107, 204–5, 226, 229 n.7
Magnetic Resonator Piano (MRP), 42, 173
Magnusson, Runar, 199
Magnusson, Thor, 4, 6, 11, 35–6, 57, 59, 64, 119, 152, 189, 190, 199
Mahillon, Victor-Charles, 63
Mahler, Gustav, 196
Mälzel, Johann Nepomuk, 97, 113
Manousakis, Stelios, 235
mapping, 7, 38, 96, 184, 189, 208–9, 236
  algorithms, 223
  engine, 36, 185
  gestural, 40–1
  layer, 37
  physical, 56, 167
  signal, 125, 140
  strategies, 64, 111, 176
Marino, Jesse, 236
Marklay, Christian, 209
marquetry, 24, 86
Marsyas, 20–1, 160–1
Marxian analysis, 52
Massive Attack, 199
material turn, 8, 12, 55, 65
mathematics, 1, 20, 22–4, 28, 31, 47, 50, 73, 75 n.3, 81, 86, 91 n.4, 112–15, 129, 145, 152–4, 160, 205–6, 226
Mathews, Max, 114, 232, 239
Matthews, Kaffe, 217–18
Max/MSP, 38, 40, 142, 142 n.4, 142 n.5, 162, 185

Mays, Tom, 39–40
mediation, 53, 170, 208–9
medieval music, xii, 6, 19–24, 31, 48, 74–5, 80–1, 109, 111, 123, 147, 196, 211, 239
Mei, Girolamo, 74
Melograph, 124, 149–51
memory
  art of, 78, 80
  computer 50
  cultural, 51, 71, 81, 83, 206
  ergomimetic, 172
  exsternalisation of, 4–6, 74, 76, 80–1, 144
  god of, 2, 75
  incorporated 10–11, 23, 76, 183
  inscribed, 72, 244
  motor, 11, 172, 175, 187
  of writing, 72, 179
Mendizabal, Alex, 181, 210
Merleau-Ponty, Maurice, 141
Mersenne, Marin, 49–50, 90, 112, 125
Merzbow, 30, 195
Messiaen, Olivier, 147
metaphor, 11, 91, 103, 115, 117, 147, 167, 185, 206, 211 n.3, 215
Metfessel, Milton, 148–9, 153
microphone, 7, 42–3, 76, 124 n.1, 131, 135–6, 158, 170, 191, 211, 236
microtonality, 40, 78, 123, 155, 186, 209
Middle Ages, 21–2, 24, 47, 72, 80, 89, 109, 225
MIDI, 10, 19, 37, 40, 44, 100, 131, 159, 202–3, 232
Miller, Daniel, 225
mimetic, 1, 5, 7, 75, 89, 242
Minecraft, 214
Minton, Phil, 60
mirror neurons, 172
Mithen, Steven, 1, 56 n.3, 72
Miwa, Masahiro, 208, 211
Mnemosyne, 2, 74, 75
mnemotechnic, 2, 4, 72, 80, 93, 143–4
Mobin, Anton, 236
modernism, ix–xi, 6, 42, 57, 102, 142, 154, 195, 98 n.2
Moholy-Nagy, László, 137–8, 140–1, 143
Molitor, Claudia, 191–2, 235
Monahan, Gordon, 216
Montes de Oca, Alejandro, 236
Monteverdi, Claudio, 90
Moog, 30, 56 n.1

Moor Mother, 236
Moorer, James Anderson, 197
Moran, Robert, 100
moveable type, xi, 50, 57, 81, 86–7, 100
Mozart, Wolfgang Amadeus, 97, 113, 118, 197, 229
MP3, 8, 94, 210
Mudd, Tom, 170 n.6
Mumma, Gordon, 57, 181
Murail, Tristan, 30, 153
Muses, Greek, 2, 20, 74–5, 91, 207, 215, 233, 244
music
  algorithmic, 96, 98, 100, 107, 109–12, 110 n.2, 114–15, 138–9, 142 n.5, 163, 180, 197–8, 200–3, 206, 252 n.3
  experimental, 13, 84, 107, 168, 176, 195, 206–7, 211, 215, 226, 229, 229 n.7, 232, 236, 239
  generative, 104 n.3, 109–11, 113–14, 117, 197–200, 216, 232, 234, 236
  information retrieval (MIR), 157, 159–61, 202
musical organics, 57, 62, 64–5
musicking, xii, 2, 161, 193
musicology, 8, 10, 90, 92, 105, 110–12, 148, 154–5, 172, 200, 233
  of code, 115, 117, 205
  ethnomusicology, 8, 123, 148, 150, 161–2
  of instruments, 12
  medieval, 22
  systematic, 200
musique concrète, 66, 131, 141–2, 151
myth, 2, 13, 19–21, 31, 74

Nakamura, Toshimaru, 235
Nancarrow, Conlon, 98
Native Instruments, 44, 107, 226–8, 233 n.3
Natyasastra, 63
Neoplatonic, 21–2
Nettl, Bruno, 72, 95, 123–4
neumes, 76–8, 80, 123
neural networks, 159, 162, 198, 200–1, 204, 208, 230
Newton, Isaac, 13, 48, 125, 147
Niblock, Phill, 30
Nicolai, Carsten, 217
Nicolls, Sarah, 209, 236
Nietzsche, Friedrich, xii, 172

Nikkal, 72
NIME (New Interfaces for Musical Expression), 30, 36, 39, 39 n.4, 42, 44–5, 107, 171, 181, 215
notation, x, xi, 4, 6–7, 12, 13–4, 22, 28, 30, 66, 68, 71–81, 83–4, 86–9, 92–102, 120, 124–5, 133, 135, 139–40, 142–51, 153–4, 163, 174, 179–89, 188 n.3, 191, 197, 203, 205, 210, 214, 219, 223, 231–2, 235, 238–9
  action, 66, 96, 99–101, 125
  animated, 99, 175, 183, 187–8, 191
  choreographic, 236
  circular, 84, 84 n.1
  cithara, 78 n.4
  cuneiform, 72
  Daseian, 77, 123
  descriptive, 95–6, 148–51, 153
  Egyptian, 123
  extended, 102
  gestural, 40, 99, 130, 185
  graphic, x, 13, 102, 104, 153
  Greek musical, 73–4, 77
  Guidonian, 79, 80, 100, 110
  idiosyncratic, 43
  Japanese, 123
  machine, 13, 109, 112
  modal, 79
  multimedia, 102
  neume, 76–8, 80
  new, 39–40, 100, 106–7, 124, 157, 179, 208–9
  opto-acoustic, 138, 141
  prescriptive, 95–6, 99, 144, 148–9, 151, 185
  printed, 12, 86–7, 90, 92–3, 105, 210, 219
  S-, 99
  seismographic, x
  synthesis, 140
  Tibetan, 123
  verbal, 99
NSynth (Neural Synthesizer), 204–5, 204 n.9
Nyman, Michael, 237 n.5

Ocarina, 43, 213
Odin, 2, 2 n.1
Odo of Cluny, 23
Oliveros, Pauline, 30, 62, 100
Ono, Yoko, 99, 101–2, 217, 236
ontology
  of music, xii, 29, 91, 93, 106, 112, 144–6, 218
  object-oriented, 8, 10, 34, 35 n.1
  software, 33–4
open work, 98–9, 106
oral culture/tradition, 72, 75, 80–1, 115, 144, 144 n.7, 148, 150, 179–80
Oram, Daphne, 100, 140, 140 n.3
Oramics Machine, 100, 140, 140 n.3
organ
  artificial, 65, 243
  bodily, 65, 109, 244
  church, 36, 44
  human, 3, 12 n.4, 18, 19, 34
  light, 147
  Moon King, 131
  pipe, 1, 18, 24–5, 47, 59, 89, 112, 219
  of rationality, 21, 47
  social, 65
  speech, 130
  water, 23, 97, 196
organology, 8, 13, 18–19, 45–6, 59, 61–5, 112, 169
organon, 11, 18, 48–9, 79
organum, 18, 47–8, 79, 80
O'Riordan, Kate, 9
Orpheus, 20, 198, 236
OSC (Open Sound Control), 41
Ouzounian, Gascia, 218
Owl Project, 64, 209

Pachet, François, 161, 198 n.3, 203
Paik, Nam June, 216
Paine, Garth, 39 n.4
Pan, Stephanie, 60
Panharmonicon, 113
Papalexandri-Alexandri, Marianthi, 67, 211, 235
Parikka, Jussi, 2, 7
Parker, Charlie, 144
Parker, Evan, 60
Parmegiani, Bernard, 140
Parmenides, 9
Partch, Harry, 209
Pascal, Blaise, 48
Pask, Gordon, 62, 175 n.10
Patel, Aniruddh, 62 n.2, 72
Patterson, Ben, 100–2
Patteson, Thomas, 17, 30, 97, 137, 140
Peirce, Charles, 79, 184
percussion, 19, 23, 25, 28, 176, 191

Perich, Tristan, 209, 235
Perotinus, 79
Perry, Lee Scratch, 145
Petrucci, Ottaviano, 86
Pfenninger, Rudolf, 102, 138, 140, 147, 188
phenomenology, 34–5, 48, 62, 141–2, 158–9, 169–70
   alien, 34–5, 55
   post-, 57
Philipsz, Susan, 217
philosophy, x, 9, 28, 47, 62, 75, 91, 93, 102, 109, 154, 196
   ancient Greek, 2, 19–20, 22, 31
   of creativity, 196
   of technology, 4, 8, 52, 55
phonograph, 13, 130, 132–3, 136–9, 218
phonography, xi, 4, 14, 95, 124–5, 130–1, 133, 133 n.6, 135, 137, 139–41, 143–5, 147, 150, 180, 226, 229
piano, 6–7, 19, 28, 42, 44, 71, 99–100, 103, 138, 160, 173–4, 176, 181, 188, 191–2, 196, 219, 233 n.1, 241
   digital, 33
   mechanical, 1, 112
   player, 7, 24, 30, 97, 140, 143, 147, 196, 226
   prepared, 236
pianola, 30, 112, 140, 196–7
Pickering, Andrew, 52, 59
Pinch, Trevor, 8, 38, 56, 56 n.1, 181, 199
plainchant, 78–9, 163
Plato, 2, 9, 14, 20–5, 72, 74, 81, 83, 91, 93, 171, 244
Polychronopoulos, Spyros, 199
Polymeneas-Liontiris, Thanos, 43
polyphony, 71 n.4, 78–80, 83–4, 89, 94, 159
Popp, Marcus, 209
postdigital, 12, 107 n.5, 118, 172, 208, 238
postmodern, 12, 195
post-recording, 12, 67, 91
Pousseur, Henri, 98, 207
Praetorius, Michael, 27–8, 50
pre-digital, xi, 118, 186
print, x–xi, 12, 25, 50, 57, 73, 80–1, 85–90, 92–3, 100–2, 128, 133, 172 n.9, 179, 210, 218–19, 226, 232
prosthesis, 6, 18, 31, 48, 51, 65, 71, 225
protocols, xii, 37, 40, 46, 50, 52, 59, 61, 63, 120, 159, 186, 198, 200, 208–10, 223, 228, 232, 238

Puckette, Miller S., 142 n.4
punk, 232 n.1
Pure Data, 38, 199
Pythagoras, 21–2, 54, 67, 73–4, 76, 126

qanun, 22, 68

Raaijmakers, Dick, 236
Rachmaninoff, Sergei, 140, 196
Radigue, Eliane, 30, 57
radio, xi, 19, 44, 133, 141, 146, 195, 213–14, 210, 223, 226, 230, 239
Radiohead, 199
Raes, Godfried-Willem, 171–2, 172 n.7, 211
raga, 78, 159
Raspberry Pi, 38
Reactable, 38–9, 56 n.2, 236
Rebelo, Pedro, 288–9
Redgate, Christopher, 135, 135 n.1
Reed, Lou, 30, 236
Regino of Prüm, 23
Reich, Steve, 103–4
remediation, 30, 129, 133, 168, 184, 215
Renaissance, 10, 21, 24–5, 27, 76, 87, 89, 147, 186, 192, 229
Rheinberger, Hans-Jörg, 167–8
ricercare, 91, 118, 183, 244
Richards, John, 212, 212 n.4
Ritchie, Graeme, 196, 198
Roads, Curtis, 114, 153, 197
robots, 24, 109, 171–2, 172 n.7, 186, 189, 191, 196, 211, 226
rock music, 8, 41, 136, 144–5, 215, 229, 233 n.2, 241
Roli Blocks, 41
Roman de Fauvel, 88, 88 n.3
Romantic music, 29, 31, 50, 57, 92–3, 105, 107, 113, 182, 191, 195, 204
Ross, Daniel, 83, 243
Rousseau, Jean-Jacques, 1
Russolo, Luigi, 30, 98

Saariaho, Kaija, 30, 153
Sachs, Curt, 17 n.1, 63, 64 n.3
Sagan, Carl, 2
Sartre, Jean-Paul, 141
Saussure, Ferdinand de, 117
Sax, Adolphe, 39
saxophone, 6, 10, 28, 30, 39, 162, 172

Scaletti, Carla, 236
Schaeffer, Pierre, 140–3, 154, 158, 239 n.1
Schedel, Margaret, 140
Schlemmer, Oskar, 97
Schmidt Horning, Susan, 133 n.6, 136
Schmitt, Antoine, 198
Schnell, Norbert, 214, 214 n.6
Schoenberg, Arnold, 98, 113, 147
Schroeder, Franziska, 189
Schwab, Michael, 168
Schwarz, Diemo, 161–2
Scipio, Agostino Di, 208, 236
scordatura, 99
score
    action, 100
    animated, xiv, 188, 236
    code as, 8
    computer-generated, 187
    databases, 162
    descriptive, 96, 151
    dynamic, 188–9
    event, 99
    following, 161
    generative, 91, 103, 116
    graphic, x, 100, 101, 106, 117, 153, 188, 209, 236
    listening, 153
    musical, x, xiv, 6, 33, 43, 50, 66, 73, 78–80, 83–7, 89, 92–6, 98–103, 105–7, 115, 117–18, 129, 133, 143–6, 154, 159, 179, 181–2, 185, 188, 191, 192, 198, 201, 212, 214, 218–19, 231, 234–6, 238, 240
    open, 85, 92, 99, 101–2, 105–6
    prescriptive, 84 n.1, 96, 151, 203 n.7
    real-time, 186
    realisation, 99
    softwarwe as, 208
    tablature, 86
    tactile, 191
    tangible, 187, 191, 236
    typology of, 96
    verbal, 183
SCOT (Social Construction of Technology), 8, 141
Scott, Edouard-Léon, 128–32, 148, 153
Scriabin, Alexander, 147, 196
Scripture, Edward Wheeler, 130
Second Life, 214

Seeger, Charles, 95–6, 123, 135, 148–50, 153, 161
Semegen, Daria, 100
semiotic, 55, 77, 171–2, 184–5
Sen, Jónas, 219
Sender, Ramon, 100
Serafin, Stefania, 215
serialism, 163, 229
seventeenth century, xii, 25, 28–9, 47, 49–50, 68, 79, 91, 91 n.4, 97, 112–13, 125–6, 174, 196
Shakespeare, William, 25
shakuhachi, 2
Shinto, 24, 24 n.4
Sholpo, Evgeny, 140
Siccio, Kate, 236
signal, 7, 29, 33–4, 42, 56–7, 59, 66, 125, 128, 139, 141–3, 149, 152, 162–3, 167, 169, 172, 202, 204
    analysis, 157
    audio, 124, 133, 147, 153, 159, 161, 167, 179, 202, 205
    control 43
    data, 55
    electronic, 141
    inscription, 6, 13, 148, 155, 179
    musical, 14
    processing, 29 n.5, 34, 56, 160, 161 n.4, 210 n.2
    vs. symbol, xi, xii, 7, 56, 67, 125, 128, 133, 139, 143, 146, 148, 154, 159
    writing, 140, 231, 238
Silverman, Robert, 48
Simondon, Gilbert, 9–10, 19, 46, 59, 62, 173
Simone, Nina, 60
Singer, Eric, 211
sketching, 54, 61, 94, 117, 129, 201, 201 n.5, 227, 241
SLÁTUR, 188, 236
Small, Christopher, xii, 2, 62 n.2, 161
Smalley, Dennis, 11, 155, 155 n.3
Smith, Harry, 102
Smith, Ryan Ross, 188, 236
Smith, Sylvia, 101
Socrates, 72
software, 1, 8, 19, 33–5, 39–40, 44, 51–4, 59, 61, 66, 91, 99–100, 109, 117, 145, 152, 154 n.2, 153–7, 159–60, 162, 168, 180–3, 186, 189, 197–9, 201, 203–5,

Index

207–8, 212–14, 223–30, 230 n.8, 232–4, 236, 238, 242
Sonami, Laetitia, 38, 183, 229 n.7
Sonic Pi, 67, 190
sonic writing, x–xii, xiv, 3–4, 13, 68, 100, 123, 131, 135, 141, 144, 151, 167, 180, 207 n.1, 238
sound
  art, 107, 216–18
  engine, 36–7, 41, 56, 185
Souza, Jonathan De, 11
SoundCloud, 233, 233 n.3
soundspotting, 161–2
speaker (loudspeaker), 7, 34–5, 42–3, 56, 67, 131, 185, 214, 218, 236
spectromorphology, 155, 155 n.3
Spiegel, Laurie, xiii, 2, 114, 236, 239
Spotify, 107, 133 n.6, 199, 203
standardisation, 6, 19, 28–30, 92–3, 95–6, 135 n.1, 182
Stapleton, Paul, 236
Star, Susan Leigh, 45, 51–2
Steen-Andersen, Simon, 236
STEIM, 37–8, 226
Sterne, Jonathan, 8, 130–1, 133 n.6, 208–9, 218
Stewart, Andrew, 40 n.6
Stiegler, Bernard, 3–6, 10, 47, 62, 65, 72, 83, 144–5, 172 n.9, 243
Stockhausen, Karlheinz, 57, 98, 100, 104, 106, 229 n.7
Stradivarius, 29, 29 n.5, 53
Stravinsky, Igor, ix, 2, 98, 136
streaming music, 131, 133 n.6, 179, 199, 207, 223, 226, 231, 233
Stroppa, Marco, 117
STS (Science and Technology Studies), 8, 62
Sturm, Bob, 202, 204
Subotnick, Morton, 57
Sun Kim, Christine, 217
SuperCollider, 38, 185, 190, 199, 201
surrealism, 95
symbolic, 6, 53, 56, 59, 71, 139–41, 142 n.4, 142 n.5, 155, 167, 175, 185, 202–3, 231, 242n
  AI, 200, 226
  art, 25
  data, 159, 163, 203
  grammatisation, 83

instructions, 2, 51, 139, 154
inscriptions, 6, 13, 147, 179, 210 n.2
language 124, 135 n.2, 150, 154
logic, 168
meaning, 24
notation, 43, 79, 146, 148, 154, 186, 238
representation, 128, 143
sign, 184
systems, 1, 4
syntax, 96
vs. signal, xi, xii, 56, 67, 125, 128, 133, 143, 146, 148, 154, 159
thought, 50
writing, 57, 139, 144, 148, 151, 238
synthesis, 7, 39, 56, 59, 114, 130, 137–40, 142–3, 145, 151, 153–5, 157, 160, 161–2, 167, 180, 186, 200, 201, 203–5, 209, 223, 226, 233, 239, 242
synthesizer, 7, 18, 24, 30, 35, 37, 39, 65, 98, 100, 109 n.1, 118–19, 130, 137, 139, 143, 163, 181, 204, 211–12, 212 n.4, 215, 226

tablature notation, 40, 43, 79, 86, 99, 109, 130, 185
Takemitsu, Toru, 100
Tanaka, Atau, 236
tape, xi, 33, 62, 98, 103–4, 107 n.5, 135, 137, 141, 143–5, 148, 154, 158, 167, 180, 205, 207, 209, 232, 239, 239 n.1
techne, xii, 9, 24, 47, 55, 73, 74, 90, 93, 142
Telharmonium, 131
Tenney, James, 84 n.1, 239
theremin, 30, 38
TidalCycles, 190
timbre, x, 30, 41–2, 55, 66–7, 77, 92, 102, 123, 142, 151–2, 159–60, 204
Tomás, Enrique, 38, 187, 191, 236
Tomlinson, Gary, 1–3, 5, 62 n.2, 71, 72, 210
Tonkin, Mike, 236
Toop, David, 115, 232, 238
transduction, xii, 7, 10–11, 17, 34, 71, 129, 133, 137, 139, 147, 167, 172, 176, 208, 225
transindividuation, 9, 62 (*see also* individuation)
Trimpin, 211, 216
Trocco, Frank, 8, 56, 56 n.1, 199
troubadour, 48, 87, 91, 239

Truax, Barry, 155
trumpet, 19, 25, 102, 128, 160
Tudor, David, 57, 71, 71 n.1, 99
tuning, 3–4, 21, 23–4, 33, 39, 41, 47, 52, 68, 71, 71 n.4, 73, 86, 99, 123, 171, 180–1, 209
  fork, 34

Uitti, Frances-Marie, 106
Úlfarsson, Halldór, 42, 43, 174
Underwood, Sam, 211

Varela, Francisco, 55, 65, 169
Varèse Edgard, x, xi, 130, 163, 229
Vasulka, Steina and Woody, 102, 147, 216
Verrando, Giovanni, 66–7, 209
Vickery, Lindsay, 188
Vidolin, Alvisi, 109
VIME (Virtual Interfaces for Musical Expression), 216
Vinci, Leonardo da, 84, 126
vinyl, 72, 141, 143, 180, 200, 209–10, 218, 223, 231, 234
violin, 6, 12 n.5, 19, 24, 27, 29, 29 n.5, 53, 127, 130, 151, 170, 176, 196, 206, 228 n.4, 235
virtual reality (VR), 14, 48, 131, 179, 183, 186–7, 192, 199, 214–16, 223, 233
virtuosity, 6–8, 29, 31, 84, 93, 141, 196, 169 n.3, 210, 241
VoCo (Adobe), 162, 204–5
Vogel, Christian, 216
Vogelin, Salome, 218
voice, ix, xi, 1, 3, 19, 21–2, 24, 47, 60, 71–2, 76, 78–9, 78 n.4, 89, 99, 124, 126 n.2, 129, 149, 159, 162, 205, 235–6
Voinov, Nikolai, 104
VRMI (Virtual Reality Musical Instruments), 215

Wagner, Richard, 102
Waisvisz, Michel, 36–8, 51
Wang, Ge, 43–4, 213
Waters, Muddy, 170
Waters, Simon, 63, 161, 170
WaveNet, 204–5
wax cylinder, 130, 132, 135, 137, 180
Weniger, Rainer, 153
Weinberg, Gil, 211
Werktreue, 13, 93, 98
Westercamp, Hildegard, 155
Whitehead, Alfred North, 9
Whitelaw, Mitchell, 115
Whitney, James and John, 102, 140, 147, 188
Wiggins, Geraint, 196, 198
Winkel, Dietrich Nicholas, 113–14
Winkler, Peter, 124
Winner, Langton, 224
wire recording, 141, 148, 180
Wishart, Trevor, 60, 155, 216
Wittgenstein, Ludwig, 17, 124
Wolff, Christian, 100
Wood, Ronnie, 233 n.2
work-concept, 13, 92–4, 104–7, 112, 120, 143, 146, 181–2
Wu Tang Clan, 243

Xenakis, Iannis, 57, 101, 106, 231

Yamaha DX7, 37
Young, La Monte, 71, 181, 216
Young, Samson, 217

Z, Pamela, 236
zairja, 111
Zarlino, Gioseffo, 27, 89–90, 111–12
Zeus, 2, 74, 75
Zorn, John, 188, 236

www.ingramcontent.com/pod-product-compliance
Lightning Source LLC
Chambersburg PA
CBHW052152300426
44115CB00011B/1636